Engineering Optical Networks

Engineering Optical Networks

Sudhir Warier

ARTECH
HOUSE

BOSTON | LONDON
artechhouse.com

Library of Congress Cataloging-in-Publication Data
A catalog record for this book is available from the U.S. Library of Congress.

British Library Cataloguing in Publication Data
A catalogue record for this book is available from the British Library.

Cover design by John Gomes

ISBN 13: 978-1-63081-447-2

© 2018 ARTECH HOUSE
685 Canton Street
Norwood, MA 02062

All rights reserved. Printed and bound in the United States of America. No part of this book may be reproduced or utilized in any form or by any means, electronic or mechanical, including photocopying, recording, or by any information storage and retrieval system, without permission in writing from the publisher.

All terms mentioned in this book that are known to be trademarks or service marks have been appropriately capitalized. Artech House cannot attest to the accuracy of this information. Use of a term in this book should not be regarded as affecting the validity of any trademark or service mark.

10 9 8 7 6 5 4 3 2 1

In the ever loving memory of my father
Late C. V. M. Warier

To my Sons
Vihaan & Neil

Contents

	Acknowledgments	xix
	Part I **Introduction to Telecom Networks: The Conceptual Framework**	**1**
1	**Telecommunication Networks: Evolution, Key Components, and Techniques**	**3**
1.1	Chapter Objectives	3
1.2	Evolution of Telecommunication Networks	4
1.3	PSTN	4
1.3.1	Basic Terminology	6
1.4	PSTN Components	6
1.4.1	Class 1 Switch	8
1.4.2	Class 2 Switch	8
1.4.3	Class 3 Switch	9
1.4.4	Class 4 Switch	9
1.4.5	Class 5 Switch	9
1.4.6	Transmission Facilities/Circuits	11
1.4.7	Network	11
1.4.8	Transmission Media and Signals	12

1.5	PSTNs: Advancements	19
1.6	Summary	20
1.7	Review	21
1.7.1	Review Questions	21
1.7.2	Exercises	23
1.7.3	Research Activities	23
1.8	Referred Standards	24
1.9	Recommended Reading	24
1.9.1	Books	24
1.9.2	URLs	24
	References	25

2 Voice and Data Transmission 27

2.1	Chapter Objectives	27
2.2	Introduction	28
2.3	Multiplexing	29
2.4	PCM	35
2.5	Quantization Noise	41
2.6	Companding	44
2.7	Differential PCM	45
2.8	Adaptive DPCM	46
2.9	Summary	47
2.10	Review	47
2.10.1	Review Questions	47
2.10.2	Exercises	49
2.10.3	Research Activities	49
2.11	Referred Standards	50
2.12	Recommended Reading	50
2.12.1	Books	50

2.12.2	URLs	50
	References	51

3 Fiber-Optic Communication Fundamentals — 53

3.1	Chapter Objectives	53
3.2	Introduction	54
3.3	Basics of Light Transmission	55
3.3.1	Modal Propagation	58
3.4	Optical Fiber Classification	59
3.5	Standard Optical Fiber Designs	60
3.6	Types of Optical Fibers	62
3.7	Fiber Design Specifications	65
3.8	Fiber-Laying Techniques	71
3.9	Cable Preparation, Splicing, and Termination	73
3.9.1	Cable Preparation	74
3.9.2	Splicing	74
3.9.3	Fiber Termination	75
3.10	Summary	76
3.11	Review	77
3.11.1	Review Questions	77
3.11.2	Exercises	78
3.11.3	Research Activities	79
3.12	Referred Standards	79
3.13	Recommended Reading	80
3.13.1	Books	80
3.13.2	URLs	80
	References	81
	Selected Bibliography	81

4 Optical Link Design — 83

4.1	Chapter Objectives	83

4.2	Optical Transmitters	84
4.2.1	Optical Sources	85
4.2.2	Modular Optical Interfaces	86
4.2.3	Transmitter Design Parameters	89
4.3	Optical Receivers	92
4.4	Optical Modulation Techniques	93
4.5	Optical Connectors	96
4.6	Link Budgeting	100
4.7	Safety Guidelines	107
4.7.1	Causes of Injury	108
4.7.2	Maximum Permissible Exposure	108
4.7.3	AEL	110
4.7.4	Fiber-Handling Techniques	110
4.8	Summary	112
4.9	Review	113
4.9.1	Review Questions	113
4.9.2	Exercises	114
4.9.3	Research Activities	115
4.10	Referred Standards	115
4.11	Recommended Reading	116
4.11.1	Books	116
4.11.2	URLs	117
	References	117

Part II
Photonic Network Architecture: The Immediate Past and Present — 119

5 Transport Networks: Prologue — 121

5.1	Chapter Objectives	121
5.2	Introduction	122
5.3	Asynchronous Networks	124
5.3.1	Bit Stuffing	127

5.4	PDH	127
5.4.1	CCITT Recommendations	129
5.4.2	Interfaces	130
5.4.3	Basic Frame Structures	134
5.5	Ethernet Over PDH	144
5.6	Asynchronous Network Limitations	146
5.7	SONET: Evolution	147
5.8	Summary	148
5.9	Review	150
5.9.1	Review Questions	150
5.9.2	Exercises	151
5.9.3	Research Activities	152
5.10	Referred Standards	152
5.11	Recommended Reading	153
5.11.1	Books	153
5.11.2	URLs	153

6 Synchronous Optical Networks — 155

6.1	Chapter Objectives	155
6.2	Introduction	156
6.3	Transport Network: An Overview	156
6.4	Transport Network: Need, Benefits, and Function	157
6.4.1	Evolution of Synchronous Optical Networks	159
6.4.2	Optical Transport Networks	160
6.5	Transport Network: Architecture	161
6.6	Transport Network: Components	164
6.7	Summary	176
6.8	Review	177
6.8.1	Review Questions	177
6.8.2	Exercises	178

6.8.3	Research Activities	179
6.9	Case Study: SDH Network Architecture	179
6.9.1	Background	179
6.9.2	Requirements	179
6.9.3	Key Challenge	181
6.9.4	Proposed Solution	181
6.9.5	Recommendation	182
6.10	Referred Standards	184
6.11	Recommended Reading	185
6.11.1	Books	185
6.11.2	URLs	185
	References	186

7 Optical Transport Network — 187

7.1	Chapter Objectives	187
7.2	Introduction to TDM Core Networks	188
7.3	Business Imperatives	189
7.4	Network Organization	192
7.4.1	Standards	192
7.4.2	Network Elements	192
7.4.3	OTN Transport Hierarchy	193
7.4.4	Network Architecture	197
7.4.5	Physical Interfaces	200
7.4.6	Logical Interfaces	201
7.4.7	Connection Management	201
7.4.8	Automatic Protection Switching	202
7.5	Optical Control Plane	202
7.5.1	Standards	204
7.5.2	Logical Interfaces	205
7.5.3	Control Plane Architecture	208
7.5.4	Control Plane Functions	209
7.5.5	Protection Switching and Restoration	210
7.6	FEC	211
7.6.1	Reed-Solomon Codes	212
7.6.2	Bose-Chaudhuri-Hocquenghem Codes	213

7.6.3	Low-Density Parity Check Block Codes	214
7.6.4	Advantages of FEC	214
7.7	Tandem Connection Monitoring	216
7.8	Switching Architecture	217
7.9	Key Features	218
7.10	Summary	220
7.11	Review	221
7.11.1	Review Questions	221
7.11.2	Exercises	222
7.11.3	Research Activities	222
7.12	Case Study: OTN Deployment	223
7.12.1	Background	223
7.12.2	Challenges	223
7.12.3	Requirements	225
7.12.4	Proposed Solution	225
7.12.5	Recommendations	227
7.13	Referred Standards	228
7.14	Recommended Reading	230
7.14.1	Books	230
7.14.2	URLs	230
7.14.3	Journals	231
	References	231
8	**DWDM**	**233**
8.1	Chapter Objectives	233
8.2	Introduction	234
8.3	What is WDM?	236
8.4	Standardization	237
8.5	WDM Fundamentals	238
8.6	Evolution of WDM Networks to Meet Future Challenges: Flexible WDM Grids	246

8.7	Optical Transmission Challenges	251
8.8	WDM Network Components	255
8.9	DWDM Links	260
8.10	Case Study: DWDM Deployment	262
8.10.1	Background	262
8.10.2	Challenges	264
8.10.3	Requirements	264
8.10.4	Proposed Solution	264
8.11	Summary	267
8.12	Review	268
8.12.1	Review Questions	268
8.12.2	Exercises	269
8.12.3	Research Activities	270
8.13	Referred Standards	270
8.14	Recommended Reading	270
8.14.1	Books	270
8.14.2	URLs	271
8.14.3	Journals	271
	References	271

Part III
Next-Generation Photonic Networks: Applications and Architecture—The Future — 273

9 Photonic Circuit-Switched Network Architecture — 275

9.1	Chapter Objectives	275
9.2	Introduction	276
9.3	Optical Access Networks	277
9.3.1	Broadband Access Networks (Cable)	278
9.3.2	Optical Fiber Access Networks	280
9.3.3	5G Mobile Wireless Networks	282
9.4	FTTx Networks	288
9.4.1	PON: Architecture and Functioning	289

9.4.2	WDM Infrastructure Integration	292
9.4.3	NG-PONs	292
9.5	Photonic Core Networks	293
9.5.1	Packet Optical Evolution or the Packet Optical Transport Service	293
9.5.2	IP/MPLS Optical Core Networks	294
9.5.3	NG-POTN: The Network of the Future	296
9.6	Summary	301
9.7	Review	302
9.7.1	Review Questions	302
9.7.2	Exercises	305
9.7.3	Research Activities	305
9.8	Case Study: Next-Generation Access Networks	305
9.8.1	Challenges	305
9.8.2	Solution	305
9.8.3	Network Architecture	305
9.8.4	ODN	306
9.8.5	Wavelength Provisioning	306
9.9	Referred Standards	306
9.10	Recommended Reading	308
9.10.1	Books	308
9.10.2	URLs	308
9.10.3	Journals	308
	References	308
10	**Packet-Switched Photonic Networks**	**311**
10.1	Chapter Objectives	311
10.2	Imperatives	312
10.3	Packet-Switched Network Architecture	314
10.3.1	Network Elements	314
10.3.2	Physical Topology	318
10.4	ISPs	319
10.5	Design Considerations	319

10.6	MPLS	322
10.6.1	Packet Forwarding Through the MPLS Domain	323
10.6.2	Label Switching Process	324
10.6.3	LSP Types	326
10.6.4	Penultimate Hop Popping	327
10.6.5	Traffic Engineering	327
10.7	IP over MPLS	334
10.7.1	Next-Hop MPLS	335
10.7.2	Next-Hop MPLS: Comparing LSP Metrics	336
10.8	Virtual Routing	336
10.9	Optical Switching	338
10.9.1	Optical/Electronic Circuit Switching	339
10.9.2	Optical Packet Switching Techniques	340
10.9.3	Optical Burst Switching	343
10.10	GMPLS	345
10.11	Packet Transport Networks	346
10.12	MPLS-TP	346
10.13	Packet Versus Data Flows	346
10.14	Summary	347
10.15	Review	347
10.15.1	Review Questions	347
10.15.2	Exercises	349
10.15.3	Research Activities	349
10.16	Referred Standards	349
10.17	Recommended Reading	350
10.17.1	Books	350
10.17.2	URLs	350
10.17.3	Journals	351
	References	351
11	**Virtualization and the SDN Ecosystem**	**353**
11.1	Chapter Objectives	353

11.2	Introduction	354
11.3	NFV Business Case	357
11.4	NFV Use Cases	357
11.5	NVF Framework and Services	362
11.5.1	Functional Blocks	363
11.5.2	VNF Services	364
11.6	Design Considerations	365
11.7	Distributed NFV Architecture	367
11.8	NFV and SDN: The Linkage	368
11.9	Network Management and Orchestration	369
11.10	Open-Source NFV Platforms	370
11.11	SDN: Making a Compelling Business Case	370
11.12	SDN: Conceptual Basis	371
11.13	Architectural Framework for SDN	372
11.13.1	Architectural Components	373
11.13.2	Protocol support	375
11.13.3	Design and Deployment Considerations	376
11.14	Management	377
11.15	OpenFlow	378
11.16	SDN Use Cases	379
11.17	Summary	382
11.18	Review	383
11.18.1	Review Questions	383
11.18.2	Exercises	385
11.18.3	Research Activities	385
11.19	Case Study—Wireless Broadband Networks	386
11.20	Referred Standards	387
11.21	Recommended Reading	390

11.21.1	Books	390
11.21.2	URLs	390
11.21.3	Journals	390
	References	391

About the Author — 393

Index — 395

Acknowledgments

This book builds on the work of numerous researchers and doctoral, postgraduate, and graduate student theses; articles by corporate planners; discussions with my colleagues; and a legion of other publications—in print and online. I wish to express my gratitude to the many creators of this work; they have provided me with the contours of this book.

I would also like to thank my publisher Artech House, where Aileen Storry, Holly Smith-Young, and their team provided me with the space I needed to complete this work, while gently nudging me to stay on course. In addition, the reviewers of this book have been spot on and must be credited for the focus that they have brought to each of the chapters. It has also been a pleasure working with the production and marketing teams and the cover designer, John Gomes. The credit for the success of the book goes to the team at Artech.

I wish to thank, from the bottom of my heart, my family: my mother, my wife, and especially my sons, who had to bear with my grueling schedules, my tantrums, and my time away from them as I strove to complete this book. It has been hard on my children to be deprived of the happy family weekends they look forward to. My elder son Neil Warier has shown maturity beyond his age in taking care of my little one, Vihaan, while encouraging me to complete the book on schedule. Words cannot express my gratitude to my family for having given me the space to write this book.

Part I

Introduction to Telecom Networks: The Conceptual Framework

1

Telecommunication Networks: Evolution, Key Components, and Techniques

1.1 Chapter Objectives

This chapter provides the foundation for understanding the design, deployment, and functioning of modern-day telecommunication networks, which the subsequent chapters detail. The chapter is devoid of any complexities and mathematical nuances, instead presenting the basic architecture of a telecommunication network, using the example of the old public switched telephone networks (PSTNs). The chapter presents and links the basic building blocks of a telecom network. The chapter also presents the conceptual framework required for understanding the architecture of modern-day photonic networks, presenting information in a manner that facilitates easy assimilation, while allowing readers to understand and appreciate their significance and application. The reader is, however, insulated from mathematical constructs and other technicalities that are not relevant in the context of this book. The concepts introduced in the chapter are reinforced with suitable examples as required.

Key Topics
- The evolution of telecommunication networks;
- The architecture, terminology, components, and functioning of PSTN;
- The developments in the PSTN architecture in conjunction with technology obsolescence and customer expectations.

1.2 Evolution of Telecommunication Networks

The word telecommunication is a loose adaptation of the French word télécommunication. It is a combination of the Greek prefix tele- ($τηλε$), meaning "far off" and communication, meaning "to communicate or exchange information" [1].

Telecommunication refers to the transmission of signals over a distance for the purpose of communication. In the past, our ancestors relied on ingenious methods for communication including the use of smoke signals, drums, and beacons. In modern times this process involves transmitting rays of light through a waveguide—fiber-optic cable (FOC). In the current technology-driven era, telecommunication has become all pervasive and permeates into the realms of data networks, radio, and television. Telecommunication involves interconnection of a vast array of networks connecting a myriad of devices, a converged network, capable of providing multiple services including voice, data, and video—in the form of triple play services. This has led to the development of a multitude of applications from simple electronic mails to video chats to video on demand services (VoD) and more complex medical imaging applications.

The advent of telecommunication, as we know it, began in the year 1876 with the development of the telephone by Alexander Graham Bell. The invention, which was patented by Bell, even though there are other claimants to the development, led to the formation of the Bell telephone company in the United States. The company's operations spread across the length and breadth of the country and emerged as a near monopoly supplier of telephone services within the country. The company, subsequently renamed American Bell, deployed the American Bell Telephone System, which provided local connections within the cities and towns across the United States. However, since the networks in the cities and towns were not interconnected customers could make only local calls. To overcome this limitation, a new subsidiary of American Bell was unveiled and christened American Telephone and Telegraph (AT&T), with the sole objective of building and operating long-distance communication facilities. The expiry of the sole licensing agreement of the original Bell telephone company in the year 1893 fostered the emergence and the resultant competition of new telecommunication service providers leading to the rapid evolution and deployment of telecommunication networks.

1.3 PSTN

PSTN [2] refers to the vast array of networks—government-controlled and commercial—linking cities, countries, and continents. Figure 1.1 illustrates the basic components of a PSTN.

Telecommunication Networks: Evolution, Key Components, and Techniques 5

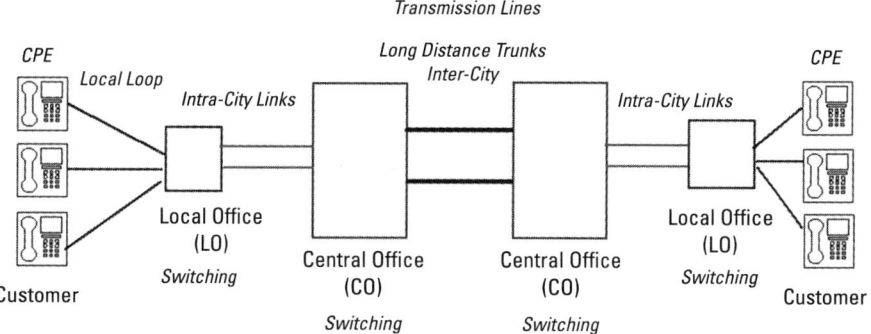

Figure 1.1 PSTN.

The name public switched is indicative of the generic communication framework provided to the masses by employing a network of switching equipment. The network, also referred to as plain old telephone service (POTS) [3], was based on analog technology with copper wires (strung on overhead poles and later buried underground) as the preferred media. However, the advent of digital technologies witnessed changes in the network architecture and the use of cutting-edge optical technologies with optical fiber emerging as the media of choice. Telecommunication networks link telephone and computer equipment, thereby facilitating voice, data, text, and video transmission. There are several interconnected components in a network including switching and transmission equipment. The primary components and their functions are outlined in the following section. Figure 1.2 illustrates the major building blocks of a PSTN.

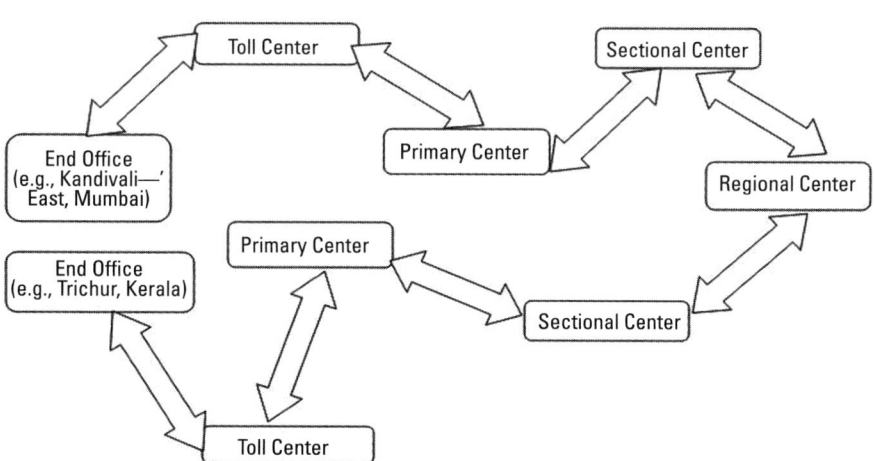

Figure 1.2 Building blocks of a PSTN.

1.3.1 Basic Terminology

- *Network:* The network is an interconnection of devices using suitable media, with the objective of providing specified services (e.g., voice services, data services, and file sharing).
- *Topology:* The method of interconnecting devices to form a network (e.g., bus or linear, star, ring, and mesh).
- *PSTN:* PSTN refers to the worldwide telecommunications network accessible to every individual and institution.
- *Customer premise equipment* (CPE): The equipment at the customer end used to interface with the PSTN.
- *Switching offices:* Switching offices interconnect transmission facilities at key locations and switch traffic through the PSTN. A switch facilitates the interconnection of multiple end users through limited transmission facilities by setting up temporary paths or circuits. Two types of switching techniques that are employed:
  - *Circuit switching:* The technique, used in telecommunication networks, wherein network paths are set up for use by two or more entities for a certain duration of time before being switched for use by another set of entities;
  - *Packet switching:* This technique is employed in data networks where all the network paths are shared by users. The messages between users are broken down into smaller units, referred to as packets, and sent through the best path through the internetwork. The receiving end reassembles the packets to recreate the original message.
- *Transmission facilities:* The communication paths that carry the traffic between the PSTN nodes. They can be classified into two main categories:
  - *Lines:* Communication paths from the CPE to the switching equipment (also referred to as customer loops);
  - *Trunks:* Communication paths between two switching systems.

Figure 1.3 illustrates the PSTN system hierarchy and traces the call flow through the network.

1.4 PSTN Components

A network is formed by interconnecting equipment through a suitable media and designed to provide services to customers or allied networks. The components of a PSTN are outlined as follows:

Telecommunication Networks: Evolution, Key Components, and Techniques 7

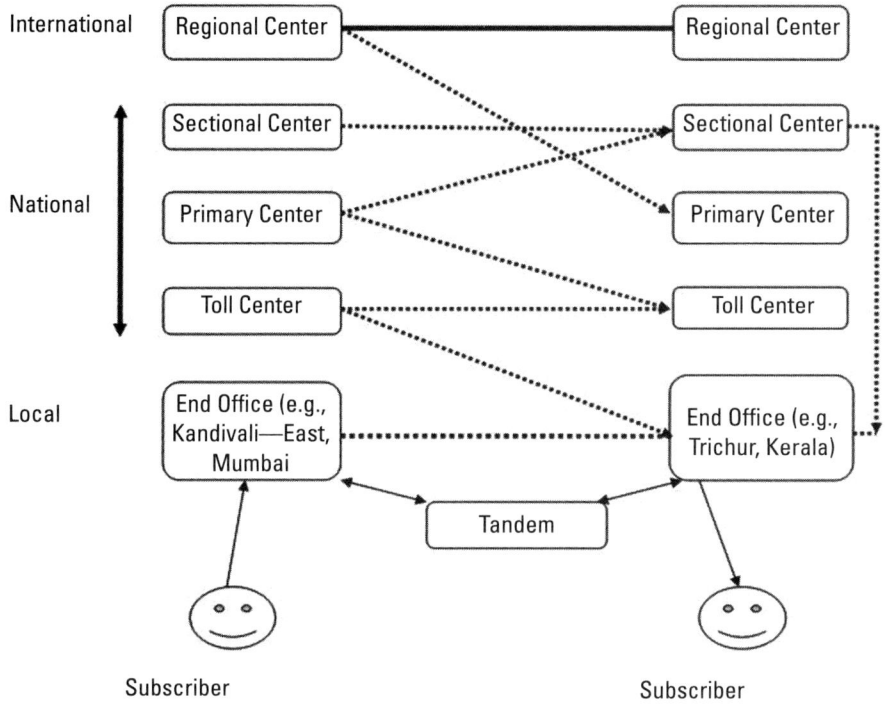

Figure 1.3 PSTN system hierarchy and call flow.

1. *Switching equipment:* For communication between two customers, a dedicated pair of transmission lines is required. The number of the required lines increases exponentially with the number of customers. To overcome this limitation, a piece of equipment that permits a number of customers to communicate with each other through a limited number of lines was devised. This equipment is known as a switch and connects network transmission facilities while routing calls through the network. Switches can also create call detail records (CDRs), which are used to bill customers for calls made. The initial switches were built using electromechanical devices that were subsequently replaced by electronic components based on digital technologies. The latest development is the emergence of the soft switch, which is essentially a computer with the necessary software to handle call switching. The purpose of a switch is to provide the path for a call. To process a call the switch performs three main functions:

a. Identifying the customer;
  b. Setting up the communication path;
  c. Supervising the call.

The switches are grouped, as per the functions provided, into different categories. In the early days of the PSTN, the switching of calls between two subscribers was done manually, by an operator. To facilitate automated operator dialing and later direct distance dialing (DDD), AT&T organized the various switches in its network in a hierarchy containing five levels (or classes), as illustrated in Figure 1.4. The term class stands for custom local area system.

1.4.1 Class 1 Switch

A class 1 switch is a part of the regional center (RC) whose connections are the last alternative for the final setup of calls when other routes between centers lower in the hierarchy are not available.

1.4.2 Class 2 Switch

Class 2 switches form a part of the sectional center (SC) and connect major interstate or intercity toll centers (TCs) to make available connections for long-distance calls.

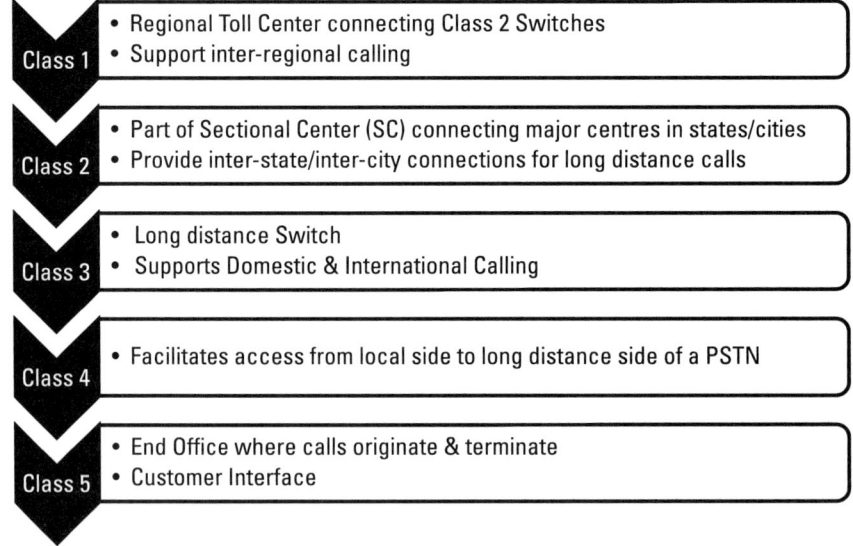

Figure 1.4 Types of switches.

1.4.3 Class 3 Switch

The class 3 switch represents the primary center (PC). Calls outside the confines of a small geographical area connected directly by class 4 toll offices are passed from the TC to the primary center using high-capacity trunks.

1.4.4 Class 4 Switch

A class 4 switch is the TC, toll point (TP), or intermediate point that routes calls between two end offices not directly connected. The TC is also used to connect to the long-distance network for charged calls like operator-handled services. This TC may also be called the tandem office because calls have to be routed through it to get to another part of the network.

1.4.5 Class 5 Switch

The class 5 switch is the local exchange or end office (EO)—also referred to as a branch exchange—that serves as the nearest connection to the end customer and provides basic dialing services.

Modern-day networks generally employ class 4 and class 5 switches. A class 4 switch—referred to as a tandem switch, a backbone switch, and a core switch—is a high-capacity switch positioned in the physical core or backbone of the PSTN that interconnects CO switches. As per the traditional PSTN hierarchy, a tandem switch may be a class 1 regional TC, a class 2 sectional TC, a class 3 primary TC, or a class 4 tandem TC. Local exchange carriers (LECs) are connected to the interexchange carriers (IXCs) over dedicated interoffice trunks (also referred to as access trunks) by the access tandem switch. The PSTN hierarchy, based on the functions performed by the switch, is illustrated in Figure 1.5.

The architecture (Figure 1.5) as well as the functions of the switch evolves in response to technological advancements and new service requirements. The important changes are detailed as follows:

- The primary function of the class 5 switch remains the same: to establish a path connecting a caller to the intended receiver utilizing available network resources.
- The new generation switches are capable of supporting value-added services (VASs), which include special features such as call waiting, call forwarding, and conferencing.
- Special calling features, such as caller identification [also known as caller line identification (CLI) and number portability] are handled by special

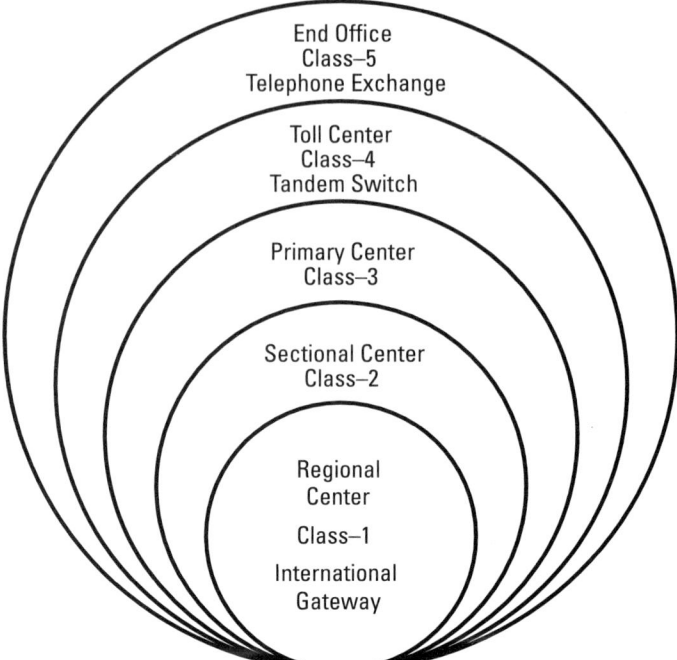

Figure 1.5 PSTN switching hierarchy.

applications servers dedicated for the purpose. This is done so that basic switching functions are not affected due to non-traffic-handling tasks.

- The advanced intelligent network (AIN), developed by Bell Communications Research and recognized as an industry standard in North America, was a modern-day telephone network architecture that separates service logic from switching equipment. This development facilitates the addition of new services without the need to redesign existing switches.
  - Capability set-1 (CS-1) is the ITU-T–equivalent version of AIN.
  - Number portability is one of the features of the AIN.
- The class 5 switch is capable of forwarding calls within a local exchange as well as to other local exchanges located anywhere in the world.
- The class 5 switch itself does not handle call routing to other class 5 switches. This function is performed by a class 4 or tandem switch that handles intercity as well as international calls.
- Class 4 switches are located on core or transport networks and are responsible for sending traffic over long distances or across long-distance calling areas (LDCAs).

- For providing voice services as well as voice-over-IP (VoIP) services, an end user would need to connect to a class 5 switch via an internet protocol (IP) voice gateway.

1.4.6 Transmission Facilities/Circuits

Transmission facilities, commonly referred to as circuits, provide a communication path carrying traffic between two points in a network. As illustrated in Figure 1.1, the two types of circuits are local loops and trunks, described as follows.

- *Local loop:* Refers to the communication link, either copper (e.g., cable televisions and cable modems using coaxial cables) or optical fibers (optical access networks) or the air interface [e.g., wireless in local loop (WiLL), WiMax, Wi-Fi, LTE], between the service provider's local or company office and the customer. The local loop is designed for a particular service, such as voice or data, and can be a 2- to 8-wire connection or an air interface depending on its use. For example, a telephone line with a 2-wire interface provides half-duplex transmission capability wherein both the calling and the called party can talk and listen, but not simultaneously. Two-wire transmissions have a high degree of signal loss over distance and hence require the deployment of amplifiers.
- *Trunk:* A circuit that connects two switches. Most trunks connections in the PSTN network are 4-wire circuits employing a pair of lines for each direction (i.e., transmission and reception). The ability to transmit at the same time in both directions is referred to as a full-duplex connection. An analog 4-wire circuit overcomes resistance to a greater extent thus reducing the need to deploy amplifier stages, resulting in better signal reproduction. The legacy, overhead copper cables and/or underground (buried), transmission networks have long been replaced by new generation optical networks.

1.4.7 Network

The PSTN consists of two interdependent parts—the local network and the long-distance network. Their features are as outlined as follows:

- *Local network*: The local exchange network, provides local calling services, and is a part of the local service provider (LSP) network. It consists of a central office (CO), providing services within designated local calling areas, and EOs, which provide associated interfaces, including dial tones and basic signaling, to the end customer. The CO contains class 5 switching equipment connecting subscribers to each other locally as well as over long-distance lines. A CO switch can be connected directly to another similar switch or it can be connected through a class 4 switch for switching local interoffice trunks and accessing long distance networks.

- *Long-distance network:* These networks provide the communications paths for carrying long-distance calls through a telecommunication network. They are also referred to as intercity networks and in modern parlance as core, transport, or backbone networks.

1.4.8 Transmission Media and Signals

Transmission media [4] refers to the facilities over which information is transmitted from one point to another. A telecommunication network may include just one medium or a combination of wireline and wireless media. Figure 1.6 illustrates the different types of media that can be deployed for telecommunication purposes. The various types of media are described as follows.

- *Wired media:* Wired media provides a physical pathway for interconnecting the different components of the network and establishing a communication channel between two or more entities. The legacy networks used copper wires as the transmission media. The advent of digital communication and the consequent increase in teledensity has brought in an explosion in bandwidth requirements, mandating the deployment of FOCs as the medium of choice. An optical fiber supports very high transmission bandwidths and is less susceptible to interference and crosstalk. The different types of wired media and their salient features are described as follows:

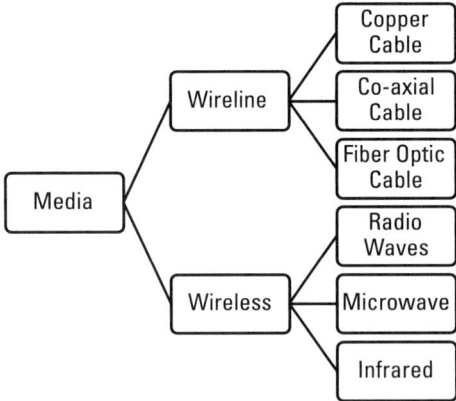

Figure 1.6 Types of media.

- *Copper wires:* In the early days of telecommunication, both analog and digital transmission was over copper cables that were strung on poles (open-wire/single-wire circuits). This was followed by the use of a twisted pair of copper wires, to reduce interference. Initially unshielded twisted-pair (UTP) cables were used (illustrated in Figure 1.7); however, they were susceptible to electromagnetic interference (EMI). This limitation was overcome by the usage of shielded twisted-pair (STP) cables (illustrated in Figure 1.8). Twisted-pair cables were also used extensively in local area network (LAN) deployments. It was designed to support speeds of up to 100 Mbps but can support speeds of 1,000 Mbps or 1 Gbps. Twisted-pair cabling is made up of pairs of solid or stranded copper 22-AWG or 24-AWG cables; the number of pairs in the cable depends on the cable type, twisted around each other. The twists reduce the vulnerability of these cables to EMI and cross-talk.

  The modern-day network architecture and its building blocks are vastly different from the legacy networks of the past; however, copper wire is still employed to link the customer's premises, within buildings and commercial complexes, and the service provider's connection points or junction boxes. Copper cable is a low-efficiency medium because of its narrow bandwidth and its inability to maintain signal strength, paired with the fact that it is expensive.

- *Coaxial cable:* Coaxial, or coax, cables consist of an inner copper wire that is held in place with an insulated material. Coax is a shielded cable and hence is less susceptible to EMI. The term coaxial comes from the fact that the inner conductor as well as the outer shield

Figure 1.7 UTP cable [5].

Figure 1.8 STP cable [5].

shares the same axis. The coaxial cable can be used to transmit low-frequency audio signals as well as high-frequency broadband signals. As illustrated in Figure 1.9, a metallic shield and an insulator covers the core. Thus coaxial cables have high resistance to noise interference facilitating the transfer of signals over larger distances without any amplification. It can also carry signals farther than twisted-pair cabling without frequent amplification. Coaxial cables were used extensively for building LANs—thin Ethernet until a decade ago. It has also seen widespread use in the cable television industry. Now, how-

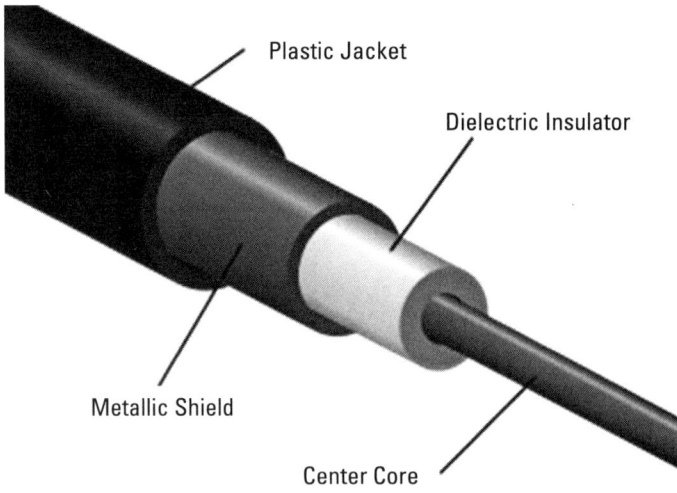

Figure 1.9 Coaxial cable.

ever, coaxial cable usage is decreasing with the advent of optical fibers, direct-to-home (DTH), and other technology developments [6].

- *Optical fibers:* Optical fibers (illustrated in Figure 1.10) are made of high-purity silica or glass and are currently being used extensively in core, metro, and access networks. The optical fiber is the medium of choice for short- as well as long-distance communication, and it is also being used for connecting customer premises, buildings, campuses, and LANs, as well as huge terrestrial networks. Fiber-optic transmission involves the conversion of electrical signals into optical signals that are transmitted over the fiber, by using the principle of total internal reflection (TIR), and their reconversion at the receiving end.

    An optical fiber consists of an inner conductor known as the core surrounded by a layer known as the cladding. The core and the cladding may be glass (most commonly used) or plastic or a combination of both. The refractive index (RI) of the core is greater than that of the cladding. This helps in guiding the ray of light within the core over a large distance. The source of light can be a light-emitting diode (LED) or laser amplification by simulated emission of radiation (LASER). The receiver is a normally a positive intrinsic negative (PIN) or an avalanche photodiode (APD).

    FOCs are weatherproof and small in size as compared to other media, and they have high bandwidth-carrying capacity, are immune from EMI, and can be deployed using a variety of means (underground as well as overhead).

- *Wireless media:* Wireless media, as the name suggests, does not require physical connectivity between two network elements (NEs). The atmosphere (referred to as ether in the past) or the air (air interface) is used as a media for communication between NEs. Wireless technologies are being used increasingly to provide voice, initially low-speed data, and data

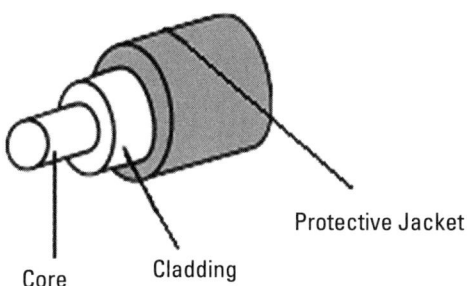

Figure 1.10 FOC.

networking services (including those for internet-enabled devices like palmtops and net books). The fourth generation (4G) networks with LTE as the radio access network (RAN) can currently (under release 9) support 100–300 Mbps [with the use of multiple input multiple output (MIMO) technology]. This speed will be enhanced to 1 Gbps with the release 13 of the 3GPP standards.

The rapid advancements in the field of semiconductor technology have led to the proliferation of PC devices and have supported the enhanced mobility of these devices, with technologies like Wi-Fi gaining prominence. The different types of wireless media, illustrated in Figure 1.6, are described as follows.

- *Radio waves:* The spectrum refers to a band, series, or broad range of frequencies that can be used for communication. The electromagnetic spectrum refers to the entire frequency range of electromagnetic waves. Radio waves are electromagnetic waves that map to the radio frequency portion of the electromagnetic spectrum. The radio wave frequency falls between 9 KHz and 300 GHz. Table 1.1 presents the radio frequency spectrum.

    There are certain radio bands that are internationally reserved for use by the industrial, scientific, and medical (ISM) communities. These are referred to as ISM bands, and all communications equipment must be capable of tolerating interference generated by the equipment using these bands. The most common ISM device is the microwave oven that operates at 2.45 GHz. These bands are also used by wireless LANs and cordless phones that operate at 2.45 GHz and 5.8 GHz. These unregulated frequencies are easily saturated, thereby limiting the broadcast power and consequentially the coverage of the communication equipment.

    Radio wave transmissions can be further divided into three categories based on their power levels, described as follows:

    1. *Single-frequency low-power:* The systems that fall in this category operate in the range of 20–25m only and transmit over a single frequency, with a low-power output. These, generally inexpensive, systems offer throughputs ranging from 1 Mbps to 10 Mbps. Since the systems operating in this range have lower power output, they are highly susceptible to attenuation and EMI from other systems operating in the same frequency range.
    2. *Single-frequency high-power:* The systems operating in this category can communicate over larger distances owing to their higher power output. The communication is effected over line-of-sight

Table 1.1
Radio Waves: Frequency Spectrum [7]

Classification	Starting Frequency	Units	Ending Frequency	Units
Extremely low frequency (ELF)	0	-	0	
Very low frequency (VLF)	3	KHz	30	KHz
Radio navigation and maritime/aeronautical mobile	9	KHz	540	KHz
Low frequency (LF)	30	KHz	300	KHz
Medium frequency (MF)	300	KHz	3,000	KHz
AM radio broadcast	540	KHz	1,630	KHz
Traveling information service	1.610	KHz		-
High frequency (HF)	3	MHz	30	MHz
Shortwave broadcast radio	5.95	MHz	26.1	MHz
Very high frequency (VHF)	30	MHz	300	MHz
Low band—TV band 1, channels 2–6	54	MHz	88	MHz
Midband FM radio broadcast	88	MHz	174	MHz
High band—TV band 2, channels 7–13	174	MHz	216	MHz
Super band (TV and mobile fixed service)	216	MHz	600	MHz
Ultra high frequency (UHF)	300	MHz	3,000	MHz
Channels 14–70	470	MHz	806	MHz
L band	500	MHz	1,500	MHz
Personal communication services (PCSs)	1,850	MHz	1,990	MHz
Unlicensed PCS services	1,910	MHz	1,930	MHz
Super-high (microwave) frequencies (SHF)	3	GHz	30	GHz
C band	3,600	MHz	7,025	MHz
X band	7.25	GHz	8.4	GHz
Ku band	10.7	GHz	14.5	GHz
Ka band	17.3	GHz	31.0	GHz
Extremely high frequencies (EHF)—millimeter-wave signals	30	GHz	300	GHz
Additional fixed satellite	38.6	GHz	275	GHz
Infrared radiation	300	GHz	430	THz
Visible light	430	THz	750	THz
Ultraviolet (UV) radiation	1.62	PHz	30	PHz
X-rays	30	PHz	30	EHz
Gamma rays	30	EHz	3,000	EHz

(LOS) or reflections through the atmosphere. The systems or networks operating in this category are less affected by EMI and are used to serve the needs of the mobile population.

3. *Spread spectrum:* As the name suggests, systems that employ spread spectrum use several frequencies to provide reliable interference-

resistant data transmissions. There are two techniques employed for spread spectrum communications, described as follows.

1. *Direct sequence modulation*: This method breaks the data to be transmitted into chips that are sent over several frequencies. The frequency pattern is known only to the receiver that reassembles the data.

2. *Frequency hopping:* As opposed the direct sequence modulation technique, frequency hopping uses a set of frequencies for specific time duration. In this method the transmitter and receiver are programmed to switch frequencies at preset time intervals. A burst of data is transmitted on one frequency, the next burst on a different frequency. These systems require that the exact frequencies be known for proper reception of data and are therefore less susceptible to eavesdropping. These systems, which were earlier also used for military purposes, can achieve a maximum throughput of 2 Mbps.

- *Microwaves:* Microwaves are electromagnetic waves that operate at higher frequencies than radio waves and hence have better throughput than radio waves. Microwave systems are based on LOS transmission, which mandates that the transmitter and receiver be in sight of each other. Microwave frequencies are regulated internationally and therefore licensed, increasing the setup and operational costs in comparison to systems based on radio waves. The distance between the transmitting and receiving stations is limited due to the necessity of LOS communications and is typically in the range of 50–70 km. This limitation is due to the curvature of the Earth, and repeater stations are required to increase the distance if necessary. Microwave systems can be broadly classified into the following two types:

1. *Terrestrial microwave:* Most of the present day core networks are built using optical fibers. Terrestrial microwave systems are generally used in areas that are inaccessible or where the terrain makes conventional cable-laying difficult. For transmission between two places to occur there must be direct LOS between the parabolic antennas or repeaters configured over relay towers in between. Terrestrial microwave systems uses licensed frequencies and operate in the low gigahertz band. The installation of microwave stations is moderately difficult due to the alignment required for meeting the stringent LOS requirements.

2. *Satellite microwave:* As the name suggests this method of communication involves the use of a satellite, generally a geosynchronous

satellite, to relay signals between different locations on Earth. A satellite dish is employed to send the transmit signal to the satellite, which redirects the same to the receiver's dish. A major drawback of this method is the introduction of latency due to the propagation delay. This propagation delay may vary from less than a second to several seconds depending upon the distance involved. The other disadvantages of this method of communication are the high deployment costs, average throughput, low security, and susceptibility to interference.

- *Infrared:* Infrared refers to the high frequencies that are located below that of the visible light spectrum. Infrared communications are affected by obstructions in transmit or receive paths, but they can provide a high throughput due to their high frequencies. Infrared transmission is immune to EMI and is used generally for short-range communication between computing devices (e.g., PCs, PDAs, and handhelds). There are generally two types of infrared systems, listed as follows:
  1. *Point-to-point:* Point-to-point infrared systems are used by computing devices to communicate with each other directly or through a LAN. This necessitates the use of highly focused infrared beams to transfer data between two systems directly. The major advantage of these systems is the high levels of security afforded and the immunity from EMI. However, these frequencies are generally unlicensed, and the radiation levels are high. Also point-to-point infrared systems require direct alignment between devices.
  2. *Broadcast:* A broadcast infrared system, unlike point-to-point systems, broadcasts signals, thereby facilitating the use of multiple receivers. This method also eliminates the problems associated with the direct alignment and unobstructed transmit and receive paths that plague point-to-point systems. Broadcast systems operate in the same frequency band as point-to-point systems but can offer throughputs of only 1 Mbps.

1.5 PSTNs: Advancements

The PSTN has evolved over its existence. The modern-day PSTN architecture incorporates several changes; its evolution is summarized as follows.

1. At first, PSTN was essentially a wireline network that employed analog transmission over copper cables.

2. The initial topology on the access side (i.e., customer interfacing and service delivery) was point-to-multipoint, LO to customer, and sometimes LO to LO and CO, and linear between CO (trunks).

3. The modern-day access networks are interconnected by core, backbone, or transport networks. Earlier, these networks were based on synchronous digital hierarchy (SDH) or synchronous optical network (SONET) standards, but these have been replaced by optical transport network (OTN) standards and IP/MPLS-based core networks. These networks are based on optical fibers with ring and mesh topologies.

4. In the past, mechanical (manual) or electromechanical switching was employed in exchanges. This has been replaced with digital switches, which are now being complemented with or replaced by next-generation network (NGN) switches and photonic (optical) switches.

5. The local loop (customer interface) is still predominantly copper-based. However, optical fibers are increasingly being deployed for connectivity to the outdoor boxes as well as household connections. These outdoor (or indoor) boxes are referred to as digital loop carriers (DLCs).

6. DLCs, mounted indoors or outdoors, serve as local aggregation points for customer voice and data connections and digitize the analog voice signals carried by the copper lines before transmission to the CO over an optical fiber.

7. Digital data transmission is now possible over the local telephone network (local loop) by employing digital subscriber lines (DSL) technology. This is now commercially available as ADSL (asymmetric DSL). These lines terminate at a DSL access multiplexer (DSLAM) that is further connected over fiber to the service provider's CO. The DLC as well as a DSLAM can be colocated in a single enclosure.

8. In addition, cable television networks based on data over cable service interface specification (DOCSIS) standards and fiber-to-the-x (FTTx) networks using packet optical network (PON) technologies are increasingly being used for access level connectivity.

1.6 Summary

- Telecommunication refers to the transmission of signals over a distance for the purposes of communication.

- The PSTN refers to the vast array of networks, including public or government-controlled networks and commercial networks that link cities, countries, and continents.
- A switch is a piece of equipment that permits a number of customers to communicate with each other through a limited number of transmission lines. The switch connects network transmission facilities and routes calls.
- Transmission facilities, commonly referred to as circuits, provide a communication path carrying traffic between two points in the network.
- Local loop refers to the communication link, either copper optical fiber or an air interface, between the service provider's local or CO and the customer.
- Wireless media, as the name suggests, does not require physical connectivity between two NEs. The atmosphere or the air is employed as the media for communication between NEs.
- The human ear can only discern between frequencies in the range of 300–400 Hz, which is also referred to as the audio spectrum.
- Cable television networks based on DOCSIS standards and FTTx networks using PON technologies are being used increasingly for access-level connectivity.
- The modern-day access networks are interconnected by core, backbone, or transport networks earlier based on SDH or SONET standards, which have been replaced by OTN standards and IP/MPLS-based core networks. These networks are based on optical fibers with ring and mesh topologies.
- The 4G networks with LTE as their RAN can currently (under release 9) support 100–300 Mbps (with the use of MIMO technology). This speed will be enhanced to 1 Gbps with release 13 of the 3GPP standards.

1.7 Review

1.7.1 Review Questions

1. The radio wave frequency falls between _____.
    a. 9 KHz and 300 GHz
    b. 7 KHz and 700 GHz
    c. 3 KHz and 300 GHz
    d. 6 KHz and 300 GHz

2. The refractive index of the core is _____ than that of the cladding.
   a. Less
   b. Equal to
   c. Greater
   d. Cannot predict
3. The _____ is immune from EMI.
   a. STP
   b. OFC
   c. UTP
   d. CAT5
4. The _____ switch is capable of forwarding calls within a local exchange as well as to other local exchanges located anywhere in the world.
   a. Class 5
   b. Class 4
   c. Class 3
   d. Class 1
5. _____ switches are located on core or transport networks and are responsible for sending traffic over longs distances or across LDCAs.
   a. Class 5
   b. Class 4
   c. Class 3
   d. Class 1
6. Number portability is one of the features of _____.
   a. PCM
   b. WDM
   c. AIN
   d. AIM
7. A _____ is also referred to as tandem switch, also known as a backbone switch or a core switch.
   a. Class 2
   b. Class 4
   c. Class 3

d. Class 1

8. A coaxial cable provides better resistance to EMI than does a STP cable.

   a. True

   b. False

9. The maximum throughput using broadcast infrared is _____.

   a. 10 Mbps

   b. 100 Mbps

   c. 1 Mbps

   d. 2 Mbps

10. The most commonly employed light source in communication systems is _____.

    a. LED

    b. Laser

    c. VCSEL

    d. CFL

1.7.2 Exercises

1. Using a simple diagram explain the major blocks of a PSTN.

2. List the different types of media that can be employed in a telecommunication network with their relative merits and demerits. Explain in greater detail the media most commonly employed in modern-day telecom networks.

3. Write a brief note on AIN.

4. Explain the function of a DLC in the network.

1.7.3 Research Activities

1. Draw a block diagram of a PSTN and briefly explain the major components.

2. List the commonly used circuit-switched networks along with their salient features.

3. Explain briefly the standards referred to in this chapter.

4. Explain what is meant by the term DOCSIS? Briefly describe the application context.

    5. Write a short note on the LTE network? What is the significance of the term LTE?

1.8 Referred Standards

ITU-T G.711: Pulse code modulation (PCM) of voice frequencies.

ITU-T G.726: 40, 32, 24, 16 kbit/s Adaptive Differential Pulse Code

TBR-21: Terminal Equipment (TE); Attachment requirements for pan-European approval for connection to the analogue Public Switched Telephone Networks (PSTNs) of TE (excluding TE supporting the voice telephony service) in which network addressing, if provided, is by means of Dual Tone Multi Frequency (DTMF) Signalling

1.9 Recommended Reading

1.9.1 Books

Harte, L., *Telecom Basics*, Third Edition, Althos Publishing.

Freeman, L. R., *Fundamentals of Telecommunication*, Second Edition, John Wiley & Sons Inc., 2005.

Fischer, R. F., *Optical System Design*, Second Edition, SPIE Press, 2008.

1.9.2 URLs

http://en.wikipedia.org/wiki/List_of_basic_telecommunication_topics.

http://en.wikipedia.org/wiki/Class_5_telephone_switches#Telephone_switch_hierarchy.

http://en.wikipedia.org/wiki/Direct_Distance_Dialing.

http://en.wikipedia.org/wiki/Subscriber_trunk_dialling.

http://en.wikipedia.org/wiki/Pulse-code_modulation.

http://www.btplc.com/Thegroup/BTsHistory/1912to1968/1958.htm.

http://www.btplc.com/thegroup/BTsHistory/Eventsintelecommunicationshistory/Eventsintelecommunicationshistory.htm.

http://www.stanford.edu/class/ee368b/Handouts/15-DPCM.pdf.

http://www.cs.wustl.edu/~jain/bnr/ftp/p_2tel.pdf.

References

[1] Haykin, S., *Communication Systems*, Fourth Edition, John Wiley & Sons, 2001, pp. 1–13.

[2] Martin, M., *Understanding the Network (The Networker's Guide to AppleTalk, IPX, and NetBIOS)*, SAMS Publishing, 2000.

[3] Haykin, S., *Communication Systems*, Fourth Edition, John Wiley & Sons, 2001, pp. 67–113.

[4] Agrawal, M., *Business Data Communications*, John Wiley & Sons, 2010, pp. 54–67.

[5] Baran, I., UTP cable.jpg (22 Oct 2008). Retrieved March 17, 2008 [WWW document]. Source URL < http://en.wikipedia.org/wiki/File:UTP_cable.jpg>.

[6] K. Tim (Tkgd2007) (31 May 2008), A cutaway diagram of a coaxial cable, retrieved March 09, 2008 [WWW document]. Source URL http://en.wikipedia.org/wiki/File:Coaxial_cable_cutaway.svg.

[7] Spectrum chart [n.d.], retrieved March 21, 2008 [WWW Document]—Source URL < http://min.midco.net/pk/charts/radio%20Spectrum.html>.

2

Voice and Data Transmission

2.1 Chapter Objectives

This chapter introduces the basic building blocks of a telecom network along with the key concepts of multiplexing. The chapter includes brief coverage of wavelength-division multiplexing (WDM), which is detailed in Chapter 8. The chapter also includes a brief discussion of space-division multiplexing (SDM), which is a more recent development. The chapter begins with an explanation of the ubiquitous pulse-code modulation (PCM) technique, which is essential to understand the foundations of time-division multiplexing (TDM). The topic has been presented in a simple manner so that readers can imbibe the basic concepts required to navigate subsequent chapters. The chapter lays the framework required for understanding the detailed chapter covering synchronous network architecture and the detailed concepts of synchronous frame transmission and overhead processing, which are covered in Part 2. The concepts included in this chapter are introduced in a simplified manner so that readers can understand them and appreciate their significance and application, while avoiding unnecessary discussion of the technicalities involved. These concepts have been reinforced with suitable examples, wherever relevant.

Key Topics
- The audio spectrum;
- The concepts of multiplexing;
- The different types of multiplexing techniques and their application context;

- The pulse code modulation technique and its flavors;
- The companding technique and its application.

2.2 Introduction

POTS was based on analog transmission and circuits. The transformation to newer-generation technologies commenced in the early eighties (much later in developing countries including the Indian subcontinent). When a human being speaks or communicates, the output or voice is in the form of sound waves. The microphone in the telephone receiver converts this sound wave into an electrical representation in the form of an electrical wave whose intensity (current/voltage) varies with respect to time. These electrical signals approximately mirror the variations of the sound wave and hence are also referred to as analog signals (since they are analogous to each other). Figure 2.1 illustrates this concept.

The electrical wave amplitude varies in conjunction with the loudness of the speech, and the frequency varies with respect to the pitch. This electrical signal is represented by a sine wave whose frequency is denoted by hertz (Hz) or cycles per second.

While speaking, a human being generates frequencies ranging from the lower band to the higher band, referred to as audio or audible frequencies. The accepted standard range for these frequencies is 20 to 20,000 Hz. However, the range of frequencies required for transmission of speech of commercial quality fall in the range 300 to 3,400 Hz. These ranges of frequencies are therefore referred to as the voice frequencies or the voice band. The human ear sensitivity is highest between the ranges of 315 Hz to 2.5 KHz also referred to as the middle frequency band. Thus, the band of frequencies supported by the telecommunication network ranges from 0 to 4 KHz.

Since the strength of the signal varies in relation to the distance traveled, amplification needs to be done at appropriate points in the network. The fallback of this technique is that, when a signal gets amplified, the distortions in

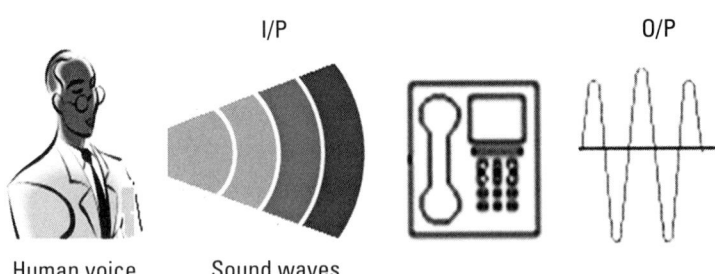

Figure 2.1 Analog transmission.

the line (caused by a variety of factors) or the noise are also amplified. If the distance between the transmitter and the receiver is very high, a large number of amplifier stages needs to be included. However, this may cause high amplification of the noise components. A number of amplifier stages will cause the line noise component to be larger than the actual voice signal being transmitted, making intelligible conversation impossible.

To overcome this limitation, another form of signal transmission, based on the binary number system, has been employed. This technique, referred to as digital transmission, involves transmission of two discrete voltage levels (generally 0 volts and 5 volts) corresponding to the 0 and 1 in the binary number system. Thus the digital signal consists of a series of discrete and discontinuous pulses that can be predicted easily. Digital signals can therefore withstand effects of noise on the transmission channel better than analog signals. Figure 2.2 illustrates the concept of digital transmission between two network points.

However, like analog signals, digital signals also lose their strength after traveling over a significant distance. To counter this, digital systems employ devices referred to as repeaters or regenerators to regenerate the signal at periodic intervals. A regenerator reproduces the original signal, at its output, thereby overcoming the problems stemming from amplitude and timing distortions suffered by amplifiers. Subsequent chapters will explain these concepts in detail and throw sufficient light on the issues regarding the deployment of this equipment.

2.3 Multiplexing

The rapid growth of telecommunication networks (higher teledensity) has led to the development of newer services, based on new applications, fueling an exponential need for transmission bandwidth. Telecom operators and service providers realized that it is not possible to link subscribers together by using dedicated lines or circuits. Accordingly, telecom companies and researchers explored newer ideas or techniques to increase bandwidth. The result was the development of multiplexing techniques. Multiplexing refers to the technique of combining multiple input informational channels onto a common output or

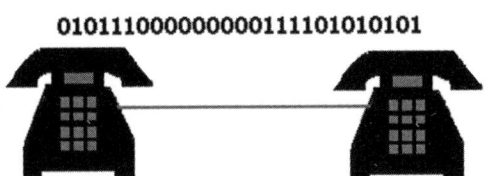

Figure 2.2 Digital signal transmission.

transmission channel. This technique facilitates the transmission of multiple information signals simultaneously over a communication path (medium, circuit, or channel). At the receiving end, these signals can be individually recovered. Multiplexing techniques have evolved over a period of time in response to technological developments, as well as need. These multiplexing techniques are as listed as follows:

1. Frequency-division multiplexing (FDM);
2. Time division multiplexing (TDM);
3. Wavelength division multiplexing (WDM);
4. Space division multiplexing (SDM).

Example 2.1

The concept of FDM is best explained by the following example: Every organization has a separate area or a room to entertain visitors. Assume that at a given point of time there are seven company officials entertaining their visitors/guests. The members of these seven groups are now attempting to initiate conversations within their group. The result would be utter chaos with no meaningful information being exchanged. In order to facilitate exchange of information within these groups, we could employ two techniques (although there are other possibilities). These two techniques are described as follows:

1. Each of the seven group members could be given a set of earplugs that filters out all sounds except their own conversation. This is an example of FDM.
2. Ten minutes of exclusive communication time could be allotted to each group in a round robin fashion until their information exchange is complete. This technique is an example of TDM.

FDM

FDM [1] standards for telecommunication purposes were defined by the erstwhile Comité Consultatif International Téléphonique et Télégraphique or Consultative Committee for International Telephony and Telegraphy (CCITT). CCITT has since been revamped and renamed the International Telecommunication Union-Telecommunication Standardization Bureau (ITU-T). Table 2.1 outlines the ITU-T FDM hierarchy.

FDM involves dividing the total frequency bandwidth into smaller frequency channels and their exclusive and continuous assignment is to transmit individual information signals. The audio spectrum lies in the range of 300 to

Table 2.1
ITU-T FDM Hierarchy

Group	Number of Channels	Bandwidth
Channel	1	4 KHz
Group	12	48 KHz
Super group	60	240 KHz
Master group	300	1.2 MHz
Super master group	900	3.6 MHz

3,400 Hz or a bandwidth of 3.1 kHz for each channel. To prevent interference among channels, guard bands are allotted on either side of a channel, resulting in a channel spacing of 4 kHz. The frequency shifting (frequency translation) of the information signal produces two new frequency bands known as the upper sideband (USB) and the lower sideband (LSB), referred to as side frequencies. This structure facilitates voice calls to be combined into blocks of appropriate sizes for transmission through a national or international telephone network. The standardized frequency assignments ensure that systems using these in various countries are compatible. Further, ITU-T recommended that all multichannel systems should be based on two alternative 12-channel groups or that one group should utilize the USB and the other should use the LSB.

TDM

As explained in Example 2.1, TDM employs a simple concept of time-sharing to facilitate the simultaneous communication over multiple channels over limited transmission facilities. Each user or subscriber is allocated a time slot or a channel that occupies the entire transmission line bandwidth for a specific period of time.

All the input channels are connected to the output transmission channel for a specific duration in a round robin fashion. (Durations can also be based on specific user-defined criteria.) This technique of TDM is detailed in Section 2.2. The resultant samples from each of the inputs (sampling) are combined to form a composite signal that is sent over the transmission facility. It is important to note that, as explained earlier, the sampled signals are converted to digital prior to line transmission. This basic function of a time division multiplexer is pictorially illustrated in Figure 2.3. Section 2.4 provides the details of the TDM techniques used in telecommunication networks.

WDM

Optical networks transmit information in the form of light beam(s) over an optical fiber. The fiber-laying cost forms a chunk of the total cost of building a network. This is especially true in countries where the time and cost of ob-

Figure 2.3 TDM.

taining the necessary permissions for standard trenching and ducting (T&D) referred to as right-of-way (RoW) is high. The RoW cost pushes up the overall fiber deployment costs.

There are a various research groups [study groups (SGs)], sponsored by industries and academia associated with ITU-T, engaged in establishing innovative techniques to improve network bandwidth while lowering deployment costs. These groups realized that increasing teledensity, especially in developing countries coupled with the deployment of newer services, would exhaust the bandwidth provided by the synchronous networks SONET and SDH (10G). Further they had to ensure that the new technology being developed would not call for a significant change in the transmission media and the network architecture.

The above needs and consequent research led to the development of WDM. As the name suggests, WDM involves multiplexing multiple rays of light or wavelengths through a single fiber. WDM enables the utilization of a significant portion of the available fiber bandwidth by allowing multiple independent signals to be transmitted simultaneously on one fiber, with each signal located at a different wavelength. The WDM concept is illustrated in Figure 2.4.

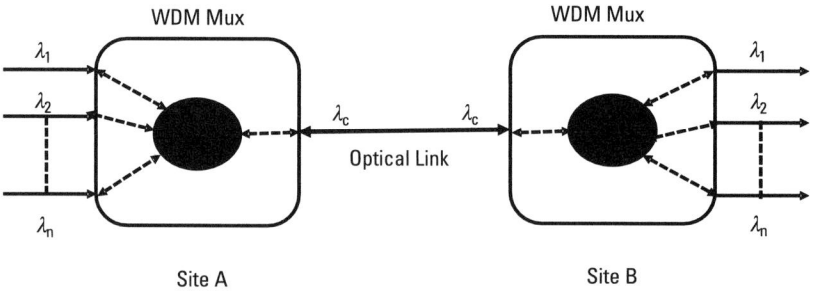

λ_c = Composite Wavelength

Figure 2.4 Example of a WDM link.

The initial demonstration of the WDM technique involved the use of two conventional wavelengths (normally used for optical transmission): 1,310 and 1,550 nm. The WDM system based on these wavelengths was referred to as normal WDM. WDM transmission was also demonstrated using 1,553 and 1,557 nm in a system referred to as narrowband WDM. Subsequently developed were two techniques referred to as coarse WDM (CWDM) and dense WDM (DWDM). CWDM systems allow the simultaneous transmission of 18 wavelengths as specified by ITU-T standard G.671 [2]. The spectral grid, based on a channel spacing of 20 nm, for CWDM wavelengths is specified by ITU-T G.694.2 [3]. The ITU-T standard G.671 defines DWDM technology with a narrow channel spacing as compared to CWDM technology. The ITU-T standard G.694.1 [4] defines the frequency grid with fixed channel spacing ranging from 12.5 GHz to 100 GHz within the C and L bands. Commercial systems facilitate the transmission of 80 wavelengths (over the C band using a 50-GHz grid). The classification of WDM systems is illustrated in Figure 2.5.

Example 2.2

As explained in the previous section long-distance data transfer (intercity) is through optical links. The information-carrying capacity of a fiber is very high and is only limited by the technology employed to transport the signals. In the case of legacy SDH networks the maximum information carrying capacity is 10 Gbps. This implies that an optical link between two cities can carry 10 Gbps of data. In cases of higher traffic requirements the telecom operator would be forced to provision an additional pair of fibers between the cities. The cost

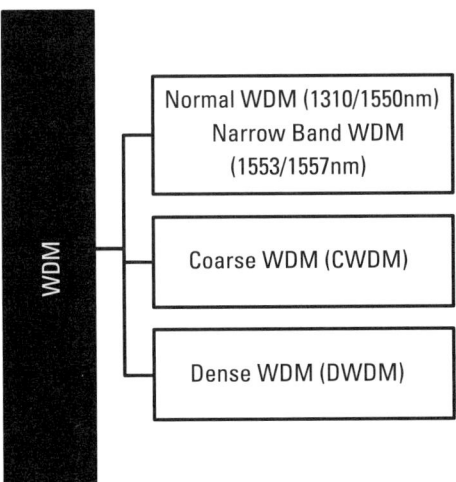

Figure 2.5 WDM classification.

of laying fibers is still very high, and the permissions required for trenching (RoW[1]) is cumbersome.

WDM permits the transmission of an additional ray of light (or wavelength) through the optical fiber (in a manner that prevents the two signals from overlapping). The capacity of the link is thus doubled. The WDM multiplexer is a passive device that contains a complex array of a lens/mirror assembly and a prism. Chapter 7 details WDM networks. The concept of WDM is similar to that of FDM, but in the optical domain. The individual rays of light (or wavelengths) can carry TDM signals, and the bandwidth is limited by the technology used.

As mentioned earlier, SDH networks support a maximum transmission capacity of 10 Gbps. The payload can contain lower order TDM signals or Ethernet frames. A DWDM network, which supports the simultaneous transmission of 80 wavelengths (80λ), will have a transmission capacity of:

$$80\lambda * 10 \text{ Gbps}(\text{per } \lambda) = 800 \text{ Gbps}$$

SDM

SDM [5] (see Figure 2.6) is a radical new approach to meet the exponential bandwidth requirements over fiber in the coming years. Multimode fibers (MMFs) have a higher core diameter that permits the propagation of several

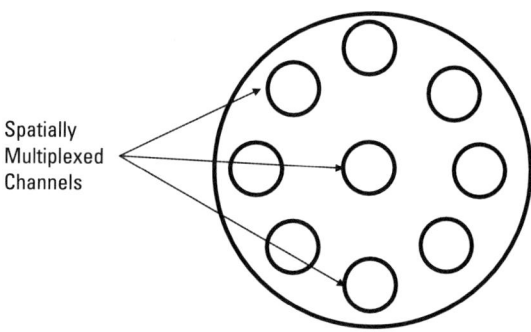

Few Mode Fiber (No. of Modes = 9)

Figure 2.6 Example of SDM.

---

1. The right-of-way (RoW) is country and state specific framework for coordination between various federal and municipal bodies to provide time-bound approvals and dispute settlements for the laying of telecom infrastructure including, but not limited to, the laying of optical fiber cables.

independent modes within the same fiber. The numbers of modes are dependent on the diameter of the core and the RI profile of the fiber. The multiple transmission modes in a MMF can be suppressed by modifying the fiber characteristics, allowing only a few modes (typically 10 modes or fewer) to propagate. These fibers, referred to as few-mode fibers, represent a fundamentally new approach to increasing the information density over fiber. The additional number of modes or space channels provides additional bandwidth capabilities.

Example 2.3

With a reference to the Example 2.2, the use of SDM enhances the transmission capacity of the optical network to:

$$80\lambda * 10 \text{ Gbps}(\text{per } \lambda) = *10(\text{spatial modes}) = 8,000 \text{ Gbps or 8 Tbps}$$

2.4 PCM

PCM [6] is a signal-coding technique defined by the ITU-T standard G.711. PCM, one of the most important basic building blocks of a telecommunication network, represents the process in which multiple analog signals are multiplexed and transmitted in a digital form over a communication link with successful reproduction at the receiving end. Figure 2.7 shows a block diagram of a PCM system.

The analog signal is represented by a series of pulses signifying binary 1 and 0. This stream of pulses of 1's and 0's are not easily affected by interference and noise. Even with a manifestation of noise, the presence or absence of a pulse

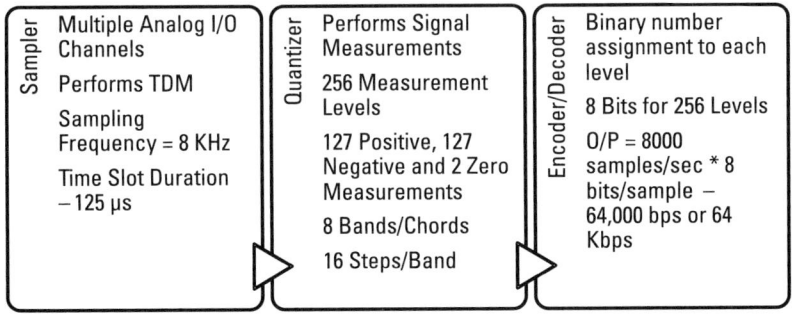

Figure 2.7 Block diagram of a PCM system.

can be easily determined. The main stages in a PCM system are described as follows:

1. *Sampling:* The process begins with the filtering of the higher frequency component of the input signal with a view to simplifying the reconversion of the signal at the receiver. The frequency of sound generated by human beings roughly lies in a 3-KHz bandwidth between 300 Hz and 3,400 Hz. The filtering process also helps in avoiding the usage of expensive precision filters.

   A constant sampling frequency is employed to sample the filtered input and acts as a precursor to the analog-to-digital (A/D) conversion. This process is referred to as pulse amplitude modulation (PAM). The instantaneous amplitude of the modulating signal, at sampling intervals, will cause variation in the amplitude of the individual pulses of the pulse train with the width and position of the pulses remaining constant (Figure 2.9). The transmitter design is simple since sampling the modulating signal at regular intervals produces PAM. The sampling frequency is determined based on the Nyquist theorem or Nyquist criteria. As per this criterion the original analog signal can be recreated at the receiving end if the sampling frequency is greater than twice the highest frequency of the original input analog voice signal. As mentioned earlier we are interested in the audio spectrum whose highest frequency component is 3.3 KHz, and the spectrum used for telecommunication purposes is 0 to 4 KHz. The process of sampling is followed by TDM, illustrated in Figure 2.8.

   A TDM multiplexer is analogous to a simple switch that connects multiple inputs to a single output channel in a round robin fashion. The derived samples are digitized in order to transmit the same over the communication network. This is achieved by deploying an A/D converter on the transmitting end and a digital-to-analog (D/A) converter on the receiving end. PCM employs a technique referred to as quantization for encoding/decoding the samples (see Figure 2.9).

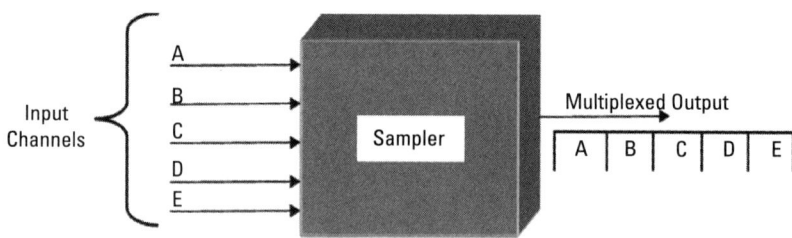

Figure 2.8 TDM.

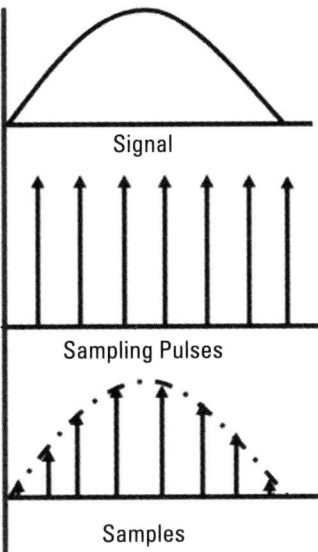

Figure 2.9 Sampling.

Example 2.4

As per Nyquist theorem the sampling frequency has to be greater than twice the maximum frequency component of the spectrum being sampled (audio spectrum). Accordingly:

$$f_s > 2(f_m)$$

$$f_s = 2*4 \text{ KHz} = 8 \text{ KHz or } 8{,}000 \text{ samples/sec}$$

where:

f_s = sampling frequency;
f_m = maximum frequency component of the spectrum under consideration.

The duration of each time slot is:

$$T_s = 1/f_s = 1/8{,}000 = 125\mu s$$

Example 2.5

ESW, Inc., needs to build a communication system that operates in the frequency range of 2 to 3 KHz. The signal spectrum falls within this range. With the assumption that the number of quantization levels is 128, calculate the basic system parameters including the sampling frequency, time slot duration, and the channel bandwidth for the system.

Solution

The maximum frequency that the system would handle is 3 MHz. Accordingly,

$$f_m = 3 \text{ KHz}$$

Therefore $f_s = 2 * 3$ KHz = 6 KHz or 6,000 samples/sec.
Thus $T_s = 1/f_s = 1/6 = 167 \mu s$.

Since the number of quantization levels is 128, we require seven bits to represent the values (i.e., $2^7 = 128$).

Therefore the channel bandwidth = number of samples/sec * number of bits/sample.

$$= 6,000 * 7 = 42,000 \text{ bps or 42 Kbps.}$$

2. *Quantization:* Quantization [7] refers to the process of assigning a discrete value to each of the analog samples. The assigned a digital sequence is a unique binary number. In the quantization phase the input signals (samples) are assigned a quantization interval. The quantization intervals are equally spaced (uniform quantization) throughout the dynamic range of the input analog signal. The process is as follows:
   - For voice communication the standard assignment is 256 levels. These are assigned as 127 positive encoding levels, 127 negative encoding levels, and two values for the zero level.
   - These levels are divided into eight bands, also referred to as chords.
   - This implies that there are 16 steps within each chord. This concept is illustrated in Figure 2.10.
   - Each PAM sample's peak would correspond to a specific chord and a step, thereby assigning it a numerical value.
   - This value is assigned a corresponding binary number, which becomes its PCM value.
   - As the input voice signal amplitude increases, the quantization levels also increase uniformly.

The process of quantization is pictorially illustrated in Figure 2.11.

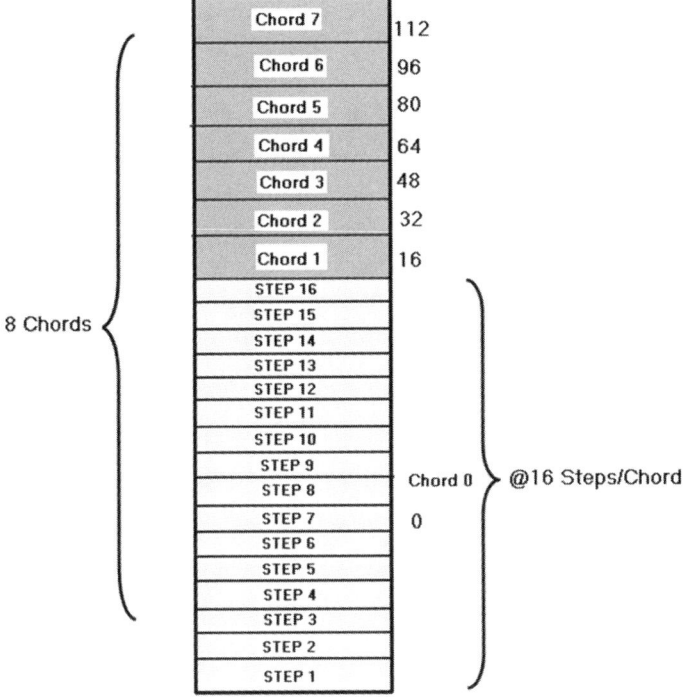

Figure 2.10 Quantization process.

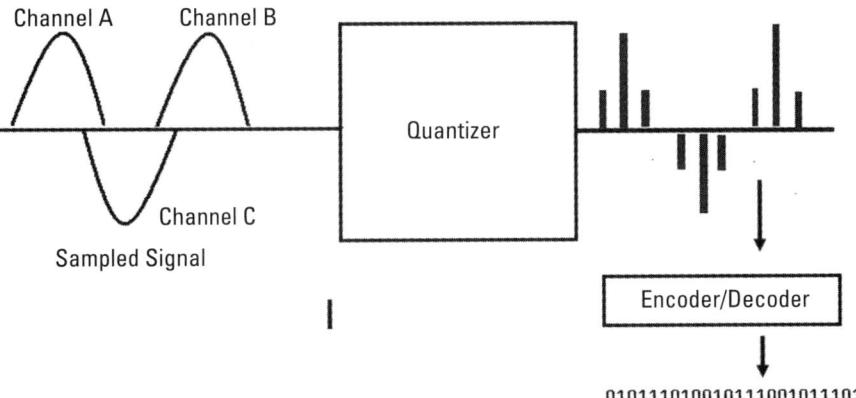

Figure 2.11 Quantization and encoding/decoding stages.

Example 2.6

As described in Section 2.3, the input signal is assigned 256 quantization levels. To represent 256 levels in binary, eight bits are required. Hence the maximum transmission bit rate or channel bandwidth is computed as:

$$\text{maximum transmission bandwidth} = \text{number of samples/sec} *$$
$$\text{number of bits/sample}$$

= 8,000 (samples/sec) * 8 (bits/sample)
= 64,000 bps
= 64 Kbps

3. Encoding: Each quantization interval is assigned a discrete value or a binary code during transmission in a process referred to as encoding. At the receiving end, the reverse process referred to as decoding is employed. Figure 2.7 presents the outputs of the quantization as well as the encoding/decoding stages.

Example 2.7

The following example depicts the PCM technique of quantizing and encoding at the transmitting end and the consequent signal reproduction through decoding at the receiver. Figure 2.7 presents the block diagram of a PCM system. In this example, eight quantization segments are employed, and the signal is quantized in nine points of time. The values that fall into a specific segment are approximated with the corresponding quantization level lying in the middle of the segment. The encoding is done as per Table 2.2.

Figure 2.12 illustrates the process of signal quantizing and digitizing. Thus the signal would be represented by the following binary sequence using 27 bits to encode the signal:

```
101 111 110 010 010 111 100 011 010 101
```

The samples shown are approximated to the nearest quantization level and coded as indicated in Table 2.2.

Figure 2.13 illustrates the process of signal restoration as per the received sample values. It may be noted that the restored signal differs from the original transmitted signal. This divergence is a consequence of the quantization noise and is independent of the signal intensity.

Example 2.7 presents a simplified process of quantization and encoding. The process can be further detailed with reference to Figure 2.13. As described there are eight bits assigned to each sample. These bits are used as described in the following. (Refer to Figure 2.12 and Tables 2.3–2.5.)

As per Example 2.7, a quantized value of +105 would be assigned a binary value of 11101001. The first 1 (MSB) indicates the polarity. The value 105 falls in between 96 and 112 or the seventh chord thus accounting for the remaining three bytes in the nibble 110. Further, each cord corresponds to 16 steps, and

Table 2.2
Quantization Levels and Equivalent Binary Codes

Quantization Levels	Binary Code
0	000
1	001
2	010
3	011
4	100
5	101
6	110
7	111

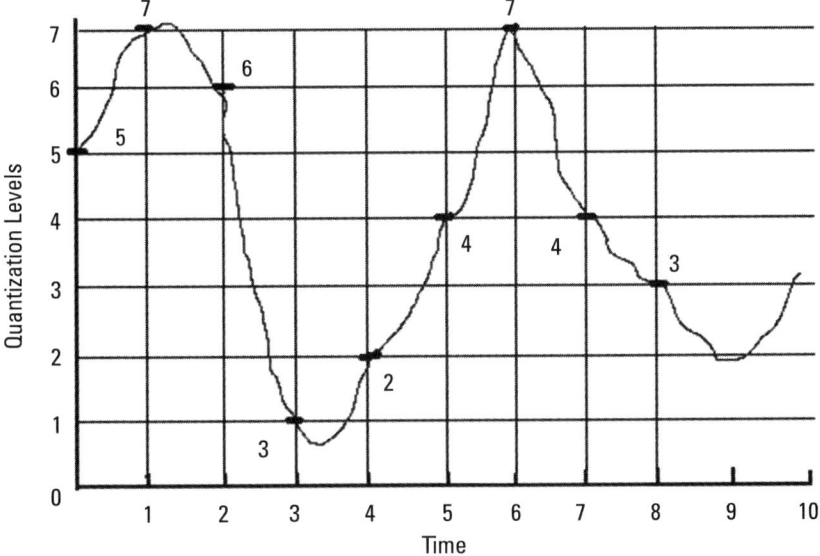

Figure 2.12 Input signal quantization and digitalization.

105 would be the tenth step (96–105), and hence the last nibble corresponds to 1001. This is graphically represented in Figure 2.15.

2.5 Quantization Noise

In a PCM system, each input sample is assigned a quantization interval during the quantization process. This interval would be the closest to the amplitude of

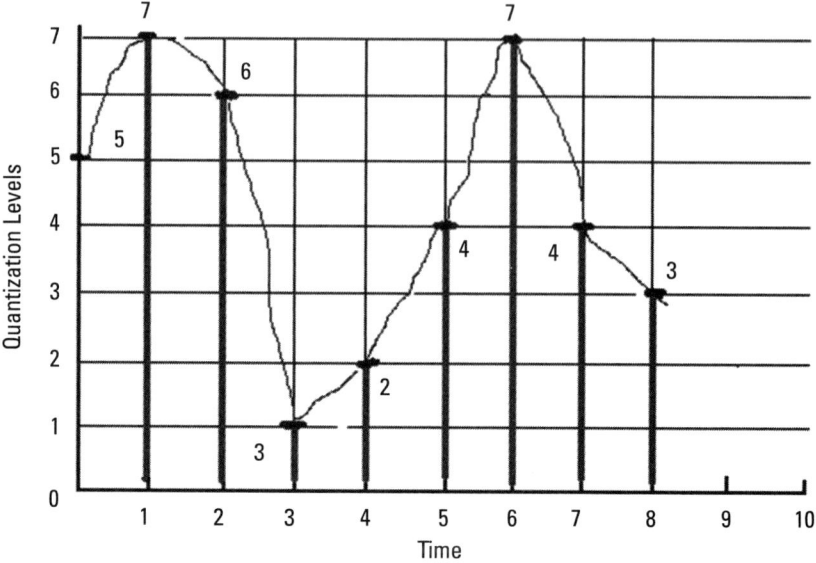

Figure 2.13 Signal restoration at receiver.

Table 2.3
Polarity Coding

Polarity	Binary Code
Negative	0
Positive	1

Table 2.4
Chord Coding

Quantization Levels	Binary Code	Decimal Value
0	000	0
1	001	16
2	010	32
3	011	48
4	100	64
5	101	80
6	110	96
7	111	112

Table 2.5
Steps Coding

Steps	Binary Code
0	0000
1	0001
2	0010
3	0011
4	0100
5	0101
6	0110
7	0111
8	1000
9	1001
10	1010
11	1011
12	1100
13	1101
14	1110
15	1111

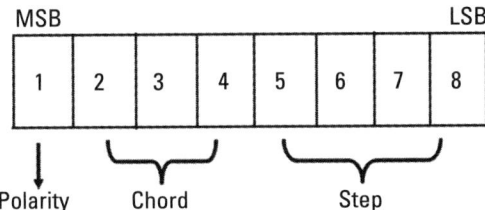

Figure 2.14 PCM value assignment.

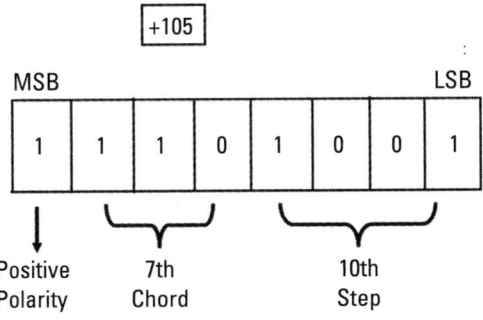

Figure 2.15 Example of the quantization process.

the input sample. In case a suitable interval cannot be assigned, an error referred to as quantization noise is introduced into the PCM process.

An increase in the amount of quantization intervals can lead to a reduction in the quantization noise. This increase in quantization intervals also leads to the reduction in the difference between the input signal amplitude height and the quantization interval. However the drawback of this method is that the amount of codes would need to be proportionately increased in relation to the quantization intervals. This could lead to problems associated with the capacity of the system. The method is summarized as follows:

- The uniform quantization method employs quantization levels that are equal throughout the entire dynamic range of an analog input signal.
- As a voice signal sample increases in amplitude, the quantization levels increase uniformly.
- The linear quantization process assigns an even spread of the 127 quantization levels over the input voice signals range.
- Thus the process suffers from a disadvantage that assigns high-amplitude or loud voice signals the same degree of resolution (step size) as soft or low-amplitude voice signals.
- The encoding technique though simple to implement does not provide optimized fidelity in the reconstruction of human voice.
- A quantization process known as companding is employed to improve voice quality at lower signal levels.

2.6 Companding

Companding, or compressing and expanding, employs a technique that involves compressing an analog signal at the source and then expanding this signal back to its original size at the destination. At the time of the companding process, input analog signal samples are compressed into logarithmic segments. Each segment is then quantized and coded using the uniform quantization technique. The compression process is logarithmic as opposed to the linear one in the case of PCM. People communicating over a telephone tend to generate low-amplitude signals (unless two individuals are communication in an extremely loud pitch). This implies that most of the energy in a human voice is concentrated in the lower end of the voice's dynamic range. Distributing quantization levels to a logarithmic, instead of linear, function gives finer resolution, or smaller quantization steps, at lower signal amplitudes. This matches the pattern of the input voice signals thereby facilitating higher-fidelity reproduction. Companding is further described in the following:

- ITU-T has recognized two standards for companding, A-law and μ-law.
- A-law and μ-law are audio compression schemes defined by the ITU-T G.711 standard, which outputs eight bits of logarithmic data from the 16-bit linear PCM data.
- A-law and μ-law use a companding technique that provides more quantizing steps at lower amplitude (volume) than at higher amplitude.
- Europe uses A-law, while North America and Japan use μ-law. An international connection employs A-law.
- The responsibility of μ-law to A-law conversion rests with the country employing μ-law.
- These laws employ linear approximations of the logarithmic input/output relationship with the dynamic signal range broken down to a total of 16 segments of which eight are positive and the remaining eight negative.
- Each segment is double the length of the preceding one and employs uniform quantization.
- The eight bit codes are structured similarly for both the laws with the most significant bit (MSB) identifying the polarity; bits 2, 3, and 4 identifying the segment; and bits 5, 6, 7, and 8 the quantization value for the segment.
- μ-law companding accounts for a better signal distortion while A-law provides a greater dynamic range.

2.7 Differential PCM

There exists a significant correlation between the successive samples of a PCM system. This makes it possible to quantize and transmit the values of the input signal at the beginning and subsequently transmit only the differences of the signals from the initial values. Differential PCM (DPCM) involves the process of A/D conversion in which an analog signal is sampled, and then the difference between the actual sample value and its predicted value (based on previous sample or samples) is quantized and subsequently encoded into a binary value. DPCM code words represent differences between samples unlike PCM where the code represents the sample value. It is obvious that the difference between the input samples would always be less than the entire input sample. Hence the number of bits required for transmission is effectively reduced. Thus the system throughput is reduced from 64 Kbps to 48 Kbps [8].

Example 2.8

$$\begin{aligned}\text{Channel bandwidth} &= \text{number of samples/sec} * \text{number of bits/sample}\\ &= 8{,}000 \text{ samples/sec} * 6 \text{ bits/sample}\\ &= 48{,}000 \text{ bps or } 48 \text{ kbps}\end{aligned}$$

In a DPCM system a constant sampling frequency is used to sample the input signal and modulate it using the PAM process, as in case of PCM system. The sampled input signal is then stored in a unit known as predictor, which passes a stored signal through another unit known as a differentiator. The previously sampled signal is compared with the currently sampled signal and the resulting difference is sent to the quantizing and coding phase of PCM (A-law or μ-law) and finally to the receiving end where the entire process is reversed.

2.8 Adaptive DPCM

Adaptive DPCM (ADPCM) is a specialized version of DPCM defined by the ITU-T G.726 specification. It uses predictive techniques to increase information coding efficiency. This technique is employed in many modern audio and video compression algorithms like video conferencing and VoIP implementations. The working of an ADPCM system is similar to that of a DPCM system. However ADPCM employs a predefined algorithm on both the transmitter as well as receiver that predicts the likely value of the next information symbol (referred to as delta). The system transmits only the difference between the two samples instead of transmitting the complete sampled values. In DPCM and ADPCM systems the quantization level is adapted to the size of the difference input signal. This process results in a signal-to-noise ratio (SNR) that is uniform throughout the dynamic range of the difference signal. The bit rate of voice transmission in an ADPCM system is reduced to 32 Kbps while retaining the same quality of speech as provided by A-law or μ-law PCM.

Example 2.9

$$\begin{aligned}\text{Channel bandwidth} &= \text{number of samples/sec} * \text{number of bits/sample}\\ &= 8{,}000 \text{ samples/sec} * 4 \text{ bits/sample}\\ &= 32{,}000 \text{ bps or } 32 \text{ kbps}\end{aligned}$$

The current networks are triple play–enabled, which means that they can carry conventional voice traffic, data, and video (multimedia). In such cases it would be extremely helpful if the network devices (switch) could reduce the channel bandwidth of multimedia transmissions in conditions of high network traffic. For example the switch could drop the LSB (or bits) of multimedia transmissions in case of high network traffic. However, conventional systems cannot support such schemes. To meet this objective, compression techniques known as embedded codes were developed. These codes facilitate the working of DPCM systems at 8/7/6/5/4/3/2 bits/sample or from 64 Kbps to 16 Kbps.

2.9 Summary

- Multiplexing is a technique of transmitting more than one information signal simultaneously over a communication path (medium, circuit, or channel) that can subsequently be recovered individually.
- WDM enables the utilization of a significant portion of the available fiber bandwidth by allowing many independent signals to be transmitted simultaneously on one fiber, with each signal located at a different wavelength.
- PCM is a signal-coding technique defined by the ITU-T standard G.711. It is one of the most important basic building blocks of a telecommunication network and performs one of the most important functions of converting an analog signal to the digital form.
- ADPCM is a specialized version of DPCM defined by the ITU-T G.726 specification. It uses predictive techniques to increase information-coding efficiency. This technique is employed in many modern audio and video compression algorithms like video conferencing and VoIP implementations.

2.10 Review

2.10.1 Review Questions

1. The _____ modulation technique is used with digital transmission.
    a. PAM
    b. PCM
    c. PMC
    d. PMA

2. The sampling frequency employed by PCM system is _____.
   a. 64 Khz
   b. 32 Khz
   c. 4 Khz
   d. 8 Khz
3. The number of quantization levels employed in a PCM system is _____.
   a. 128
   b. 224
   c. 256
   d. 255
4. In a PCM system the duration of each time slot is _____.
   a. 125 μs
   b. 120 μs
   c. 100 μs
   d. 64 μs
5. The channel bandwidth of a PCM system is _____.
   a. 48 Kbps
   b. 64 Kbps
   c. 36 Kbps
   d. 128 Kbps
6. What is the channel bandwidth of a system employing DPCM?
   a. 48 Kbps
   b. 64 Kbps
   c. 36 Kbps
   d. 128 Kbps
7. The transmission rate of a system employing ADPCM is lower than that of a system employing DPCM.
   a. True
   b. False
8. _____ is a specialized version of DPCM defined by the ITU-T G.726 specification.
   a. ADPCM
   b. DAPCM

c. MDPCM

   d. SDPCM

9. Multiplexing is a technique of transmitting more than one information signal simultaneously over a communication path.

   a. True

   b. False

10. Modern-day networks are referred to as _____ networks, which means that they can carry conventional voice traffic, data, and video (multimedia).

    a. Dual-play

    b. Multiplay

    c. Triple-play

    d. Triple-convergent

2.10.2 Exercises

1. List the key stages involved in PCM and their functions.

2. Compare ADPCM and DPCM.

3. What does the term multiplexing mean? List and briefly explain the different multiplexing techniques employed in the telecommunication arena.

4. Calculate the channel bandwidth if the sampling rate is 7 KHz and the number of bits per sample is nine.

5. With reference to Example 2.3, find out the binary equivalent if the quantized value is 85.

2.10.3 Research Activities

1. What is meant by the term concentration ratio? Explain briefly.

2. Describe the physical interface for a 64-K channel by referring to the relevant ITU-T standards.

3. Highlight the importance of SDM and its impact on the networks of the future?

4. Briefly describe few-mode fibers. Do existing fiber plants need to be replaced to accommodate SDM? List the key points to be considered for deploying SDM in existing photonic networks.

2.11 Referred Standards

ITU-T G.671: Transmission characteristics of optical components and subsystems

ITU-T G.694.1: Spectral grids for WDM applications: DWDM frequency grid

ITU-T G.694.2: Spectral grids for WDM applications: CWDM wavelength grid

ITU-T G.702: Digital hierarchy bit rates

ITU-T G.711: Pulse code modulation (PCM) of voice frequencies

ITU-T G.726: 40, 32, 24, 16 kbit/s Adaptive Differential Pulse Code Modulation (ADPCM)

TBR-21: Terminal Equipment (TE); Attachment requirements for pan-European approval for connection to the analogue Public Switched Telephone Networks (PSTNs) of TE (excluding TE supporting the voice telephony service) in which network addressing, if provided, is by means of Dual Tone Multi Frequency (DTMF) Signalling

2.12 Recommended Reading

2.12.1 Books

Harte, L., *Telecom Basics*, Third Edition, Althos Publishing, 2004.

Freeman, L. R., *Fundamentals of Telecommunication*, Second Edition, John Wiley & Sons Inc., 2005.

Fischer, R. F., *Optical System Design*, Second Edition, SPIE Press, 2008.

2.12.2 URLs

http://www.btplc.com/Thegroup/BTsHistory/1912to1968/1958.htm.

http://www.btplc.com/thegroup/BTsHistory/Eventsintelecommunicationshistory/Eventsintelecommunicationshistory.htm.

http://www.stanford.edu/class/ee368b/Handouts/15-DPCM.pdf.

http://www.cs.wustl.edu/~jain/bnr/ftp/p_2tel.pdf.

References

[1] ICT Technologies [n.d.], *Modulation & Multiplexing*, retrieved March 17, 2008. Source URL http://cbdd.wsu.edu/kewlcontent/cdoutput/TR502/page19.htm.

[2] ITU-T Recommendation G.671—Transmission Characteristics of Optical Components and Subsystems.

[3] ITU-T Recommendation G.694.2—Spectral Grids for WDM Applications: CWDM Wavelength Grid.

[4] ITU-T Recommendation G.694.1—Spectral Grids for WDM Applications: DWDM Frequency Grid.

[5] Schneider, K. S., *Primer on Fiber Optic Data Communications for the Premises Environment* [n.d.]. Retrieved June 10, 2008. Source URL http://www.telebyteusa.com/foprimer/fofull.htm.

[6] *Waveform Coding Techniques,* did: 8123, retrieved March 18, 2008, from source URL http://www.cisco.com/en/US/tech/tk1077/technologiestechnote09186a00801149b3.shtml.

[7] University of Illinois (2006), retrieved March 18, 2008. Source URL http://www2.uic.edu/stud_orgs/prof/pesc/part_3_rev_F.pdf.

[8] *DPCM Overview* [n.d.], Stanford University, retrieved March 21, 2008. Source URL <http://www.stanford.edu/class/ee368b/Handouts/15-DPCM.pdf>

3

Fiber-Optic Communication Fundamentals

3.1 Chapter Objectives

The media is an important component of the telecommunication network that directly impacts throughput, transmission errors, and coverage distance. The choice of media significantly impacts the capital expenditure (CAPEX), operational expenditure (OPEX), and quality of service (QoS) of the network. An exponential increase in the number of telecom subscribers coupled with an enhanced range of user services and applications has led to explosive bandwidth demands. An optical fiber is capable of supporting very high transmission bandwidths, currently limited only by the type of technology employed for data transfer and is thus the only medium of choice (currently as well as in the near foreseeable future) for backbone and access networks. The emergence of wireless 5G technologies might reduce the dependence on fiber in the access networks; however, it will remain the media of choice for backbone networks.

The cost of laying an optical fiber link is still very high (with geographical variation) and forms a major component of the overall network cost. This chapter lays the strong conceptual foundation required to understand and appreciate the intricacies of network design, equipment installation, commissioning, and operation and maintenance techniques, presented in the subsequent chapters. The chapter provides a clear understanding of the principles governing the transmission of light through an optical fiber, along with the requisite description of the key design and operational parameters of optical fiber plants. The chapter also outlines the industry-accepted best practices for laying optical fiber cables including T&D, use of horizontal directional drilling machines, and fiber-blowing techniques. Guidelines for handling optical fibers and tips for maintaining optical plants, patch cords, and allied fiber termination points are also presented.

Key Topics

- The fundamentals of light transmission over an optical fiber;
- The different classes of optical fiber;
- Commercially available fiber plants and their key design parameters;
- Different types of optical fibers and their application context;
- Different techniques for laying fiber plants;
- Best practices in optical cable preparation, splicing techniques, and methods for terminating fiber end points;
- Basic network diagnostic techniques.

3.2 Introduction

An optical fiber consists of a strand of glass or plastic that allows the passage of light in a manner similar to the conduction of electricity through copper cables. Light travels through the fiber due to reflection from its inner surfaces. The possibility of using a ray of light to transmit data originated from visual observations and experiments originating in the late nineteenth century. The year 1880 witnessed an initial patent on a method to transmit light using glass pipes. This led to subsequent developments in this field followed by the first demonstration of a system based on optical transmission. The early twentieth century witnessed significant progress in the growth of fiber-optic technology, with the development of a device referred to as a fiberscope acting as a catalyst. The fiberscope (from which the name fiber-optics was derived) employed a glass fiber and functioned as an image-transmitting device. The major limitation of this method was the excessive optical losses of the fiber, which severely limited the transmission distance. Transmission losses were reduced through the use of two layers, the inner core and the outer cladding, with different RIs. The core facilitated the transmission of light while the cladding prevented the light from leaking out of the core by reflecting the light within the boundaries of the core.

The initial challenge was to offset the drastic losses during transmission of light through the optical fiber. These losses were primarily due to the purity of glass used in the fiber construction and the effects of dispersion. The losses, which were very high, were on the magnitude of 1,000 dB/km. Technological developments, primarily the ability to manufacture high-purity glass fibers, reduced the attenuation losses to less than 20 dB/km. Corning has been a world leader in developing cutting-edge technologies relating to the manufacture of glass and is responsible for pioneering efforts in the development of optical fiber with some early patents on the fabrication process. These technology advances brought about a continuous decrease in the attenuation characteristics of the glass fiber and, in the process, facilitated guided light communication. The development of the plastic optical fiber as well as PCS cable can be directly attributed to these developments.

Figure 3.1 presents a block diagram of a basic fiber-optic transmission/receiving system. The electrical signal, either analog or digital, representing the user signal is fed to the input of a light source (LED laser), which converts the electrical energy into a ray of light. This light is focused onto the core of the optical fiber, which carries the signal to the receiver consisting of a diode (an APD) that converts the incoming light rays to electrical pulses again.

3.3 Basics of Light Transmission

The effect on a ray of light that passes through different media is described by Snell's law. As per the law [1]:

$$\sin\Theta_1/\sin\Theta_2 = v_1/v_2 = \lambda_1/\lambda_2 = n_2/n_1$$

where

Θ is the angle measured from the normal of the boundaries of the two medium;

V is the velocity of light in mps within the respective medium;

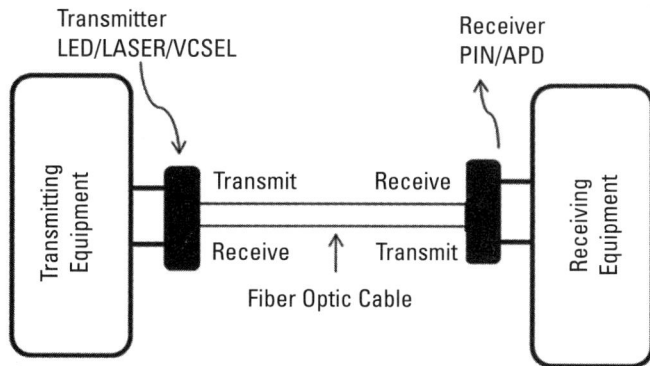

Figure 3.1 Basic block diagram of an optical system.

n refers to the refractive indices of the respective medium;

$\sin\lambda_1$ and $\sin\lambda_2$ refer to the angles of incidence and refraction, respectively.

The RI of a material describes the manner in which light propagates through a medium other than free space. Since the RI is a ratio of two variables there are no dimensions attached to it. RI can be defined by the following formula:

$$(\text{RI}) = n = c/v$$

where

n = RI of a medium;

c = speed of light in vacuum;

v represents the phase velocity of light in a specific medium [2].

This section explores the fundamental principles governing the transmission of light through a medium. The understanding of the possible effects of a ray of light incident on a media is a precursor to understanding the transmission of light. A ray of light incident on a media can undergo any of the following effects, as illustrated in Figure 3.2:

1. *Reflection:* Change in the direction of the ray of light at the junction or boundary of two different mediums;

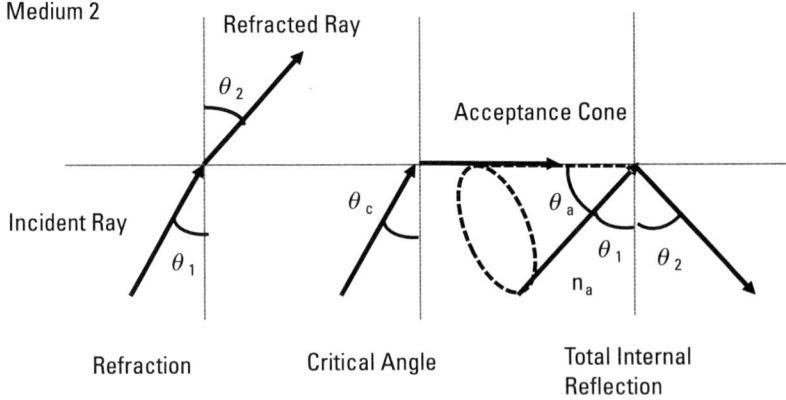

Figure 3.2 Propagation of light through different mediums.

2. *Refraction:* The bending of a ray of light as it passes from one transparent medium (medium of varying RIS) to another;
3. *Absorption:* The complete attenuation of the incoming ray of light (converted to heat);
4. Scattering: The diffusion of the incoming ray of light into multiple directions.

This behavior of a ray of light incident on a media is dependent on several factors including the following:

1. *RI of the media:* A measure for how much the speed of light is reduced as it passes through the medium. For example if a medium has an RI of 2 then a ray of light passing through it would travel at 1/2 = 0.5 times the speed in air or vacuum.
2. *Angle of incidence, $\emptyset_i$:* Angle at which the ray of light strikes the media with reference to the normal (line perpendicular to the surface of the media).

Transmission of light over a distance necessitates the use of a waveguide or a channel over which the rays of light can propagate. A FOC, also referred to as optical fiber, is media for light transmission. Optical fibers are made of glass, plastic (generally used for MMFs), or both and contain an inner conductor, referred to as core, surrounded by an outer conductor referred to as cladding. The RI of the core is greater than that of the cladding.

The transmission of a ray of light through an optical fiber or FOC is due to the principle of TIR. This is based on Snell's law, which determines the effect of light incident on a medium. The law states that the angle at which light is reflected is dependent on the RI of the two mediums under consideration. In the case of an optical fiber these are the core and the cladding. The lower RI of the cladding (with respect to the core) causes the light to be angled back into the core.

The propagation of light through an optical fiber is graphically illustrated in Figure 3.3. As highlighted in the figure, the incident ray of light cannot travel through the cladding and is confined to the core through which it propagates due to the successive internal reflections.

Another important parameter that aids in the transmission of light through an FOC is the critical angle $\emptyset_c$. The following list highlights the relationship between the angle of incidence and the critical angle:

1. If the angle of incidence is less than the critical angle, the ray of light incident on the core of a fiber will get refracted through the cladding.

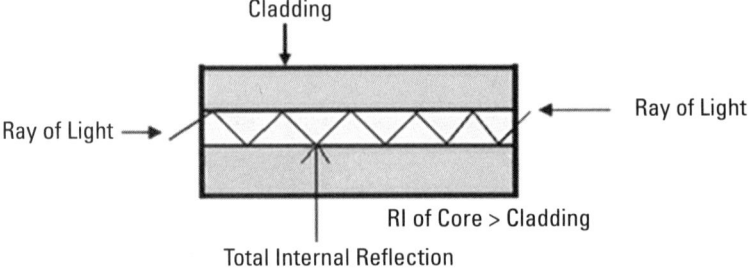

Figure 3.3 Transmission of light through an optical fiber.

2. If the angle of incidence is equal to the critical angle, the ray of light incident on the core of a fiber will travel along the boundary of the core and cladding. The ray will be weakly guided and is likely to be refracted from the fiber at some point in time.
3. If the angle of incidence is greater than the critical angle then the ray of light incident on the core of a fiber will undergo total internal reflection and propagate through the fiber.

The principle of TIR can also have other applications, including the following:

1. Medicinal (endoscopes);
2. Fiber diagnostic techniques (fiberscope);
3. Sensors;
4. Optical instruments;
5. Fingerprinting devices.

3.3.1 Modal Propagation

The ray of light traveling through the core of an optical fiber exhibits variations in its intensity as it travels down the fiber. These variations are referred to as modes. Modes can be thought of as rays of light. The number of modes (constituents) depends upon the dimensions of the core and the variation in the RIs of the core and cladding. The modes are numbered in the ascending order. As the name implies, a single-mode fiber (SMF), because of its small core diameter, allows only a single ray of light to travel along its length. In contrast, a MMF has a large core diameter, which permits multiple modes to travel down its length. Optical fibers can be classified based on their modes of propagation and RI profiles as described in the following [3].

1. *Single-mode step index:* Propagation of a single lower order mode of light through a narrow cylinder over the axis of the optical fiber.
2. *Multimode step index:* The RI of the core is greater than the cladding. The larger core diameter of the core results in multiple-mode transmission (which can have more than thousand modes based on the fiber diameter) with the higher modes leaking onto the cladding as well as leading to the conversion of optical energy to heat due to absorption. The higher levels of attenuation limit the transmission distance to a few meters to a kilometer.
3. *Multimode graded index:* The RI decreases gradually from the center of the core toward the cladding. This results in reduction in dispersion through differential mode delay[1].

The core diameter of a SMF is typically in the range of 8–12 μm, while that of a MMF-graded index (MMF-GI) is in the range of 50–100 μm and that of a MMF-step index (MMF-SI) in the range of 50–200 μm. The cladding size is usually standardized at 125 μm.

3.4 Optical Fiber Classification

As discussed in previous sections, FOCs are used extensively in building modern-day convergent telecommunication networks. Thus these fibers are deployed to link geographically distant sites (intercity networks) as well as sites within a campus or building. Based on the deployment we can classify the cables as follows [4]:

1. *Outside plant (OSP):* As the name suggests these cables are deployed extensively by telecom/internet service providers for intracity as well as intercity connections. These may be blown into ducts that are laid in trenches. (The depth may be regulated by state, central, or federal authorities or municipal corporations in order to accommodate multiple service providers as well as to manage the need to keep rodents at bay.) Depending upon the terrain these cables may also be strung up over poles or laid underwater (submarine cables) for international connectivity. Outside plant installations are all SMF (detailed in the subsequent section) and cables often have very high fiber counts. Cable designs are optimized for resisting moisture and rodent damage.

---

1. Differences in group velocity among propagating modes due to variations (imperfections) in the RI profile of the fiber.

2. *Inside plant (ISP):* As opposed to OSP cables, ISP cables involve short lengths and are installed typically within a building or campus; they may be a few hundred feet with two to 48 fibers per cable. The fiber may be single-mode, multimode, or hybrid cable with both MMFs and SMFs and may be comprised of glass or plastic.

3.5 Standard Optical Fiber Designs

The two basic cable designs commonly available are the following [5]:

1. *Loose-tube cable:* Loose-tube cable, as illustrated in Figure 3.4, has a modular design that typically holds up to 12 fibers per buffer tube with a maximum per cable fiber count of more than 200 fibers. These optical fibers are housed and protected in color-coded plastic buffer tubes. A gel filling compound effectively blocks out water infiltration. Routine bending and installation stresses are prevented by employing excess fiber length (relative to buffer tube length). A central steel member or a dielectric provides antibuckling support. The primary tensile strength member for the cable core is generally aramid yarn with an outer polyethylene jacket extruded over the core. In the case of armored cables, a corrugated steel tape is wound over a jacketed cable and an additional jacket extruded over the armor. The modular buffered tube facilitates easy drop-off of fibers without any interfer-

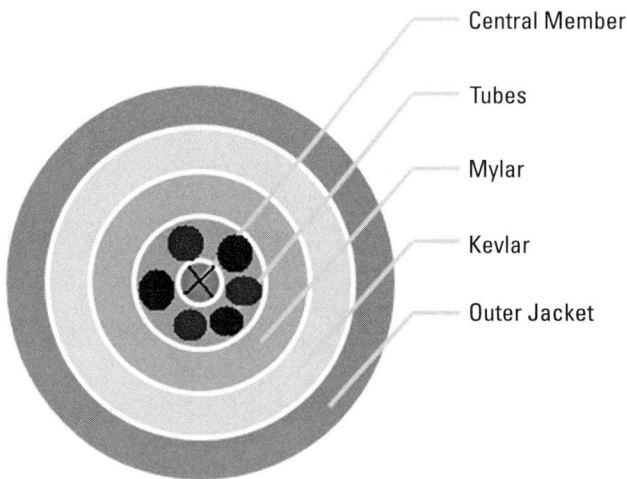

Figure 3.4 Loose-tube design.

ence to the protected buffer tubes being routed to other locations. The loose-tube design also helps in the identification and administration of fibers in the system and hence is used for OSP installation applications both overhead and underground.

2. *Tightly buffered cable:* Tightly buffered cable, illustrated in Figure 3.5, offers a rugged design with the buffering material in direct contact with the fiber. It provides a cable structure that protects individual fibers during handling and routing and during dropping connections. In order to prevent installation-related tensile stress on the cable, yam strength members are deployed. This unique design makes these cable types highly suitable for jumpering or connecting OSP cables to terminal equipment and for linking various communication devices within premises.

In addition to the cable types listed earlier, multifiber cables are available. These cables are constructed with strength members that can better resist crushing and bending forces generally associated with cable installation. The outer cable jackets are optical fiber nonconductive riser (OFNR) (riser-rated), optical fiber nonconductive plenum (OFNP) (plenum-rated), or low-smoke, zero-halogen-rated (LSZH). These specifications are explained in the next section. The primary advantage of multifiber tightly buffered cables is their routing and handling flexibility. In addition to the aforementioned cable types, ribbon cables are available. These are generally used extensively for internet service provider (ISP) and data center applications. Individual cables have a flat ribbon-like structure facilitating increased cable concentration and savings in space.

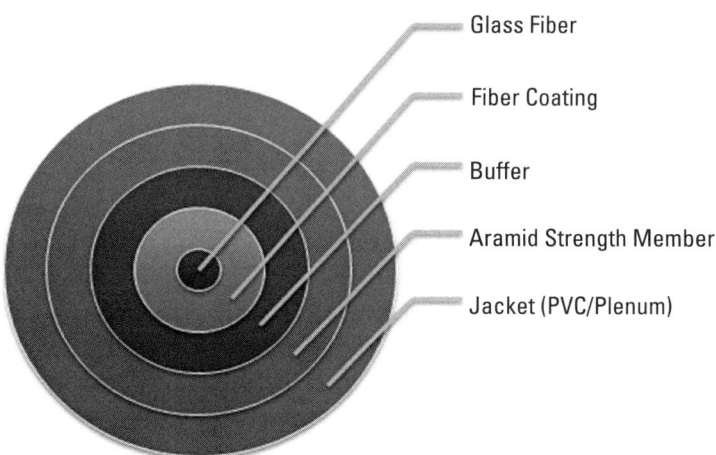

Figure 3.5 Tightly buffered cable.

3.6 Types of Optical Fibers

There are two types of FOCs that are employed in communication networks. They are single-mode and MMFs. The primary difference between the cables is with respect to bandwidth-carrying capacities due to the differences in their attenuation characteristics, material composition, and manufacturing techniques. The attenuation is principally induced due to the double effect of absorption and scattering. The attenuation of light within the fiber is dependent on the transmitted wavelength and can be minimized by choosing an optimal window within the optical spectrum. The wavelengths that are used by communication networks include 780 nm; 850 nm; 1,310 nm; 1,550 nm; and 1,625 nm. The 780- and 850-nm wavelengths correspond to the high-loss region but were used earlier due to the low equipment cost. The 1,310-nm (zero-dispersion) and the 1,550-nm wavelengths fall in the low-loss region and are used in telecom networks. Table 3.1 highlights the profiles of some of the commonly employed FOCs [4], described as follows.

1. *SMFs*: Single-mode consists of a strand of fiber made of glass with a core diameter of around 8.3 micrometers with a single transmission mode. The relatively narrow diameter of the core facilitates the propagation of only a single mode of light typically at 1,310 or 1,550 nm. These fibers can carry significantly higher bandwidths than a MMF over very large distances. However the small core size necessitates the use of a small (concentrated or intense) light source with constricted spectral width. SMFs are extensively used for most telecom/data applications[2], including WDM systems. The 1,310-nm wavelength is preferred because of the zero-dispersion characteristics; however, the losses are the least in the 1,550-nm region. SMFs can have modified RI profiles to suit different applications. The RI can include fibers with abrupt changes in the RI or a gradually reducing RI (from the outer surface of the core toward its central axis). There are three basic classes of SMF used in modern telecommunications systems [4]; they are described as follows.

    a. Standard SMF has been in existence since the beginning of optical communication and is still used extensively. These fibers are intended for use with the 1,310- and 1,550-nm wavelengths, which corresponds to the low-loss band.

    b. *Dispersion-shifted fiber* (DSF) was introduced to offset the transmission losses over nondipersion-shifted fibers (NDSF) at 1,550

---

2. Refer to note 1.

Table 3.1
MMF/SMF RI Profiles

Mode	Material	Refraction Profile	Size (micrometers)	Wavelength (nm)	Attenuation (dB/km)
Multimode	Glass	Step	62.5/125	800	5.0
Multimode	Glass	Step	62.5/125	850	4.0
Multimode	Glass	Graded	62.5/125	850	3.3
Multimode	Glass	Graded	50/125	850	2.7
Multimode	Glass	Graded	62.5/125	1,300	0.9
Multimode	Glass	Graded	50/125	1,300	0.7
Multimode	Glass	Graded	85/125	850	2.8
Multimode	Glass	Graded	85/125	1,300	0.7
Multimode	Glass	Graded	85/125	1,550	0.4
Multimode	Glass	Graded	100/140	850	3.5
Multimode	Glass	Graded	100/140	1,300	1.5
Multimode	Glass	Graded	100/140	1,550	0.9
Multimode	Plastic	Step	485/500	650	240
Multimode	Plastic	Step	735/750	650	230
Multimode	Plastic	Step	980/1000	650	220
Multimode	Plastic	Step	200/350	790	10
Single-Mode	Glass	Step	3.7/80	650	10
Single-Mode	Glass	Step	5/80 or 5/125	850	2.3
Single-Mode	Glass	Step	9.3/125	1,300	0.5
Single-Mode	Glass	Step	8.1/125	1,550	0.2

From: [7].

nm (as explained earlier). In a DSF, the zero-dispersion point is moved to the 1,550-nm region by the use of suitable dopants. These fibers however exhibit severe nonlinearities when multiple wavelengths in the 1,550-nm band are transmitted as in the case of WDM systems.

c. *Non-zero-dispersion-shifted fibers* (NZ-DSFs) were developed to address the issue of nonlinearities associated with DSF. NZ-DSF is available with positive as well as negative dispersion characteristics and is deployed extensively in modern-day telecommunication networks.

2. MMF allows multiple modes of light to propagate through it. The diameter of a MMF is greater than that of a SMF and is normally of the order of 50, 62.5, or 100 micrometers. The amount of light that can be coupled on the MMF (light-gathering capacity) is higher than the SMF, due to the higher core diameter. However it has a lower bandwidth-distance limit, as compared to SMF, due to propagation of multiple modes (modal propagation). In most short-range applications a two-fiber (2F) MMF is deployed [6]. As per ISO/IEC 11801 standards, the following classes of MMF have been defined:

   a. OM1: MMF type 62.5-μm core; minimum modal bandwidth of 200 MHz·km at 850 nm;

   b. OM2: MMF type 50-μm core; minimum modal bandwidth of 500 MHz·km at 850 nm;

   c. OM3: MMF type 50-μm core; minimum modal bandwidth of 2,000 MHz·km at 850 nm;

   d. OM4: MMF type 50-μm core; minimum modal bandwidth of 4,700 MHz·km at 850 nm.

   MMF cables can be classified into the following types:

   a. *MMF-SI:* The step-index fiber contains a core with a uniform RI and a sharp decrease in RI at the core-cladding interface. Thus the cladding has a lower RI. This type of fiber is named step-index due to the sharp change in the RI.

   b. *MMF-GI:* In MMF-GI, the RI of the core gradually reduces outward from the central axis toward the cladding. This causes the rays of light to take helical or sinusoidal paths within the core. In a MMF there can be multiple modes of propagation. Accordingly certain modes would take the helical path, which is longer but faster due the varying RI, while other modes would travel along the shorter path along the central axis at a slower rate. This characteristic ensures that all the modes arrive almost simultaneously at the receiving end. GI MMFs are more expensive than typical SMFs. SMF has lower cost and attenuation and different dispersion characteristics to MMF since MMF tends to be dominated by modal dispersion. MMF-SI is available in plastic while MMF-GI is available in glass only. The usage of MMF glass is more common.

   c. *Few-mode fibers:* MMFs have a higher core diameter that permits the propagation of several independent modes within the same fiber. The numbers of modes are dependent on the diameter of the core and the RI profile of the fiber. The multiple transmission modes in an MMF can be suppressed by modifying the fiber characteristics, allowing only a few modes (typically 10 modes or fewer) to propagate. These fibers are referred to as few-mode fibers and represent a fundamentally new approach to increasing the information density over fiber.

Table 3.1 outlines the RI profiles of the commonly used MMF and SMF fibers.

Table 3.2 highlights the application context of the different types of fibers.

3.7 Fiber Design Specifications

An optical fiber consists of two layers of different types (varying RIs) of highly pure glass or plastic arranged to form the core and cladding. The core refers to the inner layer while the cladding forms the outer layer. The cladding is usually covered by multiple layers of protective coating to help it withstand different types of stresses during deployment. The protective coating is generally comprised of a soft inner layer, designed to cushion the fiber, followed by a hard outer layer designed to withstand stresses during installation or termination. Figure 3.6 shows a sectional view of an optical fiber.

A list of important design parameters, classified under operational and geometrical parameters, is illustrated in Figure 3.7. This section introduces the parameters that influence the choice of a specific type of fiber for building optical fibers. These parameters, including attenuation characteristics, modes of light propagation, supported transmission windows, and strength [4], are described as follows.

1. *Transmission losses:* There are different types of losses encountered during the propagation of light through a medium. The primary causes of attenuation of light traveling through an optical fiber are listed as

Table 3.2
Operational Wavelength for Common Fibers

Fiber Type	Wavelength (nm)	Attenuation (dB/km)
Plastic Optical Fiber (POF)	650 850	1 dB/m
MMF-GI	850 1,300	1 3
SMF	1,310 1,490–1,625	0.2: 0.4

From [8].

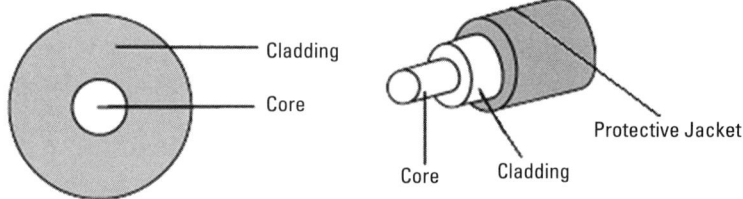

Figure 3.6 Sectional view of an optical fiber.

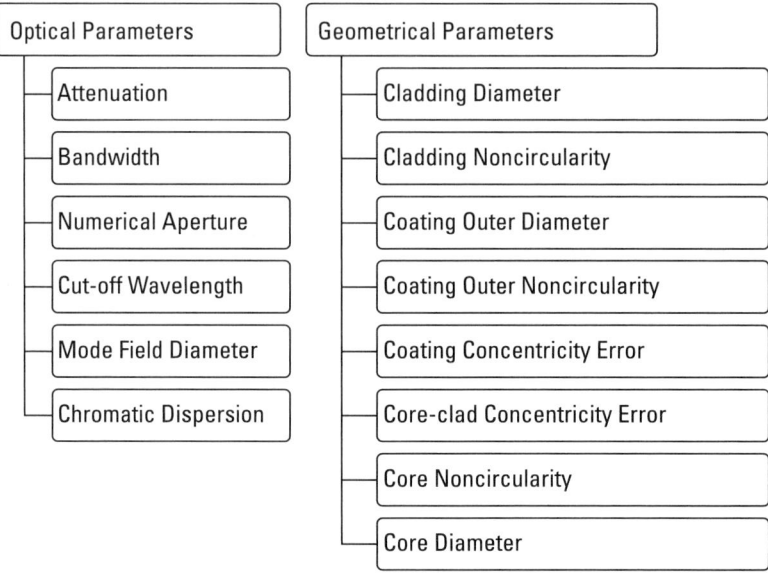

Figure 3.7 Optical fiber: key design parameters.

follows, while Figure 3.8 illustrates the attenuation characteristics of a standard (single-mode) optical fiber:

a. *Absorption:* Light consists of a stream of photons. Absorption refers to the loss of signal energy due to the absorption of the propagating photons through a media and its subsequent conversion to heat.

b. *Scattering:* Scattering results in the redirection of light through the core onto the cladding.

c. *Dispersion:* Dispersion, one of the major contributors to the losses in light transmission over an optical fiber, refers to the elongation of light waves as they travel along the fiber. Dispersion (refer to Figure 3.9) causes the propagating signal through the fiber to get

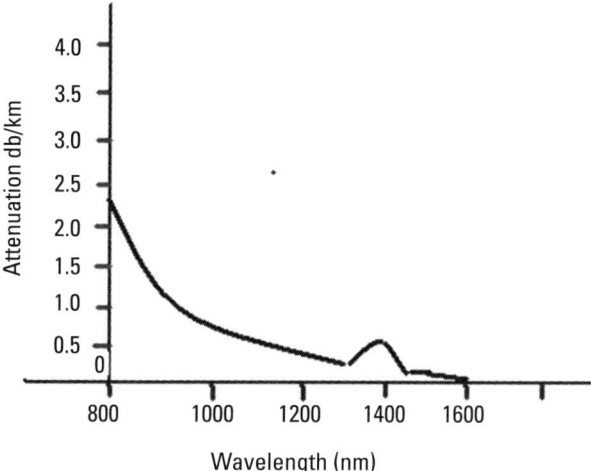

Figure 3.8 Attenuation characteristics of an optical fiber.

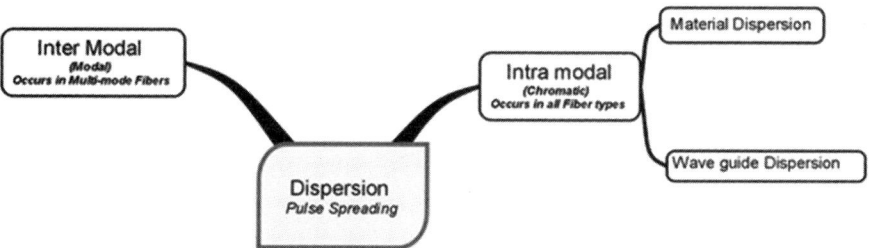

Figure 3.9 Dispersion types.

severely degraded over a distance. There are generally two sources of dispersion: intramodal dispersion and intermodal dispersion. Intramodal dispersion can occur in all types of fiber and can be further classified into either material dispersion or waveguide dispersion. Material dispersion refers to the frequency-dependent response of a media to the propagating light. Waveguide dispersion is due to the relationship between the physical dimensions of the waveguide (core diameter) and the propagating ray of light.

Material dispersion refers to the pulse spreading caused by the specific composition of the glass, while waveguide dispersion results from the light traveling in both the core and the inner cladding glasses at the same time, but at slightly different speeds. The two types can be balanced to produce a wavelength of zero dispersion anywhere within the 1,310–1,650-nm operating window.

2. *Transmission wavelengths or windows:* In the early days of fiber-optic communication the LED was employed as a light source. LEDs mostly operated at the 780-nm or the 850-nm wavelength. This region is referred to as the first transmission window. LEDs could not be employed for high-bandwidth transmissions over a long distance due to their inherent disadvantages and were replaced by LASERs. LASERs operated in two wavelength regions, namely the 1,310-nm and 1,550-nm regions, which are commonly referred to as the second and the third optical transmission windows. The effects of dispersion are zero at the 1,310-nm window, whereas the losses are the least at 1,550-nm window. The modern fiber-optic networks operate around 1,550 nm. This wavelength band is also particularly important to the WDM networks that are increasingly being deployed in networks worldwide. These networks use amplifiers to counter the effects of attenuation. Commonly deployed amplifiers are the erbium-doped fiber amplifiers (EDFAs), which provide signal amplification across a range of wavelengths around 1,550 nm and 1,625 nm. This window is commonly referred to as the EDFA window. The optical transmission windows are depicted in Figure 3.10.
3. *Propagation modes:* Light is propagated through the core of an optical fiber by the principle of total internal reflection. The optical fiber thus acts as a waveguide. Light traveling through a FOC exhibits certain modes or variations in the intensity of the light as it travels over the fiber. The number of modes that can exist is dependent on the cable

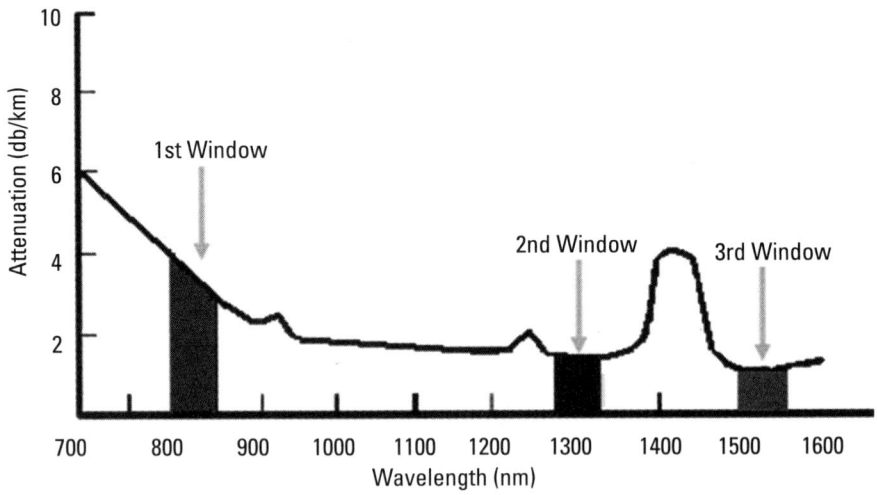

Figure 3.10 Optical transmission windows.

dimensions as well as the RIs of the core and cladding. Fibers that support only a single mode are called SMFs, while those providing multiple propagation paths are referred to as MMFs.

The commercial fibers available in the early days of optical communication had a large core diameter facilitating multiple modes of light to simultaneously propagate through the fiber. The larger core diameter of these MMFs also facilitated the use of lower-cost optical transmitters. A SMF has a small core diameter that allows only one mode of light to propagate through it. Hence the optical signal can travel over longer distances in the absence of modal dispersion. It is interesting to note that contrary to common belief, standard MMFs do not provide a higher transmission bandwidth than SMFs. Thus, SMFs are the preferred choice for deployment in telecommunication networks as well as environments requiring long-distance, high-bandwidth applications.

4. *Dimensions:* The dimensioning of an optical fiber is very crucial and governs the various losses that affect the propagation of light waves through it. A large core diameter ensures higher coupling of light energy from the source onto the fiber. This can however cause saturation problems at the receiver. Typically these dimensions are expressed in the form:

$$x/y$$

where x indicates the diameter of the core, and y indicates the diameter of the cladding.

Example 3.1

The value 40/120 indicates a core diameter of 8 microns and a cladding diameter of 120 microns.

The common dimension of a SMF is 8/125, while that of a MMF is 62.5/125. These dimensions are standardized internationally so as to maintain compatibility between connectors and allied accessories. The standardization of the communication media and its associated interfaces are an important prerequisite for universal communication. Accordingly, the standard for the outer cladding diameter of single-mode glass optical fibers is 125 microns (μm) and 245 μm for the coating. SMFs have a miniature core size, approximately 8 to 10 μm in diameter while MMFs have core sizes of 50 to 62.5 μm in diameter as illustrated in Figure 3.11.

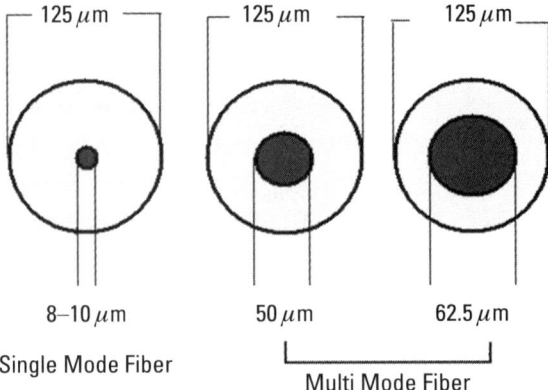

Figure 3.11 SMFs and MMFs: cross-sectional view.

5. *Strength:* The glass used for normal household use is highly brittle. However, the optical fiber is extremely durable with a high tensile strength in excess of 600+ pounds per square inch. Sometimes, however, flaws or microscopic cracks develop during the fiber manufacturing process. Thus, manufactured fiber is subjected to load tests to ensure that its tensile strength is within permissible limits.

6. *Longevity:* Optical fibers are designed to provide service for a lifetime as long as the recommended installations as well as splicing guidelines are adhered to. There are a number of tests that can be used to gauge the service life of fiber installations. Telcordia provides a comprehensive range of testing services that service providers can use. In practice the useful lifespan of a fiber depends upon its type, deployment, and operating environment. The biggest factor for obsolescence is technological advancements.

7. *Bending diameter:* One of the biggest advantages of an optical fiber is the ease of installation due its lightweight structure, small size, and flexibility. However, extreme care should be taken to ensure that tight bends (macrobending) is avoided. Bends cause high losses of the rays of light traveling through a fiber. Also a bend along an inherent flaw (as discussed earlier) can lead to breakage. Standard tests as well as experience highlights that most fibers can sustain a bend diameter of two or three inches. However it is advisable to use extra caution to avoid bends while employing fiber-handing devices like splice trays, fiber risers, or racks to minimize bending losses.

3.8 Fiber-Laying Techniques

Optical fibers differ mechanically from copper and steel cables, and hence the techniques for installation are significantly different. Fibers are elastic to a certain extent after which they are extremely brittle. Fibers must be protected from compressive forces, tensile forces (axial), and bending. The long-term transmission characteristics of an optical fiber are dependent on the installation procedures followed. Fiber-laying is thus a specialized discipline to be carried out by a trained and experienced field force. Figure 3.12 illustrates the commonly employed fiber-laying techniques [4]. They include:

1. *Trenching and Ducting:* T&D, the traditional method of laying optical fibers, is still prevalent in most developing countries. This process involves creating a trench through soil excavation either manually or through a mechanized process. This is the preferred approach in countries where manual labor is cost effective. The trench specifications are generally regulated by the local/regional authorities and specified for each operator (in multioperator countries). The trenching process requires careful supervision to ensure that trenches do not have significant bends and that the trench floor is uniform and devoid of kinks (stones). Ducts are laid in the excavated trenches and fiber is blown through the ducts using specialized equipment (e.g., either air- or water-assisted fiber blowers, air, and water). Standard blowers employ compressed air to push fiber through the ducts (i.e., air-assisted fiber-flowing). The standard practice is to lay conduits, draw inner ducts through the conduits and blow fibers through the ducts. The size of a common conduit is 100 mm (4 inches) while 150-mm (6-inch) conduits are also available. The inner duct is usually 1.25 inch. It may be

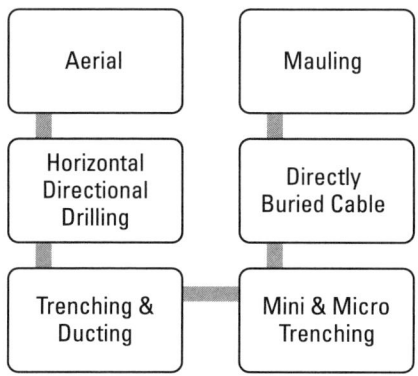

Figure 3.12 Fiber-laying techniques.

noted that the fiber cables should not occupy more than 70% of the duct space. A 1-inch fiber cable occupies 64% of a 1.25-inch duct. The ducts are made from high-density polyethylene (HDPE). HDPE is hard; it can withstand high temperatures (1,100–1,200°C); and it has excellent resistance to concentrated acids, alcohols, and bases.

2. *Minitrenching:* Minitrenching uses mechanized equipment to created precise trenches on a variety of surfaces. The minitrenching technique can be applied on routes that generally involve asphalted surfaces such as roads and sidewalks. This technique is not recommended for sandy soil or soil with gravel or cobbles. The advantage of this technique over conventional cable-laying technologies lies essentially in its speed of execution. The depth and cross-section of the trench is in accordance with the number of ducts to be laid and is generally between 30 and 40 cm, while a cross-section can vary between 7 and 15 cm. There are three methods of minitrenching:

   a. Standard;
   b. Semi-automated;
   c. Fully automated.

   In the fully automated method the process of trenching, ducting, backfill operations, and cleaning are completed simultaneously. In a standard and semi-automated method some of the operations are not done concurrently.

3. *Microtrenching:* Microtrenching creates a shallow groove in the surface (asphalt) as opposed to a conventional deep trench. The groove may be as small as 2-cm-wide and 30-cm-deep as compared to traditional approaches. A special microtube is then placed in the groove, which is filled with a suitable material, such as cold asphalt. The fibers are subsequently blown into the tubes.

4. *Aerial cables:* The installation methods for aerial FOC are very similar to those of copper cables used in legacy telecommunication networks (PSTNs). However it is important to note that the transmission characteristics of fiber are affected by tensile stress, compressive stress, and bends. It is therefore essential to ensure that the cable sag is within tolerable limits. Galvanized iron (GI) wires, also referred to as support or messenger wires, are used to provide support for the fibers strung between GIs or other suitable poles. The FOC is then lashed onto the support wire. The two methods of lashing the FOC are described as follows.

   a. Moving-reel method: The moving-reel method uses reel-carrying vehicles on the entire route of the fiber.

b. *Stationary method:* The cable is pulled into place using cable blocks. All-dielectric self-supporting (ADSS) cables are the best suited cables for aerial installation. As the name suggests ADSS does not require the use of supporting or messenger wires.

5. *Directly buried cables:* This technique involves the use of specialized mechanized equipment to make tiny incisions on any surface, including asphalt and concrete, and laying a fiber cable with protective covering and filling it up with epoxy. The protective covering can withstand the compressive and tensile strengths that a fiber would commonly be susceptible to. This is conceptually similar to micro-trenching but does not involve ducting. The technique is very useful in urban areas where RoW permissions cannot be obtained for standard ducting and trenching. The typical depth for laying these cables is 12 inches.

6. *Horizontal directional drilling (HDD):* HDD, also referred to as directional boring, is a guided (steered) and trenchless method of installing underground pipes, conduits, and cables. The boring is performed in a shallow arc along a prescribed bore path by using a surface rig that performs the drilling. There is minimal impact on the surrounding area, and hence this method is preferred when trenching or excavating is not practical, as in the case of urban areas. It is suitable for a variety of soil conditions including road, landscape, and river crossings. In some cities RoW permissions may only be allotted for HDD-based cable laying. The multistaged process involves creating an entrance pit with a receiving hole. The first stage involves drilling a pilot hole through the designed path while the second stage (reaming) enlarges the hole by passing a larger cutting tool known as the back reamer. The reamer's diameter depends on the size of the pipe to be pulled back through the bore hole. The third stage places a casing pipe in the enlarged hole by way of the drill stem. Advanced HDD machines employ a fully automatic gyro-based drilling mechanism.

7. *Mauling:* Mauling is a smaller version of HDD used for short-distance cable-laying like under roads and junctions. Mauling is a mechanized process that drills a straight hole under the surface.

3.9 Cable Preparation, Splicing, and Termination

It is important to ensure that the optical cable is properly prepared prior to splicing. This is very important and has a direct correlation with the performance of a link and adherence to link budget (LB).

3.9.1 Cable Preparation

The important steps in cable preparation are illustrated in Figure 3.13 and outlined as follows:

1. Each fiber must be cleaned thoroughly before stripping for splicing.
2. The buffer coating of the fiber(s) must be stripped off and a proper length of bare fiber exposed.
3. The exposed fiber should be cleaned using good quality standard wipes available for the purpose.
4. In order to ensure a perfectly flat end face they must be cleaved. A cleaver is a specialized tool that contains a sharp blade made of diamond, sapphire, or tungsten carbide that is used to cleave and cleanly cut the fiber. The cleaver must be suitably cleaned (depending upon the type used).

3.9.2 Splicing

Splicing refers to the process of joining optical fibers. There are several methods of splicing. The most commonly used methods are outlined as follows.

1. *Fusion-splicing:* Fusion-splicing, the process of fusing fibers together with an electric arc, is the most widely used method of splicing. It provides the lowest loss with low reflectance with strong and reliable joints between two fibers, in comparison to any other splicing technique. Fusion-splicing machines are generally classified according to the method employed to align the fibers being spliced.
    a. Active alignment fusion splicers employ a vision system to directly align the cores of the fibers, thereby reducing the splice attenuation. These systems automatically compensate for core concen-

Figure 3.13 Fiber preparation.

tricity errors and provide an estimate of splice attenuation after completion. Active alignment splicers can only splice single fibers at any given time.

b. In contrast splicers with a passive alignment system can be used for splicing single as well as ribbon fibers. These splicers use longitudinal movement to align fibers; hence the quality of the splices is dependent on the fiber geometry. After splicing a protective jacket may be used to cover the splice in order to ensure the tensile strength of the fiber.

2. *Mechanical splicing:* Mechanical splicing is an alternative method of making a permanent connection between fibers. In the past, this method of splicing suffered from the disadvantages of high losses, poor performance, and low reliability. Technological advancements, however, have contributed to the significantly increased performance of these types of splices. The advantage of this splicing method is its lower costs as well as simpler and quicker connections.

3.9.3 Fiber Termination

The rapid pace of technological change in optical transmission equipment, light sources, and interfaces necessitates the interconnection of multiple generations of transmission equipment. This could be achieved by the use of the following:

1. *Optical distribution frames (ODFs):* A fiber distribution unit or an ODF is used to provide a standard termination node for the fiber network to which transmission equipment can be interfaced. ODFs provide for the general functions of test access and organizing the cable fiber, connectors, and jumpers at termination nodes and distribution nodes in the fiber network. The distribution frame provides a connection to the backhaul/transport equipment using optical patch cables or patches or jumpers. The patch cables consist of a length of fiber with connectors at both ends. The patch cables used earlier were FC-FC and FC-SC/PC/APC while the newer equipment uses LC-LC, LC-SC/PC/APC. SC is an acronym for standard and square connectors. PC refers to physical contact connectors or polished connectors while APC refers to angled physical contact or angled polished connectors. SC/PC connectors (Figure 3.14) have a polished surface that results in a high end face, providing better core connectivity and low optical return loss. The SC/PC connectors have a blue color code. APC connectors, color-coded green, have a precise 80 cut in the ferrule and do not exhibit Fresnel reflection. APC connectors (Figure

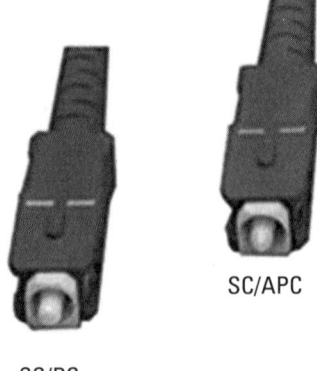

SC/APC

SC/PC

Figure 3.14 SC/PC and SC/APC connectors. (*Source:* The Fiber Optic Association, 2009, reprinted with permission. Source URL http://www.thefoa.org/.)

3.14) are best suited for high-bandwidth applications due to their very low reflectance. However APC connectors have slightly higher insertion losses as compared to SC/PC connectors.

2. *Fiber management systems (FMSs):* FMSs can be located in central exchanges or convergence nodes as well as customer premises that require fiber terminations. These systems are capable of handling fibers in a consistent and structured manner. The connectors in a termination application are sometimes replaced by fiber splices, reducing the number of connectors in the link.

FMS is the smaller version of the ODF and comes in different configurations including 12-port, 24-port, and 96-ports and occupies a small footprint (typically 1U to 4U where 1U = 1.75 inches). Network elements are connected to the FMS using optical patch cables.

3.10 Summary

The chapter provided the conceptual basis for understanding the causes and effects of propagation of light through a media. The basic concepts conform to the fundamental laws of physics and can be expressed mathematically. This book has been designed with the view of providing a comprehensive reference to field engineers, and hence the mathematical treatise has been kept to a bare minimum. There are myriad of books on the subject that readers may refer to for additional mathematical content. The chapter introduced the theory related to the propagation of light through a media and provides the background for understanding complex issues surrounding the design, deployment, and troubleshooting of modern-day next-generation optical networks

The chapter also describes modal propagation and its effects on light transmission through a fiber. A ray of light traveling through the core of an optical fiber exhibits variations in its intensity as it travels down the fiber. These variations are referred to as modes. The number of modes (constituents) depends upon the dimensions of the core and the variation in the RIs of core and cladding. The modes are numbered in the ascending order.

This chapter lays the groundwork for understanding complex concepts related to the design, deployment, and maintenance of optical networks.

3.11 Review

3.11.1 Review Questions

1. Electromagnetic radiation can be expressed using the following units:
   a. Energy: Electron volts or joules
   b. Wavelength: Meters and its subunits
   c. Frequency: Cycle per second or hertz (Hz)
   d. All of the above

2. Visible light corresponds to a frequency range of 430 to 770 THz within the electromagnetic spectrum.
   a. True
   b. False

3. The wavelength of a waveform (traveling at a constant speed) is provided by the following equation: $\lambda = v/f$.
   a. True
   b. False

4. _____ describes the manner in which light propagates through a medium other than free space.
   a. Reflection
   b. Refraction
   c. Scattering
   d. RI

5. The transmission of a ray of light through an optical fiber is due to the principle of _____.

a. TIR
  b. Reflection
  c Refraction
  d. Absorption
6. If the angle of incidence is less than the critical angle, the ray of light incident on the core of a fiber will get _____ by the cladding.
  a. Reflected
  b. Refracted
  c. Absorbed
  d. Scattered
7. _____ refers to the traditional method of laying optical fibers.
  a. Mauling
  b. T&D
  c. HDD
  d. Trenching and directional boring
8. An ODF is considerably smaller than an FMS.
  a. True
  b. False
9. Choose the correct steps in preparing an optical fiber cable for splicing.
  a. Clean-strip-cleave
  b. Clean-strip-clean
  c. Cleave-strip-clean
  d. Strip-clean-cleave
10. The effects of _____ may be highly pronounced in links with transmission speeds of 40 Gbps and above.
  a. PMD
  b. CD
  c. SBS
  d. SRS

3.11.2 Exercises

1. List the important design considerations for an optical network.
2. List and describe nonlinearities in fiber.
3. Describe the commonly used light sources in an optical fiber network. List their key parameters.
4. Which type of fiber is suited for commercial deployment? Provide detailed reasoning for your answer.
5. Describe the major differences between WDM and SDM.

3.11.3 Research Activities

1. Describe in detail the types of dispersion and illustrate with examples their effect on network design and performance.
2. Briefly describe e-dispersion compensation.
3. Describe SDM. Detail its relation with optical fiber communication and modal propagation.
4. Briefly describe few-mode fibers and the relation between FMF and SDM. List the key points to be considered for introducing SDM in existing photonic networks.

3.12 Referred Standards

EIA-440-A

EIA-455-A

EN 13501-6

EN 86 275-801

EN 86 275-802

G.651: Characteristics of a 50/125-μm multimode graded index optical fiber cable

G.652: Characteristics of a single-mode optical fiber cable

G.653: Characteristics of a dispersion shifted single-mode optical fiber cable

G.654: Characteristics of a cut-off shifted single-mode optical fiber cable

G.655: Characteristics of a non-zero dispersion shifted single-mode optical fiber cable

G.656: Characteristics of a fiber and cable with non-zero dispersion for wideband optical transport

G.657: Characteristics of a bending-loss insensitive single-mode optical fiber and cable for the access network

GR-1081

GR-1435

GR-326

GR-761-CORE: Generic criteria for chromatic dispersion test sets

IEC 60793-1-42: Measurement methods and test procedures for CD

IEC 61754-15

IEC TR 62221: Test methods for microbending

ISO/IEC 11801: Information technology—Generic cabling for customer premises

ITU-T G.650.1: Definitions and test methods for linear, deterministic attributes of single-mode fiber and cable

R-761-CORE: Generic criteria for chromatic dispersion test sets

TIA FOTP-175-B: Chromatic dispersion measurement of single-mode optical fibers

TIA FOTP-175-B: Chromatic dispersion measurement of SMF

TIA/EIA 604-16

TSB-62

3.13 Recommended Reading

3.13.1 Books

Govind, A. P., *Fiber-Optic Communication Systems,* Fourth Edition, Wiley, 2010.

Chomycz, B., *Fiber Optic Installer's Field Manual,* McGraw Hill, 2000.

Huber, J. C., *Industrial Fiber Optics Networks,* Instrument Society of America, 1995.

Senior, J., *Optical Fiber Communication: Principles & Practice,* Prentice Hall, 2008.

Keiser, G., *Optical Fiber Communication,* McGraw Hill, 2008.

Senior, J. M., *Optical Fiber Communications: Principles & Practice,* Third Edition, Pearson Education: India, 2010.

Silvello, B., *Coherent Optical Communications Systems,* John Wiley & Sons, 1995.

Warier, S., *The ABCs of Fiber Optic Communication,* Norwood, MA: Artech House, 2017.

3.13.2 URLs

http://ecmweb.com/mag/electric_basics_fiber_optics_4/

http://search.techrepublic.com.com/search/fiber-optics.html

http://www.arcelect.com/

http://www.ask.com/questions-about/Fiber-Optics

http://www.avap.ch/

http://www.ciscopress.com

http://www.fiberopticproducts.com

http://www.fibersolutionsonline.com/

http://www.hubersuhner.com/

http://www.nfpa.org

http://www.optiwave.com/

http://www.telebyteusa.com/

http://www.ul.com/telecom

References

[1] Kinsler, L. E., *Fundamentals of Acoustics*, John Wiley and Sons, 2000, p. 136.

[2] Joannopoulos, J., et al., *Photonic Crystals: Molding the Flow of Light*, Second Edition, Princeton, NJ: Princeton University Press, 2008.

[3] Alwayn, V., (2016, November 20). Propagation of Light. Retrieved from Cisco Press: http://www.ciscopress.com/articles/article.asp?p=170740&seqNum=5.

[4] Warier. S., *The ABCs of Fiber Optic Communication,* 2017, Norwood, MA: Artech House, 2017, pp. 29–114.

[5] Underwriters Laboratories, Optical Fiber Raceway, Association [n.d.], retrieved April 22, 2008. Source URL http://www.ul.com.

[6] *Multi-Mode Fibers* [n.d.], retrieved April 22, 2008. Source URL http://www.fiberoptics-4sale.com/Merchant2/multimode-fiber.php.

[7] Schneider, K. S., *Primer on Fiber Optic Data Communications for the Premises Environment* [n.d.], retrieved June 10, 2008. Source URL http://www.telebyteusa.com/foprimer/fofull.htm.

[8] Hayes, J., *Understanding Wavelengths in Fiber Optics* [n.d.], retrieved from The Fiber Optic Association: http://www.thefoa.org/tech/wavelength.htm.

Selected Bibliography

"Guide To Fiber Optics & Premises Cabling," The Fiber Optics Association, retrieved December 22, 2015.

Buzzelli, S., et al., "Optical Fiber Field Experiments in Italy: COS1, COS2 and COS3/FOSTER." International Conference on Communications, Seattle, 1980.

"Bell Labs Breaks Optical Transmission Record, 100 Petabit per Second Kilometer Barrier," Phys.org, September 29, 2009.

Fiber Technology Incorporated. (November 20, 2016). Transmission Loss. Retrieved from fiberopticstech.com. Source URLhttp://www.fiberopticstech .com/technical/transmission_loss.php.

Rigby, P., "Three Decades of Innovation," *Lightwave,* Vol. 31, No. 1, 2014, pp. 6–10.

Stone, J. A., and J. H. Zimmerman, (December 28, 2011), "Index of refraction of air," Engineering metrology toolbox. National Institute of Standards and Technology (NIST). Retrieved 2014-01-11.

Elert, G., "The Electromagnetic Spectrum, The Physics Hypertextbook," Hypertextbook.com, retrieved October 16, 2010.

Bannon, M., and F. Kaputa, "The Newton–Laplace Equation and Speed of Sound," *Thermal Jackets,* retrieved 3 May 2015.

4
Optical Link Design

4.1 Chapter Objectives

Optical fiber is the preferred medium for deployment in modern-day access, metro, and long-distance networks. However, the length of a fiber is finite, and hence fibers need to be joined together or spliced as per the network requirements. The ends of the fiber also need to be connectorized at the termination points. This is achieved through the use of splicing and connectorization techniques. These techniques are highly critical and have a on the network operational costs as well as system performance. There has been major development in the field of fiber-optic interfaces or connectors. Technology developments have resulted in a transition from the old ferrule core connectors to the recent short-form factor—plus interfaces. The size, cost, and attenuation characteristics of the interfaces have decreased drastically while the density of optical interface ports have significantly increased. This chapter, in a continuation of Chapter 3, provides the background needed to design, deploy, and troubleshoot optical links. It also introduces commonly used splicing techniques along with a detailed treatise of the different types of optical fiber interfaces. The chapter concludes with an in-depth discussion of optical link budgeting and span analysis that is particularly useful for field engineers and network planners. This chapter offers readers the opportunity to develop and cement the core concepts, presented in Chapter 3, while appreciating the practical nuances of deploying optical links.

Key Topics

- The functioning and key operational parameters of optical transmitters and receivers;
- Optical interfaces;
- Optical modulation techniques;
- The features of commonly used optical connectors;
- Link budgets and how to prepare them;
- Safety guidelines for handling optical fibers and working with optical links;
- Industry standard diagnostic techniques.

4.2 Optical Transmitters

The term transmitter refers to the light source in an optical system. It is fairly obvious that the light sources are an integral and a crucial component of a fiber-optic network. The span length and the number and placement of regenerators and amplifiers are all dependent directly or otherwise on the transmitter types. In all commercial applications the conversion from electrical to optical and vice versa is done by an integrated electro-optical (E/O) unit that functions as a transmitter and receiver. The issues of thermal dissipation, efficacy of heat sink, core diameter, and the method of coupling the optical power onto a fiber are all taken care of by the manufacturer, and in most cases the users do not have much say about them. The modern-day architecture consists of racks or skeletons that are equipped with cards also referred to as plug-in-units (PIUs) or plug-in-cards (PICs). The E/O unit is integrated with other functional components and is in the form of a PIC. The user can only decide among the various types of cards offered by the manufacturers, based on specific applications. The transmitter serves two major functions [1]:

1. Providing a source of light for transmission over the fiber;
2. Modulating the source in accordance with the transmitted data.

The former function involves emission and the subsequent coupling of light onto a waveguide, which is nothing but the fiber. The second function (modulation) involves the variation of the intensity of light being coupled onto the fiber.

4.2.1 Optical Sources

Early transmitters were fabricated using discrete electrical as well as E/O components. These were subsequently replaced by integrated circuits (ICs) of varying designs starting from small scale integration (SSI) to the current multilayer very large-scale integrated (VLSI) chipsets. Commonly employed optical sources include LEDs, laser diodes (LDs) and their variation, vertical-cavity surface-emitting lasers (VCSELs). Some of the important design considerations for a transmitter [1] are described as follows:

1. The physical dimensions must be compatible with the core diameter of the FOC being deployed.
2. This implies that the emitted light must be coupled into a cone with a cross sectional diameter of 8 to 100 microns.
3. The optical source must be able to generate enough optical power so that the desired bit error rate (BER) can be met.
4. In order to ensure the optimal usage of the high bandwidth offered by the FOC, the optical source must be capable of being modulated by a high-frequency electrical signal.
5. The coupling efficiency (the amount of power that is actually transmitted along the fiber) should be high.
6. The source should have a high coupling efficiency.
7. The linearity of the optical source should facilitate the suppression of harmonics and intermodulation distortion.
8. The source must be made available in a compact, highly reliable, small-footprint, and lightweight package.

There are two basic types of lasers:

1. Fabry-Perot (FP) lasers: FP lasers have a high SNR and are slower than distributed feedback (DFB) lasers, but they are highly economical. FP lasers emit light at a number of discrete wavelengths. There are two commonly used FP laser types, buried hetero (BH) and multi-quantum well (MQW) types. BH lasers were extensively used earlier. MQW are used in present day applications. There are several advantages of MQW lasers over BH lasers. These include the following:
   - Low threshold current;
   - High efficiency;
   - Low noise;
   - Enhanced linearity;

- Stability over a wide range of temperatures;
- Lower operating and manufacturing costs.

However, MQW lasers have higher back reflections in comparison to DFB lasers, which have an antireflection coating on one side of the optical cavity[1].

2. DFB lasers: DFB lasers are monochromatic (emit a single color of light) and are quieter and faster than MQW lasers. They are high-performance lasers that are used extensively in high-speed applications (analog as well as digital). They offer enhanced linearity and have narrow spectral widths. The cross-sectional area of the core of an optical fiber is circular while the output of the lasers is elliptical, thereby impacting the coupling efficiency of the source. This issue led to the development of another type of laser, the vertical-cavity surface-emitting lasers (VCSELs).

3. VCSELs: VCSELs are lasers with a large output aperture that emits circular low-divergence beams. This facilitates a high coupling efficiency with optical fibers (highest in comparison to LED and LD). VCSEL has a low threshold current that results in low power consumption and high intrinsic modulation bandwidths. VCSELs are also used for sensing applications (proximity sensing); high-power applications including LOS systems; atomic clocks; and magnetometers. Other applications include spectroscopy, broadband transmission (analog), laser printers, and optical mouse technology. The major advantages of VCSELS include the following:

   a. Low cost (as they can be manufactured using the setup for manufacturing conventional lasers);
   b. High reliability;
   c. Scalability (fabricated as arrays);
   d. Availability in 1,550 nm;
   e. Metro DWDM (C and L band) as well as PON application support.

4.2.2 Modular Optical Interfaces

In the days prior to the development of the small form-factor pluggable (SFP), the failure of an optical port warranted the replacement of an entire PIU or a functional card. The situation was similar in the case of equipment upgrades. The development of the SFP has facilitated an easy or faster method of recov-

---

1. Does not support hot swapping. (SFP supports hot swapping.)

ering from connector failures and a cost-effective method for card or capacity upgrades [1].

1. SFP modules: SFP modules are compact, modular, hot-swappable optical transceivers that are capable of operating at speeds up to several gigabits per second. The development of the SFP reflects the significant strides made in the field of semiconductor and electronic engineering. The benefits of an SFP cannot be understated. In most of the conventional multiplexers deployed in modern-day telecom networks a capacity quadrupling calls for the replacement of only an SFP. In fact an electrical SFP (operating at E4/STN-1e) with mini BNC connectors is also available. The development of the SFP has resulted in lower CAPEXs and OPEXs) for deploying and maintaining fiber-optic networks. The inherent benefits of an SFP have brought together several leading telecom companies as well as data organizations with the common objective of standardizing SFPs. The common goals were to have an interface with a small footprint facilitating easy installation as well as maintenance. SFPs are available at STM-1/4/16 as well as STM-1e bandwidths. However, complete portability has not yet been achieved. For example, an STM-1 SFP from vendor A, used in telecom equipment manufactured by company X may not work in a similar piece of equipment from company Y. Figure 4.1 illustrates the front view of an SFP.

Figure 4.1 SFP module, front view.

2. Ten-gigabit SFP (XFP) modules: With the X representing 10, XFP [2] is a protocol-independent, hot-swappable optical transceiver operating at 850 nm; 1,310 nm; or 1,550nm. It finds use in a wide range of applications, including the following:
   - SDH/SONET networks;
   - Fiber channel (FC);
   - Gigabit Ethernet (GigE);
   - DWDM.

   The XFP (as illustrated in Figure 4.2) supports varying interfaces ranging from very short reach (VSR) to long reach (LR). The standard offerings are short reach/intraoffice, intermediate office/short haul, and long reach/long haul. The XFP module, as in case of the SFP, includes the transmission and receiving functions integrated onto a compact, flexible, and low-cost format supporting up to16 XFP modules on a typical card (European standard). As in the case of the SFP, the XFP specifications were drawn from a multisource agreement. The family also includes an XFI or "Ziffy" high-speed serial electrical interface with a nominal baud rate of 9.95 to 11.1 Gbps.

3. SFP+ modules: SFP+ (refer to Figure 4.3) represents the next-generation transceiver module with a miniature form factor as specified by the ANSI T11 Group. The module is designed for operation at 8.5 Gbps and 10 Gbps. These bandwidths correspond to that of fiber channel and Ethernet applications. The SFP+ form factor is 30% less than that of an XFP. In fact form factors one-sixth of an SFP (with similar reduction in power dissipation as well) have also been successfully designed. To achieve this reduced size certain traditional functions, including the serial/parallel conversion, clock recovery, and signal-conditioning functions, have been relocated to the associated PIC or module. The primary benefits of an SFP+ module are as listed as follows:
   - Reduced footprint and hence higher port density (up to 24 and even 48 ports on a standard equipment card);

Figure 4.2 XFP Module [3]. (Source: The Fiber Optic Association, 2009. Reprinted with permission. Source URL http://www.thefoa.org/.)

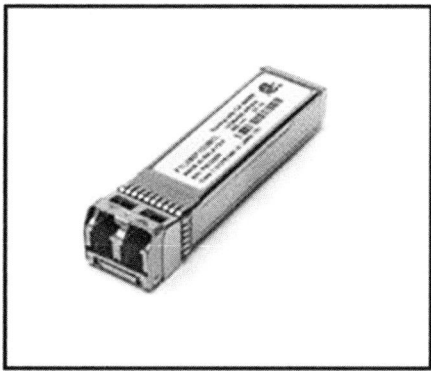

Figure 4.3 SFP+ module [4]. (Source: The Fiber Optic Association, 2009. Reprinted with permission. Source URL http://www.thefoa.org/.)

- Lower costs;
- Reduced power since many functions are relocated to the associated PIU.

Note 4.1
The SFP+ standards are laid down by the American National Standards Institute (ANSI) Technical Committee (T11) along with inputs from more than 30 fiber-optic system vendors. Table 4.1 outlines new developments in the area of physical optical interfaces.

4.2.3 Transmitter Design Parameters

The following are some of the key (and often overlooked) parameters that significantly impact the choice of an optical source and its performance:

1. Back reflection: Back reflections refer to the optical energy that is reflected back onto the waveguide especially at the spliced/terminating end. Back reflections cause distortions in the light propagating through a FOC besides increasing the effective noise. Strong back reflections can cause some lasers to be unstable and render them useless in certain applications. It can also generate nonlinearities, referred to as kinks, in the laser response. This is unsuitable for most analog applications and some digital applications.

   The importance of controlling back reflection depends on the type of information being sent and the source laser. The design may render some lasers susceptible to back reflection. A crucial factor that has significant impact on the amount of back reflections is the coupling of

Table 4.1
New Optical Interfaces

S.N	Transceivers	Description
1	AOC	Active optical cable—Employed for interconnection and short-range multilane data communication applications. AOC consists of MMF, optical transceivers, control circuitry and performs electrical-to-optical conversion at the cable ends to improve speed and distance performance of the cable while ensuring compatibility with standard electrical interfaces
2	C-BIDI SFF	Compact bidirectional small form factor1 transceiver module that supports bidirectional communication over a single fiber at rates up to 2.5 Gbps at 1310/1490 nm wavelengths. Used in access networks including fiber-in-the-loop and PON
3	CFP	C-Form factor pluggable transceiver interface with 100 Gbps support (10*10 Gbps streams or 4*25 Gbps streams)
4	CFP2	Functionally similar to CFP with half the form factor (41.5 mm as against 82 mm CFP). Supports 10*10 Gbps streams, 4*25 Gbps
5	CFP4	Functionally similar to CFP with one-fourth the form factor (21.5 mm as against 82 mm CFP)
6	CFP8	CFP8, with physical dimensions similar to CFP2 module, is a next generation optical transceiver designed for 400G Ethernet applications. It can support 16x25G and 8x50G electrical I/O along with 4x100 GigE
7	QSFP+	Quad SFP+ is a hot pluggable transceiver supporting optical as well as electrical TDM and IP interfaces. The interface was developed as a part of the multisource agreement (MSA) by the Small Form Factor Committee. The original specification supported 4*4 Gbps channels (GigE or FC or DDR Infiniband). The variations of the interface supports 4*14 Gbps FDR Infiniband or SAS-3 channels (QSFP14) and 4* 28 Gbps 100 GigE or EDR Infiniband
8	Tunable XFP	Tunable XFP 10 Gbps multiprotocol optical transceiver providing a high-speed serial link from 9.95 Gbps to 11.1 Gbps in conformance with 10 Gigabit small form factor pluggable (XFP) multisource agreement (MSA)
9	XGPON	10 Gbps PON Specification, wavelength—downstream: 1578 nm ± 3 nm, upstream: 1270 nm ± 10 nm, data rates 9.953 Gbps downstream, 2.488 Gbps upstream as per ITU-T G.987 standards

the fiber and the connector. A low-power laser generally has weak coupling to the fiber with only 5 to 10% of the laser power being directed onto its core. This implies that only 5 to 10% of the back reflection would also be coupled into the laser cavity. This makes the laser relatively immune to back reflections. On the other hand, a high-power laser may have 50 to 70% of the laser output coupled to the fiber. This would therefore imply that 50 to 70% of the back reflection would also be coupled back into the laser cavity increasing the level of back reflections.

A crucial factor that has significant impact on the amount of back reflections is the coupling of the fiber and the connector. A low-power laser generally has weak coupling to the fiber with only 5 to 10% of the laser power being directed onto its core. This implies that only 5 to 10% of the back reflection would also be coupled into the laser cavity. This makes the laser relatively immune to back reflections. On the other hand, a high-power laser may have 50 to 70% of the laser output coupled to the fiber. This would therefore imply that 50 to 70% of the back reflection would also be coupled back into the laser cavity making those lasers more susceptible to back reflections.

The losses induced when coupling light onto a fiber, due to misalignment and/or air gaps between components like connectors or due to absorption are termed as insertion losses. With advancements in manufacturing technologies the insertion loss is less than 0.2 dB. Insertion loss is expressed in decibels as:

$$IL = 10\log_{10}\left(P_{out}/P_{in}\right)$$

where P_{in} refers to the incident power and P_{out} the output power.

The losses induced due to reflectance or backscattered light are referred to as return losses. A higher value of return loss indicates a better quality of connection. Return loss is expressed in decibels as:

$$RL = 10\log_{10}\left(P_i/P_r\right)$$

where P_i refers to the incident power and P_r the reflected power.

2. Extinction ratio: Normally optical systems employ two levels of optical power, where the higher power level represents a binary 1 and the lower power level represents a binary 0. These two power levels can be represented as P_1 and P_0, where $P_1 > P_0$, the unit of representation being watts. The extinction ratio re expresses the relationship of the power used in transmitting a binary 1 to the power of transmitting a binary 0. The extinction ratio is used to describe the efficiency with which the transmitted optical power is modulated over the fiber-optic channel. It can be defined as a linear ratio—P_1/P_0 or as a power measurement—10 * log (P_1/P_0) or as a percentage, (P_0/P_1) * 100. Small changes in the extinction ratio can make a relatively large difference in the power required to maintain a constant BER. Thus variations in the extinction ratio affect the performance of an optical link.

For an ideal transmitter, P_0 would be zero and hence r_e would be infinite. Most optical transmitters have a finite amount of optical power at the low level and hence $P_0 > 0$. (Lasers must be biased so that P_0 is in the vicinity of the laser threshold.) At the receiving end two important decisions must be made:

- The instant or time to sample the received data;
- Whether the sampled value represents a binary 1 or 0.

The receiver decision circuit simply compares the sampled voltage to a reference value known as the decision threshold. The associated circuitry is generally a part of a clock and data recovery (CDR) block. P_0 is ideally equal to zero, making the optimum extinction ratio infinite. When the extinction ratio is not optimal, however, the transmitted power must be increased in order to maintain the same BER at the receiver. This increase in transmitted power due to nonideal values of the extinction ratio is called the power penalty.

Note 4.2

Small changes in the extinction ratio make a relatively large difference in the power required to maintain a constant BER. This effect is especially pronounced when the values of the extinction ratio are less than seven. A change in the value of the extinction ratio by 1 necessitates a corresponding change of 10% in the average power of the laser. This additional required power is referred to as power penalty. It is important to note that increasing the power of a laser will reduce its MTBF.

4.3 Optical Receivers

The detection of the optical signal, at the receiving end, is performed by a photodiode, and the demodulation process is carried out on the resulting electrical signal. One or more stages of amplifiers, filtration, and equalization circuits are employed to boost the weak incoming signals as well as to reshape (with amplitude distortion correction) them. The receiver component of an E/O unit primarily serves two functions:

1. The receiver must sense or detect the light from the FOC and subsequently convert the light into an electrical signal.
2. It must then demodulate this light to determine the identity of the binary data represented by the incoming signal.

In addition to its primary functions, a receiver also performs multiple secondary functions including the following:

1. Clock recovery: This is true for NEs that are part of a synchronous network like SDH or SONET and employ line timing;
2. Line decoding: 4B/5B decoding;

3. Error detection/ recovery.

The receiver must be capable of a high level of detection (sensitivity), support high data rates, and have low noise characteristics. A high-sensitivity receiver would be capable of detecting highly attenuated signals. A high bandwidth-handling capacity would allow for a swift response to state transitions, ensuring demodulation of the high-speed incoming data. In order to ensure the link BER, the receiver must have low noise characteristics.

Two types of photodiodes are commonly employed in optical equipment. They are the PIN diode and the APD. The PIN is the preferred choice in short-distance applications due to its standard power supply requirements (as it can be operated on a standard 5-V or 15-V supply) and lower costs. APD devices have much better sensitivity (5 to 10 dB more sensitivity) and can accommodate twice the bandwidth than that possible by a PIN. The major disadvantage, however, is that APDs require a stable power supply; moreover, they are considerably more expensive.

The demodulation performance of the receiver is characterized by the BER. This in turn is determined by the modulation scheme, the received optical signal power, the receiver noise, and the processing bandwidth.

Note 4.3

The receiver performance is characterized by a parameter referred to as receiver sensitivity. The receiver sensitivity is indicated by a curve on a graph representing the minimum optical power that a receiver can detect versus the data rate required in order to achieve a particular BER.

4.4 Optical Modulation Techniques

The modulation technique employed on an optical link plays a key role in realizing high data throughput. The choice of modulation schemes is especially crucial for links operating at 40 Gbps and above, as there is severe performance degradation, due to the effects of dispersion and other fiber nonlinearities. Optical modulation refers to the controlled variation of one or more of the key parameters (amplitude, frequency, phase, polarization) of a carrier signal, according to the bit pattern of the data being transmitted, prior to its propagation through the network. This concept is illustrated in Figure 4.4.

The different types of modulation techniques are illustrated in Figure 4.5 and described as follows.

1. Amplitude shift keying (ASK): ASK, also referred to as on-off keying (OOK) and intensity modulation (IM), is a technique that modulates

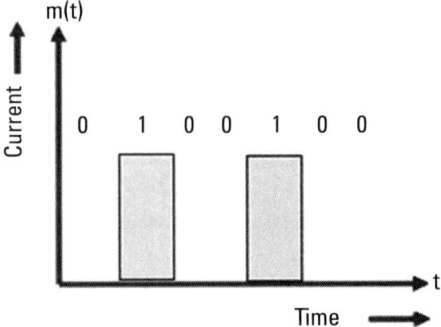

Modulation current representing binary modulation waveform

Figure 4.4 Optical signal modulation.

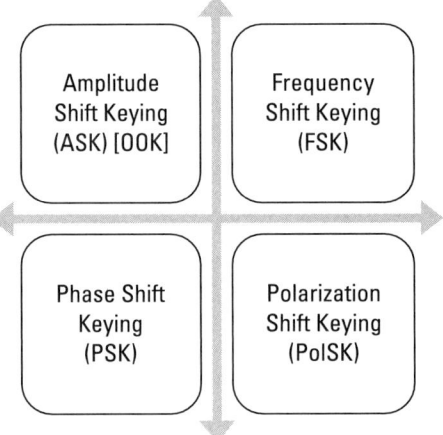

Figure 4.5 Types of modulation techniques.

the intensity of the carrier signal. The simplest form of ASK involves switching the optical source on and off depending upon the state of the digital signal being transmitted. There are several different variations of ASK used for optical modulation on links with data throughput limited to 10 Gbps. These include non-return-to-zero (NRZ) and return-to-zero (RZ). NRZ/RZ techniques are widely used in systems with transmission speeds of 10 Gbps and below. At these rates the RZ technique is more appropriate since it takes into account the nonlinear effects. It may be noted that there is a difference in the signal shape and the spectral characteristics for NRZ and RZ schemes.

2. Frequency shift keying (FSK): FSK involves using two values for frequency corresponding to the binary nature of the signal being transmitted. FSK modulation, also referred to as dispersion-supported transmission (DST), is a complex technique that requires precise alignment of transmitter and receiver parameters and is therefore not a preferred option for commercial systems.
3. Phase shift keying (PSK): PSK is a technique that uses the signal phase to encode the data being transmitted. Forward equivalence class PSK results in improved receiver sensitivity (of nearly 6 dB) [4] as compared to systems that employ the ASK technique. There are certain complexities involved with system design using PSK, hence a variation referred to as differential PSK (DPSK) is generally preferred. DPSK uses the phase change between two successive bits to encode information (the information regarding the phase difference between symbols is more important as opposed to phase information itself). Another modification of the PSK technique referred to as differential quaternary PSK (DQPSK) employs four different phases to encode information resulting in enhanced code efficiency. DPSK and DQPSK can be used in high-speed WDM long-haul links. Another variation quadrature PSK (QPSK) uses a combination of two bits and four phases of the carrier frequency to encode information being transmitted leading to an almost 50% reduction in symbol rate (as compared to the ASK technique) and consequentially a doubling of the data rate.
4. Polarization shift keying (PolSK): PolSK is a modulation format where the signal polarization switches between two orthogonal states. A binary 0 is represented as a linear polarized signal of –45° while the binary 1 is represented as a linear polarized signal of + 45°. The design of receivers based on PolSk is suited for free-space optical communication. A new class of fibers referred to as hollow core fibers may use this technique to generate a highly reliable transmission system.

Modern-day telecommunication systems are fairly complex and are capable of very high throughputs ranging from 10 to 400 Gbps. These systems have characteristics that are vastly different from their predecessors and are expected to be highly reliable over a wide range of operating conditions.

The need for higher throughput has led to advanced research in optical modulation formats, yielding an exponential increase in transmission rates that renders many of the modulation schemes mentioned in the preceding section irrelevant. As speeds move beyond 40 Gbps the effects of dispersion severely limit the distance of transmission. In addition, the effects of polarization mode dispersion (PMD) become significant. The need to groom traffic on long-distance links necessitates the deployment of optical add/drop multiplexers (OADMs) over a long-distance link that supports WDM (transmitting multiple rays of light simultaneously over a single fiber core) [5].

Example 4.1

Most long-distance networks employ DWDM, a technique that employs a 50-GHz grid (wavelength separation). This implies a 0.8 bps/Hz spectral efficiency for a link operating at 40 Gbps.

$$N = \text{link speed/channel bandwidth}$$
$$= 4,000,000,0000 \text{ bps}/5,000,000,0000 \text{ Hz}$$
$$= 0.8 \text{ bps/Hz}$$

4.5 Optical Connectors

Connectors provide remateable connections and hence are used generally at termination points where flexibility is required in routing an optical signal (for example, from equipment pigtails to the receivers, wherever reconfiguration is necessary, or for normal termination purposes [6]. These remateable connections simplify system reconfigurations to meet changing customer requirements. Several optical connectors are described as follows.

1. Epoxy and polish connectors: Epoxy and polish–style connectors were the original connectors used for termination purposes and are still deployed extensively. These connectors provide a wide range of options including ST, SC, FC, LC, D4, SMA, MU, and MTRJ. Their key advantages include the following:
   a. Robustness: Ability to withstand higher levels of mechanical/environmental stress;
   b. Cable diameter: Workable with cables of varying diameters from small to big;
   c. Multiple connectors: Supportive of single as well as multiple cables (up to 24 cables) in a single connector.
2. Preloaded epoxy or no-epoxy and polish: The primary advantage of these connectors is their simplicity of installation and the consequent reduction in skills for handling these connectors. These connectors can be classified into two types: connectors with preloaded epoxy and connectors without epoxy. These connectors employ an internal crimp

mechanism to stabilize the fiber and are available in ST, SC, and FC connector styles.

3. No-epoxy and no-polish: These connectors offer the advantages of low cost, simple design, and highly reduced installation and training costs while facilitating fast restorations. They are available in ST, SC, FC, LC, and MTRJ connector styles.

There are numerous connectors, both proprietary as well as standardized, that are employed in the field of telecom, data, and cable/television and other industrial uses. The ones included in this text include those that were used earlier as well as the ones that are in use currently. Some of the connectors that may have a significant impact in the future have also been included. These include the following [7]:

1. Subscriber connector (SC): SCs, also sometimes referred to as standard connectors (refer to Figure 4.6), are low-cost, rugged, and simple connectors that employ a ceramic ferrule for providing accurate alignment of SMF. The SC is a push-on/pull-off connector with a locking tab and a ferrule diameter of 2.5 mm.
2. Lucent connector (LC): LCs (refer to Figure 4.7), also know as little connectors, are small form-factor fiber-optic connectors that employ a 1.25-mm ferrule. These connectors have widespread use in the telecommunication industry. The different types of LC connectors are as listed as follows:
   a. Single-mode LC UPC;
   b. Single-mode LC APC;

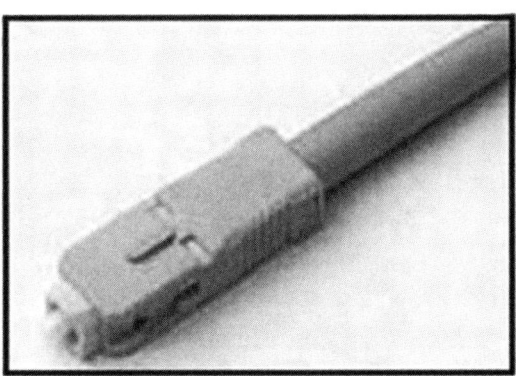

Figure 4.6 SC. (Source: The Fiber Optic Association, 2009, reprinted with permission. Source URL http://www.thefoa.org/.)

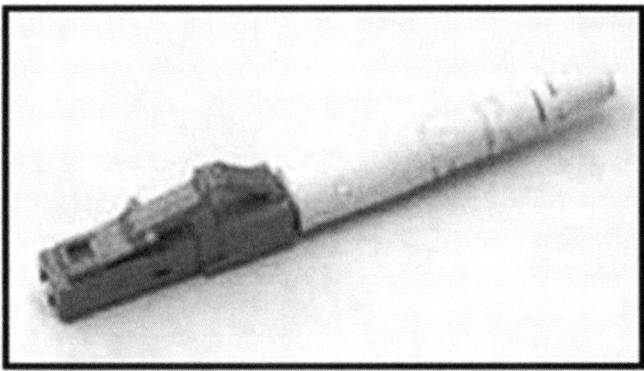

Figure 4.7 LC. (Source: The Fiber Optic Association, 2009, reprinted with permission. Source URL http://www.thefoa.org/.)

    c. Multimode LC UPC.

3. MU connector: MU connectors [8] (see Figure 4.8) are small (reduced footprint) new-generation connectors used for dense applications. They are square and feature a push-pull mating mechanism. The different variants of this connector are as listed as follows:

    a. Single-mode MU UPC fiber-optic connector;

    b. Single-mode MU APC fiber-optic connector;

    c. Multimode MU UPC fiber-optic connector.

The MU connector is used for SDH, SONET, WDM, LAN, ATM, and CATV applications.

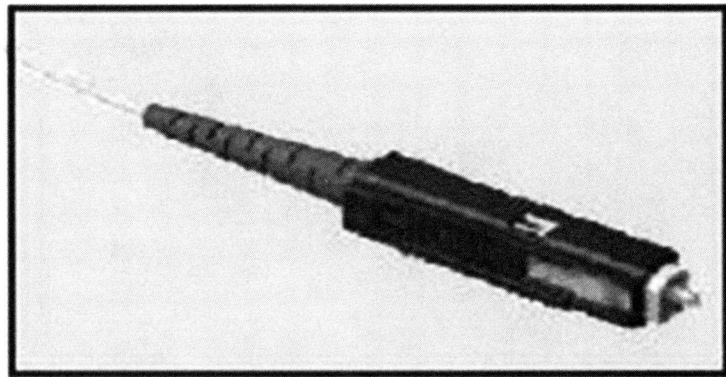

Figure 4.8 MU connector. (Source: The Fiber Optic Association, 2009, reprinted with permission. Source URL http://www.thefoa.org/.)

4. E2000 connector: The E2000 connector [9] (Figure 4.9) is increasingly being used in modern-day telecommunication networks. The unique feature of this connector is the inclusion of an integrated spring-loaded shutter that protects the ferrule from dust, dirt, and scratches. Further this connector employs a monobloc ceramic ferrule; hence, there is no problem associated with a differing coefficient of expansion. The connector is of the latched push-pull locking type. E2000 is a trademark of Diamond SA based in Losone, Switzerland. This connector is available in the following variants:

   a. Single-mode E2000 UPC connector;
   b. Single-mode E2000 APC connector;
   c. Multimode E2000 UPC connector.

   The primary advantages of this connector are the high performance and enhanced safety due the monobloc ferrule and the shutter mechanism. The return loss of the E2000 connector is one of the lowest at 0.1 dB and also supports color keying. The application of this connector includes telecommunication networks, broadband applications, fiber CATV, LAN [fiber-in-the-loop (FITL), fiber-to-the-home (FTTH), and fiber-to-the-desk (FTTD)], and data networks.

Note 4.4

Fiber connecterization results in an air gap between the mated connector end face leading to higher optical return loss (ORL), due to light being reflected back toward the optical source. In order to reduce the air gap between the mated pair of connectors the connector end face is polished to form a cone (spheri-

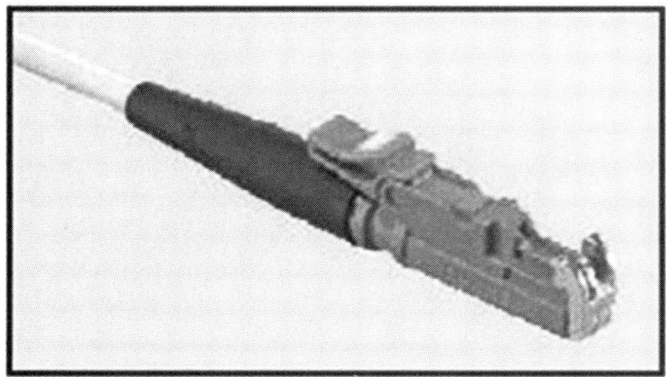

Figure 4.9 E2000 connector. (Source: The Fiber Optic Association, 2009, reprinted with permission. Source URL http://www.thefoa.org/.)

cal). The cone reduces the air gap and hence the ORL. The polished connectors are referred to as physical contact, or PC, connectors. An ultra-polished connector (UPC) is subjected to extended polishing leading to a finer surface finish. A UPC has a lower ORL as compared to PC connector. However ORL degradation occurs with repeated usage or mating/unmating cycles. This led to the development of another connector with an 8° angled end face referred to as angled physical contact (APC) connectors. APCs have a smaller end face and hence the lowest ORL among the connectors. It can also withstand repeated usage better than the other connector types.

4.6 Link Budgeting

The link loss is relative to the transmit power of a light source and represents the range of optical loss experienced by an operational fiber-optic link while meeting its operating specifications. The calculation and the subsequent verification of a fiber-optic system's operating characteristics are performed with the help of span analysis, which encompasses the following multiple functions (depicted in Figure 4.9):

1. Fiber type;
2. Fiber link parameters—length and routing;
3. Operational wavelengths;
4. Characteristics of the active components (e.g., system gain, transmitter power, receiver sensitivity, and dynamic range) and passive components (e.g., fiber loss, splice loss, and coupler/splitter loss) involved in the link;
5. Attenuation;
6. Nonlinear effects—pronounced at higher bit rates and minimized by employing dispersion compensation modules (DCMs).

The span analysis would culminate in the generation of a link loss budget that includes a safety margin to offset losses due temperature changes (1 dB) as well as component aging and any maintenance work in the future (3 dB). The actual or the true loss is measured using an optical meter and can be compared to the computed loss. Some of the key parameters to be considered when preparing a link budget (shown in Figure 4.10) are discussed as follows:

1. Transmitter launch power: The transmitter output at a specific wavelength, measured in decibels per meter. A higher power generally yields better results (higher extinction ratio and consequentially lower BER).

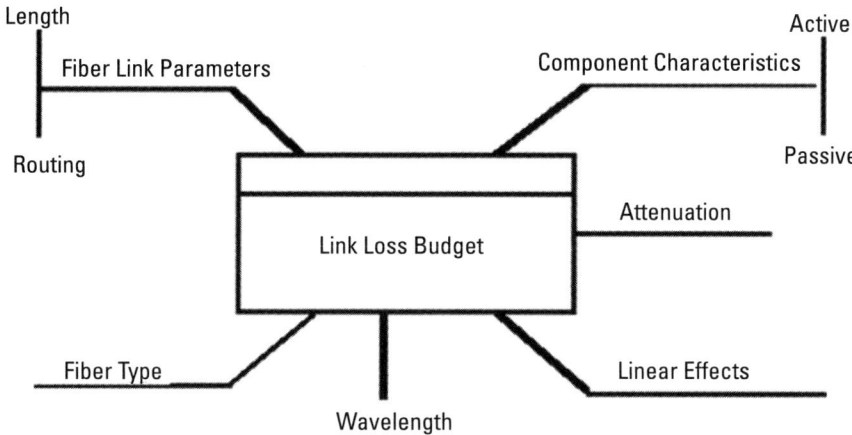

Figure 4.10 Link loss budget—key parameters.

However care should be taken that the power transmitted does not increase beyond the receiver's sensitivity or saturation level (dynamic range). A higher transmit power offsets the effects of link attenuation but can induce nonlinear effects, degrading system performance at higher bit rates.

2. Receiver sensitivity and dynamic range: The minimum acceptable value of received power that is required to obtain an acceptable BER or performance. This takes into account the power penalties caused by use of a transmitter with worst-case values of jitter, receiver connector induced distortions, extinction ratio, pulse rise and fall times, measurement tolerances, and optical return loss. However, the receiver sensitivity does not include dispersion or back-reflection power penalties. These effects are compiled separately under the heading of maximum optical path penalty. The receiver sensitivity does, however, take into account the worst-case operating and end-of-life (EOL) conditions. For example a receiver with a high optical input, 3 dBm, and a worst-case value of 29 requires an optical dynamic range of 26 dB.

3. Power budget and margin computations: A section (span) power budget that reflects the maximum transmit power is required to be computed to ensure that an optical link has adequate power for normal operation. This budget is computed on the assumption of worst-case scenarios—minimum transmitter power and minimum receiver sensitivity in order to simulate practical conditions. This ensures a margin for compensation for variations of transmitter power and receiver sensitivity levels. Accordingly

$$P_{BS}(\text{Span Power Budget}) = PT_{\min}(\text{Min. Transmitted Power}) - PR_{\min}(\text{Min Receiver Sensitivity})$$

As noted earlier span losses are the summation of the various linear and nonlinear losses and include factors like fiber attenuation, splice attenuation, connector attenuation, chromatic dispersion, and other linear and nonlinear losses. The reader is advised to refer to vendor-provided information for actual values. Single-mode connectors that are factory-made and fusion-spliced onto the optical fiber will have insertion loss values ranging from 0.15 dB to 0.25 dB while field terminations can have a higher value of up to 1 dB. Multimode connectors have insertion loss values ranging from 0.25 to 0.5 dB. The newer connectors would typically have a minimal insertion loss at 0.1 dB. In case of PON the losses due to the use of optical couplers would also need to be included. A 1:N coupler would introduce an additional loss of 10LogN dB. In case there is no coupler the value of N = 1 and the additional loss would be equal to zero. The following formula can be used to compute the span loss:

$$P_s(\text{Span Loss}) = \begin{bmatrix} (\text{Fiber Attenuation} * \text{Km}) + \\ (\text{Splice Attenuation} * \text{No. of Splices}) + \\ (\text{Connector Attenuation} * \text{No. of Connectors}) + \\ (\text{Inline Device Loss}) + (\text{Coupler Losses}) + \\ (\text{Nonlinear Losses}) + (\text{Safety Margin}) \end{bmatrix}$$

Power margin (PM) represents the amount of power existing in the system after accounting for linear and nonlinear span losses with a positive value indicating sufficient receiver power levels. The PM is computed as follows:

$$P_m(\text{Power Margin}) = P_{BS}(\text{Power Budget}) - P_S(\text{Span Loss})$$

The input power should not exceed the receiver sensitivity $PR_{\max}$ after adjusting for all span losses. In case the received power exceeds the receiver sensitivity, receiver saturation would occur. To compensate for the additional power at the input ports, attenuators would have to be used. The input signal level is denoted as P_{IN}, and the maximum launch power or transmitter power is denoted as $PT_{\max}$. The span loss P_S would, however, remain remaining constant. The receiver power is computed as follows:

$$P_{in} (\text{Input Power}) = P_{T\,\text{max}} (\text{Max. Transmitter Power})$$
$$-P_S (\text{Span Loss})$$

To prevent receiver saturation and ensure efficient system operation the following condition must be satisfied:

$$P_{in} (\text{Input Power}) = P_{R\,\text{max}} (\text{Max. Receiver Sensitivity})$$

If P_{IN} is greater than the maximum receiver sensitivity $P_{R\,\text{max}}$, passive attenuation must be considered to reduce signal level and bring it within the dynamic range of the receiver.

Now we examine a few examples of preparing optical link loss budgets.

Example 4.2

Figure 4.11 depicts an SDH link operating at STM-1 or 155 Mbps spanning a distance of 5 km with MMF fibers. The link would contain two patch panels and would employ mechanical splicing and standard values for attenuation; component and nonlinear losses are assumed.

The following are the key link parameters:

1. $PT_{\text{min}} = -13$ dBm;
2. $PT_{\text{max}} = -4$ dBm;
3. $PR_{\text{max}} = -5$ dBm;
4. $PR_{\text{min}} = -31$ dBm;
5. $\lambda = 1{,}310$ nm;
6. Link length = 5 Km;
7. Link capacity = 155 Mbps;
8. Fiber type = 50/125 μm GI.

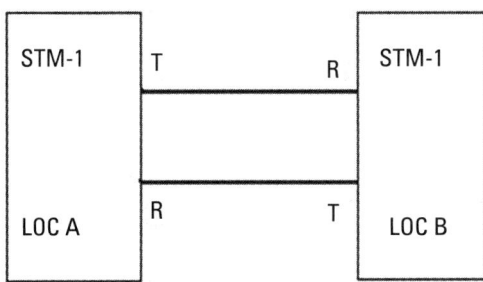

Figure 4.11 A sample STM-1 link.

Span analysis.

For a link operating at a lower throughput of STM -1 the following effects can be discounted:

- Self-phase modulation (SPM);
- Polarization mode dispersion (PMD);
- Stimulated Raman scattering (SRS)/stimulated Brillouin scattering (SBS);
- Cross-phase modulation (XPM);
- Four-wave mixing (FWM).

The effects of chromatic dispersion need to be accounted irrespective of the link speed. Table 4.2 details the span analysis.

$$P_{BS} \text{ (Span Power Budget)} = PT_{min} \text{ (Min. Transmitter Power)} -$$
$$PR_{min} \text{ (Min. Receiver Sensitivity)}$$
$$= -13 \text{ dBm} + 31 \text{ dBm} = 18 \text{ dBm}$$

The power margin is computed as follows:

$$P_m \text{ (Power Margin)} = P_{BS} \text{ (Power Budget)} - P_S \text{ (Span Loss)}$$
$$P_m = 18 \text{ dB} - 13.5 \text{ dB}$$
$$P_m = 4.5 \text{ dB}$$

Table 4.2
Span Analysis (Example 4.2)

S.N	Factor	Computation	Loss (dB)
1	MMF-GI	50/125-μm fiber (1,310 nm)—5 km at 0.7 dB/Km	3.5
2	SC connectors	2 at 0.5 dB/Km	1
3	Mechanical splice	2 at 0.5 dB/splice	1
4	Patch panel	2 at 2 dB/panel	4
5	Dispersion margin	—	1
6	Safety and repair margin	—	3
TOTAL			13.5

Optical Link Design

The total span loss is within the power budget of 18 dB (maximum allowable loss over the span)

Further, as explained in the previous section, the power at the receiver received by the receiver should not exceed the maximum receiver sensitivity ($P_{R\max}$) after adjusting for span losses. This is to prevent saturation at the receiver. This input signal level is denoted as PIN and the maximum transmitter power is represented by $P_{T\max}$. With the span loss (PS) remaining constant $P_{T\max}$ represents the launch power and accordingly the input power (P_{IN}) is computed as follows:

$$P_{in}(\text{Input Power}) = P_{T\max}(\text{Max. Transmitter Power}) - P_S(\text{Span Loss})$$

$$P_{IN} = -4 - 13.5$$

$$P_{IN} = -17.5 \text{ dBm}$$

$$-17.5 \text{ dBm}(P_{IN}) \leq -5 \text{ dBm}(PR_{\max})$$

The value of P_{IN} satisfies the receiver sensitivity constraint while ensuring the viability of the optical system operating at the rate of STM-1 rate over a span of 5 km without any amplification stages or usage of attenuators.

Example 4.3

For this example let us consider a bidirectional link with a higher throughput (STM-64 or 10 Gbps) operating over a larger distance of 30 km. The minimum optical transmitter launch power is assumed to be −9 dBm and the maximum optical transmitter launch power is + 2 dBm at 1,550 nm. The minimum receiver sensitivity is assumed to be −32 dBm and the maximum receiver sensitivity is −5 dBm at 1,550 nm. There would be a total of two patch panels involved, and the fusion splicing would be employed with a maximum of three splices spanning the 30-km distance. The fiber used is a SI 8.1/125-μm SMF cable, and standard assumptions for attenuation, component, and nonlinear losses are made.

Accordingly, the following are the key parameters for the link under consideration:

1. PT min = −9 dBm;
2. PT max = + 2 dBm;
3. PRmax = −5 dBm;
4. PRmin = −32 dBm;
5. λ = 1,550 nm;

6. Link length = 30 km;
7. Link capacity = 10 Gbps;
8. Fiber type = 8.1/125-μm SMF-SI.

The system is operating at 10 Gbps or approximately 10 GHz. At such high bit rates, SPM, PMD, and SRS/SBS margin requirements must be taken into consideration while allowing for a degree of chromatic dispersion. XPM or FWM margins are not taken into account for a single-wavelength system. The link loss and viability calculations are as follows:

$$P_{BS} \text{ (Span Power Budget)} = PT_{min} \text{ (Min. Transmitter Power)} -$$
$$PR_{min} \text{ (Min. Receiver Sensitivity)}$$
$$= -9 \text{ dBm} + 32 \text{ dBm} = 23 \text{ dBm}$$

Table 4.3 presents the span analysis for Example 4.3.

$$P_m \text{ (Power Margin)} = P_{BS} \text{ (Power Budget)} - P_S \text{ (Span Loss)}$$
$$P_m = 20 \text{ dB} - 20.62 \text{ dB}$$
$$P_m = 3.38 \text{ dB} (> 0 \text{dB})$$

Thus it is evident that the 20.62-dB span loss is well within the 23-dB power budget (maximum allowable loss over the span). The input power received by the receiver (PIN), after accommodating the span losses must not

Table 4.3
Span Analysis (Example 4.3)

S.N	Factor	Computation	Loss (dB)
1	SMF-SI 8.1/125 μm	60 km at 0.2 dB/km	12
2	LC connectors	2 at 0.5 dB/connector	1
3	Fusion splices	6 at 0.02 dB/splice	0.12
4	Patch panels	2 at 1 dB/panel	2
5	SPM margin	–	0.5
6	PMD margin	–	0.5
7	SRS/SBS margin	–	0.5
8	Dispersion margin	–	1
9	Optical safety and repair margin	–	3
	TOTAL		20.62

exceed the maximum receiver sensitivity specification (PR max) so as to prevent receiver saturation.

$$P_{in} \text{ (Input Power)} = P_{T\max} \text{ (Max. Transmitter Power)} -$$
$$P_S \text{ (Span Loss)}$$
$$P_{in} = -0 - 29.62$$
$$P_{in} = -20.62 \text{ dBm}$$
$$-20.62 \text{ dBm} (PIN) \leq -5 \text{ dBm} (PR\max)$$

This satisfies the receiver sensitivity design equation while ensuring the viability of the optical system at a throughput of STM-64 without the need for amplification or attenuation. The calculations have not considered the deployment of DCMs, which would introduce additional link losses. The margins for nonlinear effects have, however, been included in the computation

4.7 Safety Guidelines

Fiber-optic networks have completely replaced copper-based transmission networks and form the core of the transport/backhaul network. The network could have an all-IP core or traditional transport-based hierarchies like SDH or the newer OTN along with DWDM nodes. Understanding the hazards in working with optical fibers and associated systems is of paramount importance for technicians and field engineers as well as planning, installation, and commissioning personnel. Several potential safety precautions to adopt while working with optical fiber networks are described as follows.

1. Material safety: Fiber-optic splicing and termination employs various chemical cleaners and adhesives. Specified handling procedures for these substances should be observed. Cleaners like isopropyl alcohol are highly flammable and should be handled with care. Skin contact must be avoided with chemicals through the use of specified gloves and other safety accessories.

2. Fire safety: Fusion splicers use an electric arc for cleaning and splicing fibers. Care must be taken to ensure that there are no flammable gases or liquids in the splicing area. Furthermore, splicing should never be attempted in manholes where there is a possibility of flammable gas accumulation. Also, smoking should be avoided near splicing stations in order to avoid contamination of the joints due to ash or smoke particles.

3. Electrical safety: Electrical safety is an important aspect to be considered for fiber installations near electrical cables, as in the case of common utility ducts, or aerial cable installations.
4. Optical safety: Lasers are classified by their wavelength and maximum output power into a total of seven classes of which four classes find application in the telecommunication industry. These classes are based on output power of the lasers and consequentially define the extent and nature of injury to individuals upon exposure. The classification system in existence since the 1970s was last revised in 2002. High-power lasers are employed for industrial and medical usage.

4.7.1 Causes of Injury

The primary mechanism of injury when exposed to a laser beam is thermal. Lasers can cause damage in biological tissues, both to the eye and to the skin, due to several mechanisms:

1. Thermal damage or burn occurs due to heat transfer leading to protein damage or breakdown.
2. Prolonged exposure (hours) to short wavelength light can trigger chemical reactions in tissues resulting in photochemical damage of the affected area.
3. Absorption of ultraviolet light with wavelengths less than 400 nm by the cornea and lens can cause photochemical damage.
4. Direct ocular (eye) exposure to a laser beam can cause temporary blindness as well as burns in the retina depending upon the class of the laser source.
5. Infrared light may cause permanent damage to the frontal areas of the eye, cornea, and retina due to the high levels of heat generated.

The commonly resulting injuries due to exposure to a laser beam are outlined in Table 4.4.

4.7.2 Maximum Permissible Exposure

Maximum permissible exposure (MPE) is the highest optical power that can be considered safe for ocular or skin exposure. MPE is usually about 10% of the measure that has a 50% chance of creating damage under worst-case conditions. The MPE is measured at the point of contact, for a given wavelength

Table 4.4
Common Injuries due to Human Exposure to Lasers

Wavelength	Injury
180–315 nm (UV-B, UV-C)[1]	Inflammation of the cornea (photokeratitis)
315–400 nm (UV-A)	Clouding of the eye lens (photochemical cataract)
400–780 nm (visible)	Photochemical damage to the retina, retinal burn
780–1,400 nm (near-IR)	Cataract, retinal burn
1.4–3.0 μm (IR)	Cataract, corneal burn
3.0 μm–1 mm	Corneal burn

1. Ultraviolet (UV) radiation is divided into three ranges based on the wavelength (UV-A = 315 to 400 nm, UV-B = 280 to 315 nm, and UV-C = 100 to 280 nm). Over 95% of the solar energy reaching the earth falls in the UV-A range.

and exposure time [9]. It may be noted that collimated[2] laser beams are highly dangerous even at low power levels since they have a high degree of spatial coherence (convergence). Class 1 lasers are generally considered to be within MPE. The laser classes are detailed as follows:

1. Class 1M: Class-1M lasers are safe for all conditions of use except when passed through magnifying optics such as microscopes and telescopes or fiber scopes. Class-1M lasers produce divergent beams and are within MPE. Focusing or imaging optics can narrow the divergence of the laser beam and increase the power beyond MPE.

2. Class 2M: A class-2M laser is safe for naked eye viewing due to the blink reflex. This may however not be applicable with the use of any optical viewers or other accessories that may result in light amplification. The exposure limits are same as those for class 2 without the use of instruments that may result in optical amplification

3. Class 3B: A class-3B laser is considered hazardous for direct ocular exposure. The accessible emission limits (AEL) for continuous lasers in the wavelength ranges from 315 nm to far infrared is 0.5 W. An AEL of 30 mW would be applicable for pulsating lasers that operate in the 400 to 700-nm range. The use of certified protective eyewear is recommended for applications where direct viewing of the laser source is likely.

4. Class 4: The use of class-4 lasers in the telecom industry for commercial applications is a relatively newer phenomenon. Class-4 lasers are high-power lasers whose power exceeds that of class 3B AEL. Exposure to class-4 lasers (direct, indirect, and diffused) can result in permanent

---

2. Parallel rays of light with minimal spread.

eye damage as well as skin burns. Further the power of these lasers may be enough to may ignite combustible materials. Protective eyewear is mandatory for individuals exposed to class-4 lasers.

4.7.3 AEL

The classification of a laser is based on the concept of AEL [10], defined for each laser class. The measure is usually maximum power (watts) or energy (joules) emitted at specific wavelengths (range) within a specified time period and at specific distances. AEL is independent of the wave length and can be mathematically defined as the product of MPE and an area factor also referred to as the limiting aperture (LA).

$$AEL = MPE * LA\ Area$$

The LA could be the pupil of the human eye. MPE refers to the radiation received by the eye or skin directly from a laser source or a beam after reflection. In contrast AEL refers to the emission of the laser. Table 4.5 summarizes laser classification.

4.7.4 Fiber-Handling Techniques

The connectorization or termination of fibers and their splicing generates scraps, shards, or broken ends that can be extremely dangerous, if not disposed of properly. These extremely sharp ends can easily penetrate the skin and can get blown into the eyes. Adequate care in the form of disposal containers or tapes should be employed to minimize the risk involved. Protective eyewear is recommended while splicing and terminating fibers in order to eliminate the possibility of glass particles entering the eye.

When working with bare fibers it is important to understand that bare fingers are likely to be covered with glass fragments. It is imperative that hands are washed thoroughly; avoid rubbing of eyes immediately after completion of fiber installation or maintenance tasks. If fiber particles are ingested, they can cause internal hemorrhaging.

Field technicians/engineers need to do the following when working with fiber-optic systems and networks:

1. Wear protective eyewear—mandatory for equipment with class 3 and 4 lasers.
2. Avoid consuming food and beverages in fiber work areas, as ingested fiber particles can cause internal hemorrhaging.

Table 4.5
Laser Classification [11]

S.N	Class	Laser Type	Criticality	MPE	AEL[1]
1	1	Very low-power	Safe	Not exceeded[2] (long-duration exposure 100/3,0000 sec)	$40\mu W$
2	1M	Very low-power	Safe for naked eye	Not exceeded[3]	$40\mu W$
3	2	Visible low-power	Safe for momentary exposure	Not exceeded[4] (exposure restricted to 0.25 sec)	1mW
4	2M	Visible low-power	Safe for momentary exposure	Not exceeded for naked eye (exposure restricted to 0.25 sec)	1mW
5	3R	Low-power laser	Potential hazard in case of accidental exposure	Exceeds MPE for naked eye by 5 times	5mW
6	3B	Medium-Power Laser	Ocular MPE exceeded for naked-eye viewing	Exceeds Ocular MPE for naked eye by 5 times	500mW
7	4	High-power laser	Ocular and skin MPE exceeded for naked-eye viewing	Exceeds ocular and skin MPE	No limit

1. For continuous-wave lasers.
2. Safe to use even with optical instruments.
3. May exceed MPE with the use of instruments capable of optical amplification.
4. Safe to use even with optical instruments.

3. Use disposable aprons to minimize the possibility of fiber particles getting stuck to one's clothing, since such particles may be inadvertently ingested by the operator.

4. Use protective gloves when handling bare fibers.

5. Ensure that the fiber link is dark prior to undertaking diagnostic or splicing activities.

6. Ensure, if possible, that the link is isolated from the network (remove connectivity from either end).

7. If link isolation is not possible, use a power meter to verify that a link is powered down. It is important to note that most transport systems have a safety feature referred to as automatic laser shutdown (ALS), automatic laser restart (ALR), and automatic power reduction (APR). This feature ensures that a section is powered automatically after a break. ALS and ALR ensure that a continuous laser beam is converted to a pulsating laser with a larger OFF time (than ON) time. This ensures that the average power is equivalent to that of a Class 1 laser. The default laser restart interval is 2 sec. A rule of thumb is to verify

that there is no power on a link for three minutes before commencing work. This is especially true when scoping a fiber using old fiber scopes that do not have a built-in attenuating mechanism.

8. When using a continuity checker, view fiber from at least 6 inches.
9. Ensure that hands are thoroughly washed before handling contact lenses (when applicable).
10. Dispose of all cut fiber pieces in a safe place and ensure that the work areas are kept clean.

4.8 Summary

This chapter presents the essential concepts and optical fiber-splicing techniques. The treatise is kept nonmathematical, deliberately, with the objective of presenting the key concepts and technology in a simple fashion. The key points covered in this chapter are summarized as follows.

1. Optical fiber splicing and the choice of connectors has a direct impact on the network speed, accuracy, and transmission distance.
2. The coupling efficiency of an LED is very low (around 2%) owing to its divergent output while the output of a conventional laser is highly convergent and thus results in a higher coupling efficiency. The output of a VCSEL is circular, resulting in higher coupling efficiency.
3. SFP is a specification for a new generation of optical modular transceivers that offer high speed and physical compactness. These devices are designed for use with SFF connectors and are hot-swappable.
4. The transmitter launch power measured in decibels per meter refers to the transmitter output at a specific wavelength. A higher power generally yields better results (higher extinction ratio and consequentially lower BER).
5. Receiver sensitivity and dynamic range refers to the minimum acceptable value of received power that is required to obtain an acceptable BER or performance.
6. The link loss is relative to the transmit power of a light source and represents the range of optical loss for an operational fiber-optic link, while meeting its operating specifications.
7. The modulation technique is a key element in realizing large high throughput telecommunication networks. This is especially crucial for systems operating at 10/40 Gbps as there is severe performance

degradation due to the interaction between dispersion and fiber nonlinearities.

4.9 Review

4.9.1 Review Questions

1. _____ refers to the optical energy that is reflected back into the waveguide especially at the spliced/terminating end.
   a. Back reflection
   b. Transreflection
   c. TIR
   d. Raleigh backscattering

2. There are two types of photodiodes that are commonly employed in optical equipment, the PIN diode and the _____.
   a. PLD
   b. APM
   c. ALD
   d. APD

3. The link loss is relative to the _____ of a light source and represents the range of optical loss for an operational fiber-optic link, while meeting its operating specifications.
   a. Received power
   b. Dispersion
   c. Transmit power
   d. Wavelength

4. The PIN is the preferred choice in short distance applications due to its standard power supply requirements.
   a. True
   b. False

5. The transmitter launch power measured in ____ refers to the transmitter output at a specific wavelength.
   a. dB
   b. dBm
   c. ps/nm
   d. mw

6. If PIN is greater than the maximum receiver sensitivity PR max, _____ must be considered to reduce signal level and bring it within the dynamic range of the receiver.

   a. Attenuation
   b. Amplification
   c. Regeneration
   d. None of the above

7. The power at the receiver should exceed the maximum receiver sensitivity (PR max) after adjusting for span losses.

   a. True
   b. False

8. The form factor of an SPF+ connector is one-sixth that of an SFP.

   a. True
   b. False

9. The coupling efficiency of a VCSEL is higher than that of a conventional laser.

   a. True
   b. False

10. The most commonly employed light source in communication systems is _____.

    a. LED
    b. Laser
    c. VCSEL
    d. CFL

4.9.2 Exercises

1. List the commonly used connector types in modern-day SDH, SONET, and Ethernet networks along with their key features and popular vendors.

2. What is an E/O unit? Describe its function in detail.

3. Describe the meaning of the term extinction ratio. What bearing does it have on the design of a fiber-optic link?

4. List the different types of light sources used in fiber-optic networks along with a summary of their key advantages and disadvantages.

5. What is a link loss budget? Prepare a link loss budget for an SDH link carrying voice traffic only and spanning 100 km; make standard assumptions.

6. Prepare a link budget for a bidirectional link operating at 10 Gbps over a distance of 130 km. The minimum optical transmitter launch power is assumed to be –5 dBm, and the maximum optical transmitter launch power is +4 dBm at 1,550 nm. The minimum receiver sensitivity is assumed to be –29 dBm and the maximum receiver sensitivity is –4 dBm at 1,550 nm. There would be a total of two patch panels involved, and there would be five fusion splices spanning the distance. The link would employ the following fiber—single-mode, glass, step, 9.3/125, 0.5.

   Standard assumptions are to be considered for attenuation as well as nonlinear and component losses.

4.9.3 Research Activities

1. Write a brief essay on fiber enclosures and their usage.
2. Prepare a chart evaluating the various connectors listed in Section 4.5.
3. List the latest developments and likely future developments in the areas of optical light sources and optical interfaces.
4. Find and subscribe to one online newsletter listing new developments in the field of optical interfaces.

4.10 Referred Standards

ANSI Z136.1: Safe Use of Lasers—LIA

ANSI Z136.3: Safe Use of Lasers in Health Care—LIA

ANSI Z136.4: Recommended Practice for Laser Safety Measurements for Hazard Evaluations—LIA

ANSI Z136.5: Safe Use of Lasers in Educational Institutions—LIA

ANSI Z136.6: Safe Use of Lasers Outdoors—LIA

ANSI Z136.7: Testing and Labeling of Laser Protective Equipment—LIA

ANSI Z136.8: Safe Use of Lasers in Research, Development, or Testing—LIA

ANSI Z136.9: Safe Use of Lasers in Manufacturing Environments—LIA

EIA/TIA-455-13: Visual and Mechanical Inspection of Fibers, Cables, Connectors and/or Other Fiber Optic Devices

EIA/TIA-455-187: Engagement and Separation for Fiber Optic Connector Sets

EIA/TIA-4750000-B: Generic Specification for Fiber Optic Connectors

EIA/TIA-475C000: Sectional Specification for Type FSMA Connectors

EIA-455-172: Flame Resistance of Fire Wall Connectors

EIA-455-17A: Maintenance Aging of Fiber Optic Connectors and Terminated Cable Assemblies

EIA-455-1A: Cable Flexing for Fiber Optic Interconnecting Devices

EIA-455-21A: Mating Durability for Fiber Optic Interconnecting Devices

EIA-455-26A: Crush Resistance of Fiber Optic Interconnecting Devices

EIA-455-34A: Interconnection Device Insertion loss Test

EIA-455-9: Fiber Optic Test Procedure for Bundle Connector

EIA-455-A: Standard Test Procedure for Fiber Optic Fibers, Cables, Transducers, Sensors, Connecting and Terminating Devices, and Other Components

IA-440-A: Fiber Optic Connector Terminology

TIA/EIA-455-158: Measurement of Breakaway Frictional Force in Fiber Optic Connector Alignment Sleeves

TIA/EIA-475EA: Blank Detail Specification for Connector Set for Optical Fiber and Cables, Type BFOC/2.5, Environmental Category I

TIA/EIA-475EB: Blank Detail Specification for Connector Set for Optical Fiber and Cables, Type BFOC/2.5, Environmental Category II

TIA/EIA-475EC00: Blank Detail Specification for Connector Set for Optical Fiber and Cables, Type BFOC/2.5, Environmental Category III

TIA/EIA-604: Fiber Optic Connector Intermateability Standards

TSB-62: Informative Test Methods for Fiber Optic Fibers, Cable, Opto-Electronic Sources and Detectors, Sensors, Connecting and Terminating Devices, and Other Fiber Optic Components

Z136.2: Safe Use of Optical Fiber Communication Systems Utilizing Laser Diode and LED Sources—LIA

4.11 Recommended Reading

4.11.1 Books

Chomycz, B., *Fiber Optic Installer's Field Manual,* McGraw Hill, 2000.

Huber, J. C., *Industrial Fiber Optics Networks,* Instrument Society of America, 1995.

Silvello, B., *Coherent Optical Communications Systems,* John Wiley & Sons, 1995.

Senior, J., *Optical Fiber Communication: Principles & Practice,* Prentice Hall, 2008.

Keiser, G., *Optical Fiber Communication,* McGraw Hill, 2008.

Warier, S., *The ABCs of Fiber Optic Communication,* Norwood, MA: Artech House, 2017.

4.11.2 URLs

http://ecmweb.com/training/electrical_basics/electric_basics_fiber_optics_7/.

http://ecmweb.com/mag/electric_basics_fiber_optics_4.

http://www.arcelect.com/.

http://www.nfpa.org.

http://www.ul.com/telecom.

http://www.ciscopress.com.

http://www.hubersuhner.com/.

http://www.avap.ch/.

http://en.wikipedia.org/.

http://www.telebyteusa.com/.

http://search.techrepublic.com.com/search/fiber-optics.html.

http://www.fiberopticproducts.com.

http://www.ask.com/questions-about/Fiber-Optics.

http://www.optiwave.com/.

http://www.fibersolutionsonline.com/.

References

[1] Warier, S., *The ABCs of Fiber Optic Communication,* Norwood, MA: Artech House, 2017.

[2] University of Illinois (2006), retrieved March 18, 2008. Source URL http://www2.uic.edu/stud_orgs/prof/pesc/part_3_rev_F.pdf.

[3] *DPCM Overview* (n.d.), Stanford University, retrieved March 21, 2008. Source URL http://www.stanford.edu/class/ee368b/Handouts/15-DPCM.pdf.

[4] University of Illinois (2006), retrieved March 29, 2008. Source URLhttp://www2.uic.edu/stud_orgs/prof/pesc/part_3_rev_F.pdf.

[5] Gnauck, A. H., et al., "25x 40-Gb/s Copolarized DPSK Transmission Over 12x100-km NZDF with 50-GHz Channel Spacing," *IEEE Photonics Technol. Lett. 15,* 2003, pp. 467–469.

[6] Gnauck, A. H., and P. J. Winzer, "Optical Phase Shift Keyed Transmission," *J. of Lightwave Technology,* Vol. 23, No. 1, Jan. 2005.

[7]   Tim, K., (Tkgd2007) (31 May 2008), *A cutaway Diagram of a Coaxial Cable*, retrieved March 09, 2008, from Wikipedia. Source URL http://en.wikipedia.org/wiki/File:Coaxial_cable_cutaway.svg.

[8]   ICT Technologies (n.d.), *Modulation & Multiplexing*, retrieved March 17, 2008. Source URL< http://cbdd.wsu.edu/kewlcontent/cdoutput/TR502/page19.htm>.

[9]   The Fiber Optic Association (June 13, 2008), *Guide to Fiber Optics & Premises Cabling*, retrieved from www.thefoa.org. Source URL http://www.thefoa.org/tech/connID.htm.

[10] Schröder, K. (ed.), *Handbook on Industrial Laser Safety*, Technical University of Vienna, 2000.

[11] Underwriters Laboratories, Optical Fiber Raceway, Association (n.d.), retrieved April 22, 2008. Source URL http://www.ul.com/.

Part II

Photonic Network Architecture: The Immediate Past and Present

5

Transport Networks: Prologue

5.1 Chapter Objectives

The invention of the telephone was a major milestone in the evolution of our civilization. However the initial years following the invention did not witness any development in the supporting ecosystem—on the scale that we are now accustomed to—due to the monopoly regulating telecom policies and services. The legacy networks, initially based on overhead copper cables, were referred to as transmission networks and were designed to carry only voice signals. In fact, it is ironic that though the word telecommunication refers to communication over a distance, as the growth of long-distance communication networks was stunted by the lack of standardization and the resultant incompatibility between the physical interfaces and the bit rates of the various providers within and across countries. Fortunately, the subsequent establishment and growth of standardization bodies led to a uniform standardization process that ensured compatibility between the different carriers and facilitated deployment of multivendor equipment. The introduction of T1 systems heralded the beginning of digital transmission systems in North America. These new digital transmission networks were referred to as the asynchronous networks, and the resultant multiplexing hierarchy was referred to as the North American Digital Hierarchy (NADH). The plesiochronous digital hierarchy (PDH) is the European equivalent of the NADH transmission technology employed to transport voice and data over digital transmission networks. Both the hierarchies were based on TDM. This chapter provides the conceptual basis that is required to decode and understand the complex optical communication networks currently being deployed. This chapter also lays the foundation for the deciphering the next part of the book while including frame transmission, time synchronization, payload mapping/demapping, and protection schemes. The evolution of the telecom standardization bodies, and processes as well as the key standards governing the functioning of the asynchronous optical networks is also detailed in this chapter. Moreover, the chapter details the architecture and functioning of the PDH network to provide readers with the conceptual basis for understanding the SDH networks presented in Chapter 6 and to appreciate its advantages. Please note that the present TDM transport networks are based on OTN technologies.

Key Topics

- The organization and functioning of various telecom standardization bodies;
- The architecture, functioning, and limitations of asynchronous networks;
- The organization and working of PDH networks;
- The concepts related to the transport of Ethernet frames over PDH networks;
- The need and evolution of synchronous optical networks.

5.2 Introduction

Transport networks are an important component of modern-day telecommunication networks. They form the core or the backbone of a telecom network and interconnect (provide a high-speed pathway) the service-providing elements within the network. The importance of transport networks has grown over the years, and they have become the high-availability, high-performance backbone for the voice, video, and data services commonly referred to as triple-play networks. A transport network connects all access points (e.g., voice/data switches, mobile communication towers, and exchanges) while ensuring guaranteed and error-free transmission of voice, video, and data information through the use of self-healing protection schemes.

Transport networks include a set of facilities and equipment that carries data and information between the switching network elements (NEs) that are responsible for end customer connectivity. The transport network ensures reliable delivery of data and information between end points. In the absence of a common backbone, separate networks would be required for delivering voice (using different technologies like CDMA, GSM, and 4G) as well as data (using different architecture and technologies such as Ethernet, frame relay, and ATM) and video. The transport network provides a convergent backbone that can serve as a common framework supporting real-time triple-play services.

This eliminates the need for redundant hardware, software, processes, and procedures, while providing cost-effective services. In a relatively short span of time the transport networks grew geographically as well as in terms of traffic (bandwidth) handled. It therefore became important to have effective operations, administration, maintenance, and provisioning (OAM&P) systems to ensure their manageability, reliability, and cost-effectiveness. Thus, the effective management of transport networks is of paramount importance. Consider, for example, that one of the key factors that prevented the use of Ethernet as a transport mechanism was the lack of OAM&P capabilities.

The advent of transport networks followed the emergence of standard-based SONETs. The older networks (referred to as transmission networks) were essentially point-to-point copper cables with separate wires dedicated to each voice channel. In the early days, the switching was done manually by operators with patch panels. This approach was clearly unsuitable for a large-scale implementation. Multiplexing techniques were introduced to overcome the shortcomings of manual switching. Figure 5.2 lists the development stages of the transmission/transport networks.

The older transmission networks used analog signals for transmission, and hence the multiplexing technology employed was FDM. FDM was the most appropriate multiplexing technique for working with analog signals. The overhead cables strung on poles were later replaced by underground copper cables. However, these cables were difficult to lay over mountainous terrain, deserts, and bodies of water. This led to the deployment of microwave systems in the transmission network. Analog signals require frequent amplification and suffer from problems related to a decrease in the SNR. The introduction of the ICs in

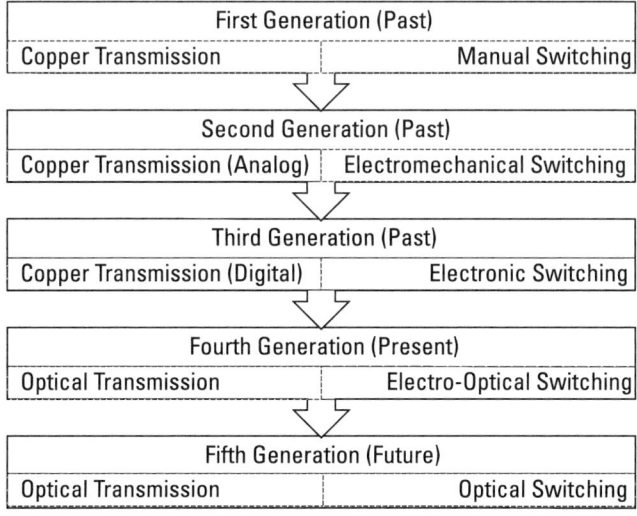

Figure 5.1 Evolution of transmission networks.

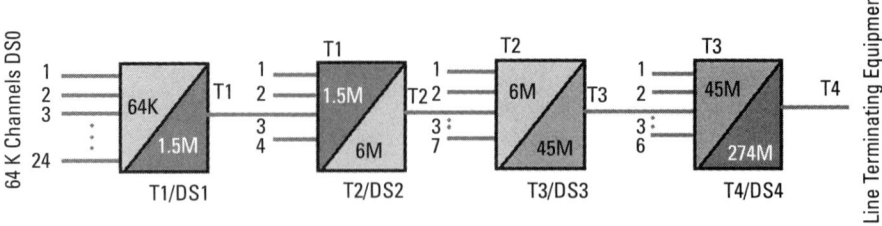

Figure 5.2 NADH Multiplexing hierarchy—simplified view.

the early 1960s paved the way for the development and deployment of digital transmission systems that revolutionized the functioning and efficiency of the transmission networks. This development also paved the way for the introduction of TDM techniques over the existing copper networks as well as microwave systems. Figure 5.2 traces the evolution of telecommunication networks.

5.3 Asynchronous Networks

The introduction of T1 systems heralded the beginning of digital transmission systems in North America. The new digital transmission networks were referred to as the asynchronous networks, and the resultant multiplexing hierarchy was referred to as the NADH. NADH was based on TDM. The input base signal for NADH is the 64K PCM signal (explained in Chapter1). The subsequent bit rates, as indicated in Table 5.1, are multiples of the first order rates, which, in turn, are multiples of the base signal of 64 Kbps. The 64-Kbps PCM signal is also referred to as Digital Signal Level 0 (DS0) in the North American parlance. The T1 also referred to as digital signal level 1 (DS1), which is the first order of NADH, has a bit rate of 1.544 Mbps. Table 5.2 presents the multiplexing rates for the North American as well as the European hierarchies. The North American rates are commonly referred to as 1.5-Mbps (1.544 Mbps), 6-Mbps (6.312 Mbps), and 45-Mbps (44.736 Mbps) signals. The bit rate calculations are discussed in Table 5.1.

Table 5.1
European PDH Multiplexing Hierarchy

1 DS0 signal has a rate of 64 Kbps.
32 such DS0s are multiplexed to give 1 E1= 2.048 Mbps.
4 E1s + 4 DS0s are multiplexed to give E2= 8.448 Mbps.
4 E2s + 9DS0s are multiplexed to give E3= 34.368 Mbps and so on…

Table 5.2
ITU-T PDH Multiplexing Hierarchy

European		North American		
Signal Designation	Bit Rate	Signal Designation	Alias	Bit Rate
E1	32 * 64 Kbps = 2.048 Mbps (2 Mbps)	T1	DS1	24 * 64 Kbps = 1.544 Mbps (1.5 Mbps)
E2	4 * E1 = 8.448 Mbps (8 Mbps)	T2	DS2	4 * DS1 = 6.312 Mbps (6 Mbps)
E3	4* E2 = 34.368 Mbps (34 Mbps)	T3	DS3	7 * DS2 = 44.736 Mbps (45 Mbps)
E4	4 * E3 = 139.264 Mbps (140 Mbps)	T4	DS4	6 * DS3 = 274.176 Mbps (275 Mbps)
E5*	4 * E4 = 557.056 Mbps	—	—	—

Input = 64-Kbps PCM channel (also referred to as DS0 in North American hierarchy).

Example 5.1 (Refer to Figure 5.2)
- A DS1 consists of 24 eight-bit PCM voice channels or DS0s.
- The size of a T1 frame is therefore 24*8 bits = 192 bits plus an additional bit added (to this group of 24 channels) for framing making it a total of 193 bits.
- In a PCM system 8,000 frames are transmitted in a second, hence the T1 bit rate works out to 8,000 frames/sec * 193 bits/frame or 1,544,000 bits/sec or 1.544 Mbps. Table 5.1 presents the standard European multiplexing hierarchy.

Note 5.1
T_x (where x = 1, 2, 3) refers to the physical interface, including cables, mechanical connectors, and repeaters, while the traffic flowing over these physical interfaces is referred to as DSx (where x = 1, 2, 3). It may be noted that DS0 refers to the 64-Kbps PCM input channel in NADH.

Example 5.2
A T1 represents the physical layer while the traffic flowing over this physical layer is referred to as DS-1. The terms, however, are used interchangeably in most literature.

Note 5.2

It is important to note that the duration of a time slot in a PCM channel containing eight bits is 125 µs. Twenty-four such channels are multiplexed together to form a DS1 signal. However, the time slot still remains 125 µs. The number of bits now occupying this time slot is 193. Hence as we go higher in the multiplexing hierarchy the number of bits, and consequently the bit rates, increases, and as a result their bit width reduces.

The second order of the hierarchy is represented by the DS2 signal, which consists of 4 * DS1 signals together that are time-division–multiplexed together. Similarly the third order of the hierarchy is represented by the DS3 signal, which consists of 7 * DS2 channels multiplexed together. The most important consideration during the development of the hierarchy was to make optimum use of the existing transmission network infrastructure (STP cables for DS1 transmission). This consideration led to choice of the multiplication factor at the different levels. The fourth-order signal rate DS4 of 274.176 Mbps was achieved by multiplexing 6 * DS2 channels together. Out of the above rates the DS1 and DS3 were the most popular and widely used rates. The major drawback in obtaining transmission speeds above the fourth-order rate of 274.176 Mbps was the bandwidth limitation of the copper cables and the lack of a common synchronization reference. A few important points regarding the NADH are summarized:

1. To form a signal of level *n*, of the NADH, a group of signals from level *n*–1 are interleaved in a round-robin fashion using bit interleaving techniques.

2. The DS1 signal is obtained by byte-interleaving 24 DS0 or 64-Kbps channels together. These 24 DS0 channels are digitized using a common clock; hence there are no problems related to time synchronization

3. However the four DS1 channels that form the input to a DS2 multiplexer would be having different clock rates as the signals are asynchronous. This would cause the frequency as well as the framing phase of these signals to be different. The DS2 multiplexer must be capable of accommodating these different frequencies within a specified range. The frequency of a DS1 signal is 1.544 MHz (corresponding to the 1.544 bit rate). The tolerance provided is ± 77.2 Hz. This would in turn imply that the number of bits coming in on the input DS1 lines would be varying. As a result these input streams cannot be directly bit-interleaved to achieve higher rates.

4. A technique known as bit stuffing is used as solution to the above problem.

Figure 5.2 presents a simplified representation of the NADH.

5.3.1 Bit Stuffing

Bit stuffing is the basic technique used in all multiplexers in asynchronous networks to counter the effects arising from the lack of a common clock. Each of the incoming bit streams (in a NADH/PDH hierarchy) is referenced to a different clock (each DS1 mux has its own local clock and the DS2 mux receives four DS1 tributaries that are timed differently) and hence has a bit rate that is variable within certain limits. In order to accommodate these clock variations the DS2 mux employs a clock rate that is higher than the combined rate of all the tributaries as well as the associated overheads. In simple terms it means that a DS2 multiplexer makes provision for more bits than the 193 it is expected to receive on a DS1 line. This allows the multiplexer to accommodate the maximum tributary rate. It is understood that if the incoming bit rate is higher (within the specified range) the extra bit positions will be filled. On the contrary if the bit rates are lower some of the bit positions will be empty or contain dummy bits referred to as stuff bits or S bits.

This implies that the receiving mux needs a mechanism wherein it can first understand the position of these extra bits (stuff bits) and second decipher whether these bit positions contain data or stuff bits. This is achieved by the use of control bits or C bits that are present within the mux overhead. By reading the value of these C bits a mux can decipher the contents of the extra bit positions. The positions of the S bits are fixed within the frame structure. The demultiplexer can therefore understand the positions of these S bits and by reading the C bits can recognize whether the extra bit positions contain stuffing bits or data.

5.4 PDH

The PDH, the European equivalent of the NADH, is a transmission technology that is employed to transport voice and data over digital transmission networks. The term plesiochronous is derived from the Greek words "plesio" meaning near, and "chromos" meaning time. These networks do not employ a common reference clock, and the receiving equipment derives its time from the incoming data on its input interfaces (the line timing clock has to be derived separately for each interface on which it is receiving data). Due to this technique the different elements of the PDH networks would not be exactly synchronized to each other.

The concept of plesiochronous timing can be explained with the help of the following example. Consider a class of five individuals, a trainer and four participants. The trainer will communicate with the participant in a round robin fashion. All the five individuals have their own watches, which are not synchronized to any common reference. The different watches follow the same timing principle, 60 seconds to a minute and 60 minutes to hour, but they may not be running at the same rate. When the trainer is communicating with participant 1, they will synchronize their watches. The same synchronization does not hold when the trainer is communicating with participant 2. They will have to resynchronize their watches. The same process continues with the other participants. This is a simple example that attempts to replicate the plesiochronous communication methodology. It should be noted that an actual mux has multiple interfaces over which it simultaneously receives and transmits data. The process, however, remains the same.

The first multiplexing order in the PDH hierarchy is formed by multiplexing 32 * 64-Kbps channels; it has a bit rate of 2.048 Mbps and is commonly referred to as the 2-Mbps channel. Out of the 32 input channels two channels are used for signaling and synchronization, and the remaining 30 channels are available for carrying traffic.

Note 5.3

The entire bandwidth of an E1 can be used for carrying data traffic. In this case there would not be any channel demarcation within the E1 frame structure (i.e., it is unchannelized).

In the PDH hierarchy the subsequent multiplexing orders use a constant multiplication factor of 4. The second order of the multiplexing hierarchy is therefore formed by multiplexing four first-order channels or E1s together. As in the case of asynchronous networks, there are variations in the bit rates of the multiplexer output. At the first order this variation is limited to ± 50 ppm (parts per million). The bit-stuffing technique is gain employed to ensure that variations in the input streams, within the tolerated range, are accommodated. The higher order signal rates are derived in a similar fashion; 4 * E2's are multiplexed together to form an E3 and 4 * E3's are multiplexed together to form an E4. Bit-interleaving techniques are employed by the multiplexers to combine the four input streams together. The receiving multiplexer uses the control bits to decode the signal, taking into account the variations in the streams. The hierarchy of signals is depicted in Table 5.2. Figure 5.3 presents a simplified pictorial representation of the PDH, while the associated signal structure is presented in Figure 5.8. Table 5.3 lists the computational details of the PDH multiplexing hierarchy, while Table 5.4 outlines tolerances and transmission voltages.

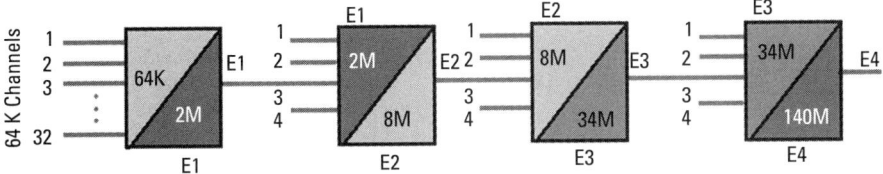

Figure 5.3 PDH Multiplexing hierarchy—simplified view.

Table 5.3
CCITT G.702 Recommendation

Digital Hierarchy Level	PDH	Signaling Channels	Service Rate	NADH	Signaling Channels (Reserved Bits)	Service Rate
1	2.048 Mbps	2*64 Kbps = 128 Kbps	1.920 Mbps	1.544 Mbps	F bit	1.536 Mbps
2	8.448 Mbps	4*64 Kbps = 256 Kbps	8.192 Mbps	6.312 Mbps	F Bit and bits in Slot 97 and 98	6.144 Mbps
3	34.368 Mbps	9*64 Kbps = 864 Kbps	33.504 Mbps	44.376 Mbps	M_i, F_0, F_{11}, F_{12}, C bits*	44.407 Mbps 44.209* Mbps
4	139.264 Mbps	28*64 Kbps = 1792 Kbps	137.472 Mbps	274.176 Mbps		

5.4.1 CCITT Recommendations

The task of drawing up international standards to facilitate the growth of telecommunication networks was entrusted to the CCITT by the ITU. The CCITT was formed by the amalgamation of the CCIF and CCIT committees. A detailed treatise of all the CCITT recommendations pertaining to the functioning of digital networks is beyond the scope of this book. However, the major recommendations of the CCITT governing the design and working of digital networks are summarized in the following sections.

5.4.1.1 G.702

The G.702 recommendation was drafted by the CCITT with a view of standardizing the hierarchical bit rates to be employed in digital networks thereby facilitating interconnections at the international level and removing the usage of proprietary bit rates and interfaces. The key considerations in arriving at these rates were the backward compatibility and the characteristics of the then

Table 5.4
G.703 Electrical Interfaces Specifications: Summary

S.N	Bit Rate	Tolerance	Interface	Impedance	Coding
1.	64 Kbps	± 100 ppm	Two symmetrical pairs of wires (data and composite timing signal–64 KHz/8 KHz)	110 Ω	AMI
2.	1,544 Kbps	± 32 ppm	Balanced twisted pair	100 Ω	AMI/B8ZS
3.	2,048 Kbps	± 50 ppm	Balanced twisted pair/coaxial cable	75Ω/120 Ω	HDB3
4.	6,312 Kbps	± 10 ppm	Symmetrical pair/coaxial Cable	110Ω/75 Ω	B6ZS/B8ZS
5.	8,448 Kbps	± 30 ppm	Coaxial cable	75Ω	HDB3
6.	32,064 Kbps	± 50 ppm	Coaxial cable	75Ω	AMI
7.	44,736 Kbps	± 20 ppm	Coaxial cable	75Ω	B3ZS
8.	139,264 Kbps	± 15 ppm	Coaxial cable	75Ω	CMI

existing analog systems, the transmission media employed, well as the ease of grouping and routing the signals and the services being provided on the network. Table 5.3 details the CCITT recommendation G.702.

5.4.1.2 G.703

The ITU-T G.703 recommendation lays down the physical and electrical characteristics of the interfaces at the bit rates specified by the G.702 recommendation. The characteristics include the general parameters, input and output port specifications, and conductor earthing requirements as well as the coding specifications. The G.703 standard facilitates the interconnection of the various multiplexing equipment as well as exchanges to set up digital links or connection internationally. The recommendation covers the functional requirements of the interfaces that include the full-duplex transmission of a 64-Kbps signal carrying traffic, a 64-KHz timing signal, and an optional 8-KHz timing signal. The 8-KHz signal is required by the controlling equipment for PCM multiplexing/demultiplexing. However it is not mandatory for the subtending equipment to use the signal provided by the controlling or the higher order equipment or to supply an 8-KHz timing signal. Table 5.4 summarizes the specifications for the G.703 electrical interfaces while Table 5.5 presents the electrical characteristics of the G.703 interfaces.

5.4.2 Interfaces

The G.703 recommendation specifies three types of interfaces, described as follows:

Table 5.5
PDH Multiplexing Hierarchy: Interface Specifications

Multiplexing Order	Standard	Bit Rate	Line Code	Amplitude (V)	Attenuation (dB)
1	G.704/732	2.048 Mbps ± 50ppm	HDB3	2.37 / 3.0	6
2	G.742	8.448 Mbps ± 30ppm	HDB3	2.37	6
3	G.751	34.368 Mbps ± 20ppm	HDB3	1.0	12
4	G.751	139.264 Mbps ± 15ppm	CMI	1.0 V	12

1. Codirectional interface: In a codirectional interface, illustrated in Figure 5.4, the timing signal or the clock is transmitted in the same direction as that of the information signal. A 64-Kbps codirectional interface has a tolerance of ± 100 ppm and employs a pair of balanced wires for each direction. A binary 1 is represented by the code 1100 and a binary 0 is represented by 1010. [Refer to Figure 5.5(a).] The binary information is converted to a three-level signal by alternating the polarity of adjacent bits [Figure 5.5(b)]. This alternation is violated after every seventh bit (eight bit) block to indicate the end of an octet [Figure 5.5(c)]. The signal pulse shape is rectangular with a voltage of 1.0V for a pulse mark and 0V ± 0.10V for a space or no pulse. The code conversion steps are summarized in Table 5.6.

2. Centralized clock interface: In a centralized clock interface, shown in Figure 5.6, the timing signals for both the directions of transmission are from a common clock. The central clock might, in turn, be synchronized to an external reference, or it may itself be the reference or

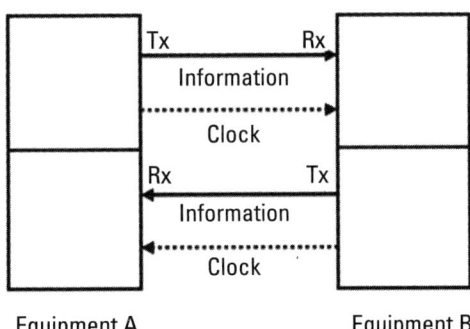

Figure 5.4 Codirectional interface.

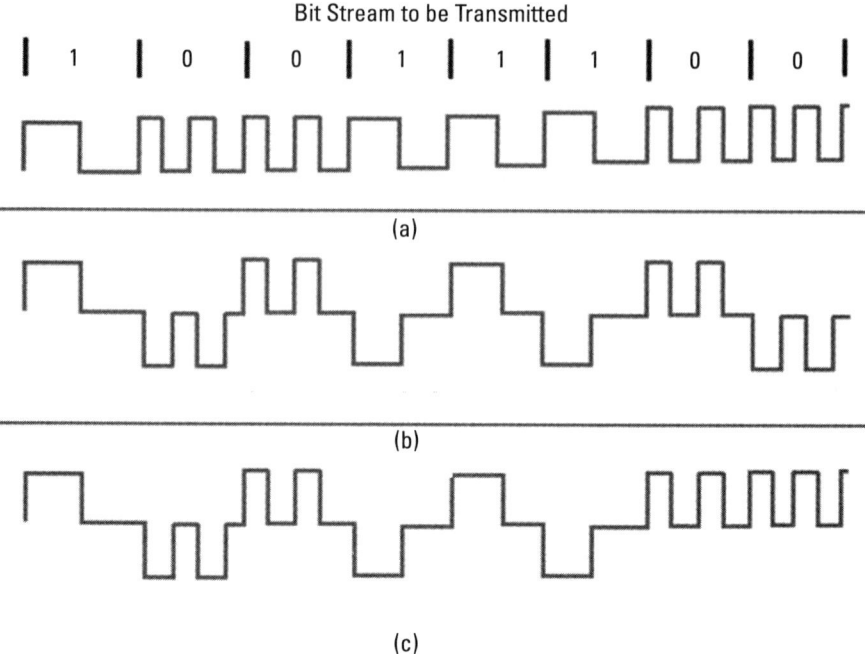

Figure 5.5 Data signal code conversion: (a) electrical representation of a bit stream, (b) polarity reversal of consecutive bits, and (c) violation bit (end of octet).

Table 5.6
Data Signal Code Conversion Steps

Step No.	Description
1	A 64-Kbps time slot is divided into two intervals of four bits each
2	A binary 0 is coded as a block with the following four-digit code assigned: 1010
3	A binary 1 is coded as a block with the following four-digit code assigned: 1100
4	The binary signal is converted into a three-level signal by alternating the polarity of the adjacent blocks
5	The polarity alternation is violated after every seven blocks (eighth block) to signify the end of an octet

employ line timing. The clock transmission is via a symmetrical pair of wires employed for each transmission direction. The line coding employed for the data signals is alternate mark inversion (AMI).

The codirectional interface and the centralized clock interface are used in synchronized networks.

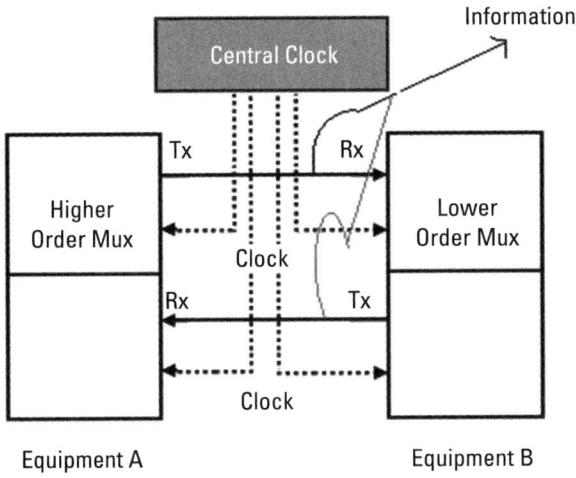

Figure 5.6 Centralized clock interface.

3. Contradirectional interface: In a contradirectional interface, shown in Figure 5.7, the timing signals for both the directions of transmission are toward the subtending equipment.

5.4.2.1 G.704

The G.704 recommendation includes the functional characteristics of the SDH NE interfaces, digital exchanges for telephony applications, and ISDN and PCM multiplexing equipment. It also provides the basic frame structures, including details of frame length, frame alignment signals, cyclic redundancy check (CRC) procedures, and other allied information. In addition, the recommendation includes information about accommodating channels at 64 Kbps as well as other bit rates within the basic frame structures defined.

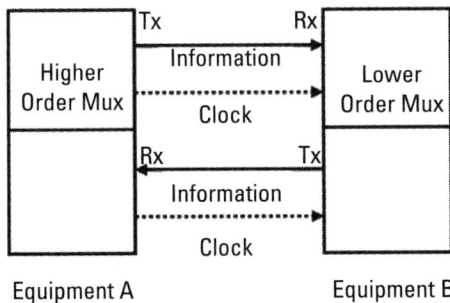

Figure 5.7 Contradirectional interface.

5.4.3 Basic Frame Structures

This section presents the basic frame structures at the various hierarchies of NADH and PDH along with the associated signaling and control mechanisms and the associated interface specifications.

5.4.3.1 NADH DS1 Frame Format

The DS1 frame format is illustrated in Figure 5.8, where the starting bit of every frame is the F bit. The F bit is used for frame alignment and performance monitoring and to provide a data link for OAM purposes.

Signaling and Control Mechanisms

The process of frame alignment and signaling at each level of the hierarchy is a complex subject that field personnel need not learn. Furthermore, the book does not detail the functioning of the legacy NADH and PDH. However, to facilitate a comparison with the SONETs as well satisfy the curiosity of a certain section of readers, the usage of the F bits at the first level of the NADH is included. Additional details can be obtained by referring to the G.704 standard document from the ITU-T website. The URLs are included in the additional reading section. Figure 5.9 presents the associated signaling at each order of the hierarchy.

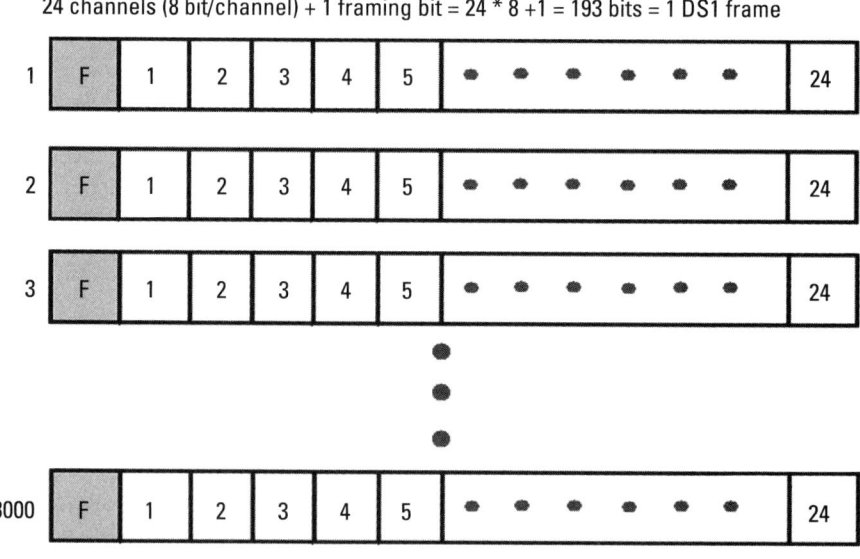

Figure 5.8 NADH DS1 frame structure.

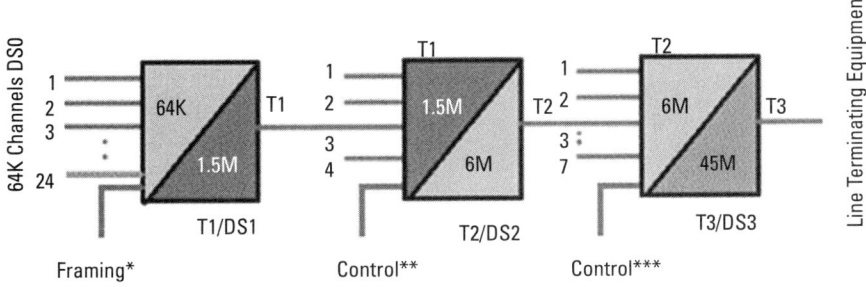

Figure 5.9 ADH multiplexing hierarchy—signaling structure.

* 1 framing bit per 24 * DS0
** 1 control bit after every 48 payload bits
*** 1 control bit after every 84 payload bits

There are two methods of using the F bits. They are outlined in Tables 5.7 and 5.8.

5.4.3.2 NADH DS2 Frame Format

A DS2 master frame consists of four * DS1 frames with additional framing bits. The alignment of the subframes (4 * DS1 frames) within the master frame is achieved by employing additional bits referred to as M bits. The payload or the information fields consist(s) of the 4 * DS1 channels that are bit-interleaved in a round robin fashion. The following are some important calculations with respect to the frame structure.

Note 5.4

- The bit rate of the DS2 frame is 6.312 Mbps.
- The tolerance range is ± 208 bps.

The DS2 signal format is shown in Figure 5.10. The F-bits provide the master frame alignment pattern while the M-bits indicate the subframe alignment within the master frame (Refer to Table 5.9). The DS2 payload consists of 4 * DS-1 payloads that are bit-interleaved together in a round robin fashion.

5.4.3.3 NADH DS3 Frame Format

A major drawback of the multiplexing technique employed in asynchronous networks is the hierarchy to be followed for multiplexing and demultiplexing. In both NADH as well as PDH networks signals can only be demultiplexed at the level at which they are multiplexed. This in turn implies that a DS3 signal can only be constructed by multiplexing 7 * DS2 signals, which, in turn, are

Table 5.7
24-Frame Multiframe Alignment

Frame Number	Bit Number	Assignment		
		Frame Alignment	Data Link	CRC*
1	1	–	M	–
2	194	–	–	C1
3	387	–	M	–
4	580	0	–	–
5	773	–	M	–
6	966	–	–	C2
7	1,159	–	M	–
8	1,352	0	–	–
9	1,545	–	M	–
10	1,738	–	–	C3
11	1,931	–	M	–
12	2,124	1	–	–
13	2,317	–	M	–
14	2,510	–	–	C4
15	2,703	–	M	–
16	2,896	0	–	–
17	3,089	–	M	–
18	3,282	–	–	C5
19	3,475	–	M	–
20	3,668	1	–	–
21	3,861	–	M	–
22	4,054	–	–	C6
23	4,247	–	M	–
24	4,440	1	–	–

*CRC: 6 (6-bit CRC)

constructed by multiplexing 4 * DS1 channels together. A DS3 signal cannot be directly be constructed by multiplexing 28 * DS1 signals together. This adds complexity in network design and deployment. A 1.5-Mbps signal drop could not be directly affected from a 45-Mbps line. The 45-Mbps would have to be demultiplexed to the 6-Mbps signal, which, in turn, has to be demultiplexed to the first order rate of 1.544 Mbps.

Table 5.8
12 Frame Multiframe Alignment

Frame Number	Frame Alignment Signal (FAS)	Multiframe Alignment Signal (MFAS)
1	1	–
2	–	S
3	0	–
4	–	S

In order to overcome the above drawback AT&T developed a different version of the DS3 signal format, referred to as the C-bit parity DS3, which allowed DS1 signals to be directly mapped on to a DS3 structure. The bit rate of the C-bit parity DS3 is unchanged, but some of the C-bits within the frame overhead are redefined to facilitate the direct multiplexing (refer to Table 5.10). The signal structure also provides for OAM&P communications between the multiplexing equipment. The DS3 was by far the most popular rate employed in the NADH networks. The ANSI proposal for SONET therefore started with a network node interface (NNI) of around 50 Mbps. However there were proprietary systems that employed rates above 45 Mbps, including 90 Mbps, 135 Mbps, 405 Mbps, and 560 Mbps. Figure 5.11 highlights the DS3 frame structure.

5.4.3.4 PDH E1 Frame Format

The E1 frame, as shown in Figure 5.12, represents the first level of the PDH. The voice (speech) signals are encoded using A-law in accordance with ITU-T G.711 specification. An E1 is formed by multiplexing 32 * 64-Kbps PCM channels (numbered 0–31) together, using A-law encoding.

The bit rate of an E1 frame is 32 * 64 Kbps or 2,048 Kbps with a sampling rate of 8,000 samples/sec. The time duration of an E1 remains 125 μs and contains 32 channels with eight bits per channel or a total of 32 * 8 = 256 bits per time slot. Out of the total of 32 channels, channel 0 is used for frame alignment and channel 16 is reserved for signaling; 30 channels are available for carrying user traffic. Figure 5.13 illustrates the PDH signaling structure, the details of which are listed in Table 5.11.

A multiframe is formed by combining 16 E1 frames. Figure 5.14 depicts a 16-frame E1 multiframe. The first octet of every alternate frame (starting from frame 0) in the multiframe is reserved for FAS. The first octet of the frames not containing the FAS is reserved for OAM&P purposes. The frame alignment pattern is 0011011. In the alternate frames not containing the FAS the second bit is set to 1 to ensure that the FAS bit pattern is not repeated.

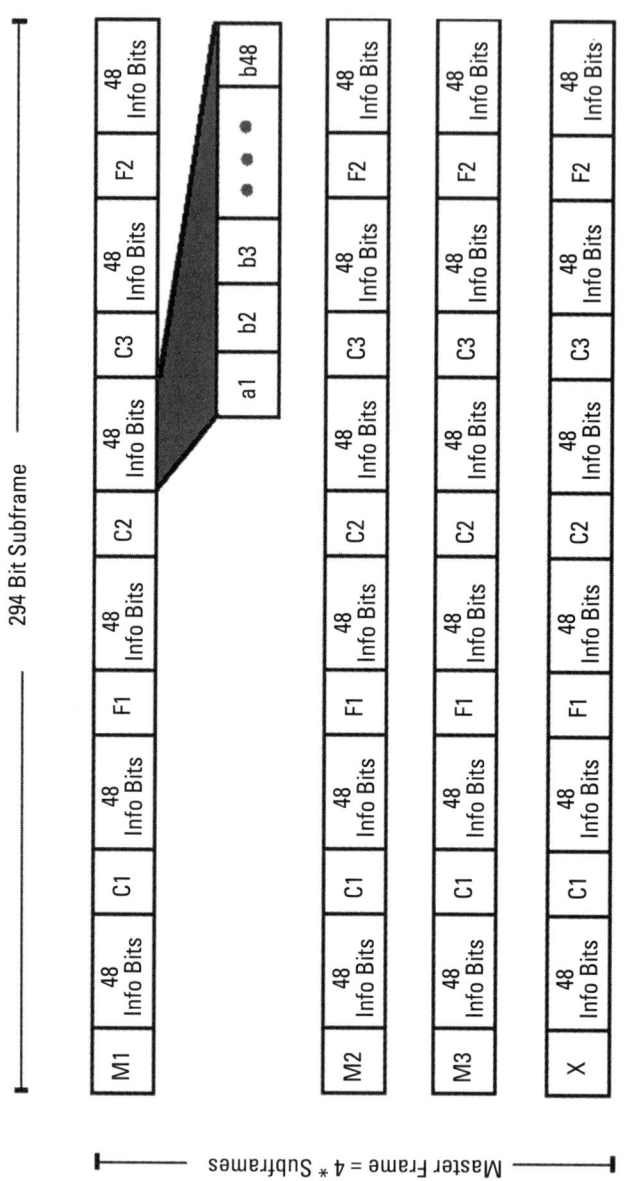

Figure 5.10 DS2 frame format.

Table 5.9
DS2 Frame Control Bit Description

Bit	Description
M1	0
M2	1
M3	1
F1	0
F2	1
X	0 (no alarm condition) 1 (alarm condition)
C1, C2, C3	0 or 1 (stuffing control bits—0 indicates no stuffing for the corresponding stuffing bit, 1 indicates stuffing for the corresponding stuffing bit)
S_n where n = 1,2,3	0 or 1 (stuffing bits—value depends upon the control bits)

Table 5.10
3 Frame Control Bit Description

Bit	Description
F0, F1	F0=0, F1=1 (FAS)
M0, M1	M0=0, M1=1 (MFAS)
P	Parity bits (corresponding to previous frame)
X	In service messages
C_{xy} where x = 1 to 7 and y = 1 to 3 (x represents the seven subframes within the multiframe and y represents the stuffing bit position within the subframe)	0 or 1 (Stuffing control bits—0 indicates no stuffing for the corresponding multiframe, 1 indicates stuffing for the corresponding multiframe)
S_n where n = 1,2,3	0 or 1 (stuffing bits—value depends upon the Control bits)

The multiframe alignment is verified by using a four-bit CRC (CRC-4). The 16-bit multiframe is divided into two halves (submultiframes) containing eight frames each. The first bit of the FAS in each submultiframe is used for the CRC-4 while the first bit of the non-FAS frames are used to carry a framing pattern and an error indication signal. Table 5.12 indicates the bit patterns employed.

5.4.3.5 PDH E2 Frame Format

The E2 frame is created by bit multiplexing four E1 signals. The E2 multiplexing is more complex than the first-order E1 multiplexing since the multiplexer must adapt to different incoming clock rates. Figure 5.15 shows an E2 frame.

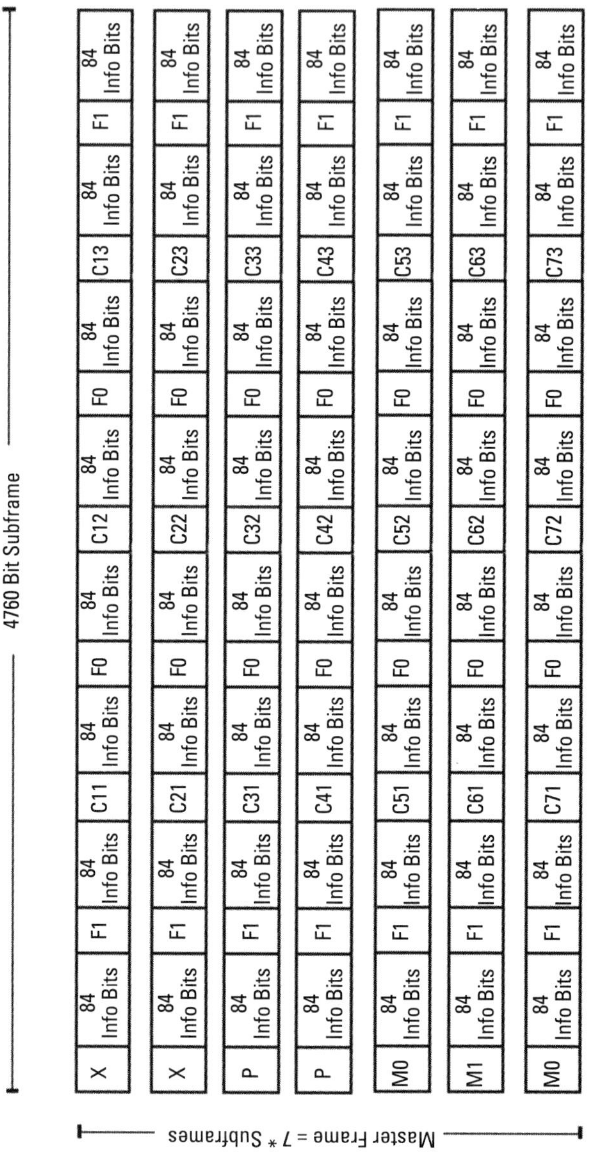

140 Engineering Optical Networks

Figure 5.11 NADH DS3 frame format.

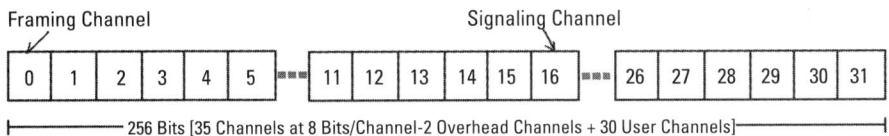

Figure 5.12 PDH E1 frame structure.

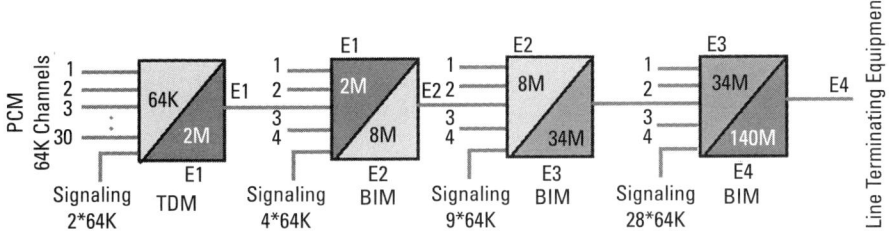

TDM: Time Division Muliplexer
BIM: Bit Inerbleaving Multiplexer

Figure 5.13 PDH multiplexing hierarchy— signaling structure.

Table 5.11
PDH Multiplexing Hierarchy—Signal Structure Details

Multiplexing Order (Level)	Computation	Bit Rate	Abbreviation
Input	PCM channel	64 Kbps	-
1	32 channels * 64 Kbps	2.048 Mbps	E1
2	4*E1+4*64K channels	8.448 Mbps	E2
3	4*E2+9*64K channels	34.368 Mbps	E3
4	4*E3+28*64K channels	139.264 Mbps	E4

To compensate for the variations in the input signal streams bit-stuffing techniques are employed. The maximum range of variations at this level of signal hierarchy is + 30 ppm.

The E2 frame is of a total of 848 bits that includes 10-bit FAS. The FAS pattern is 1111010000. This 848-bit frame is divided into four subframes of 212 bits each. With reference to Figure 5.15 the bits C1, C2, C3, and C4 are the stuffing-control bits. The three C1 bits control stuffing for the first E1 signal while the C2, C3, and C4 bits control the stuffing for the second, third, and fourth E1 signals respectively. A value of 1 in the control bit position indicates stuffing while a value of 0 indicates no stuffing. The C1, C2, C3, and C4 bits

142 Engineering Optical Networks

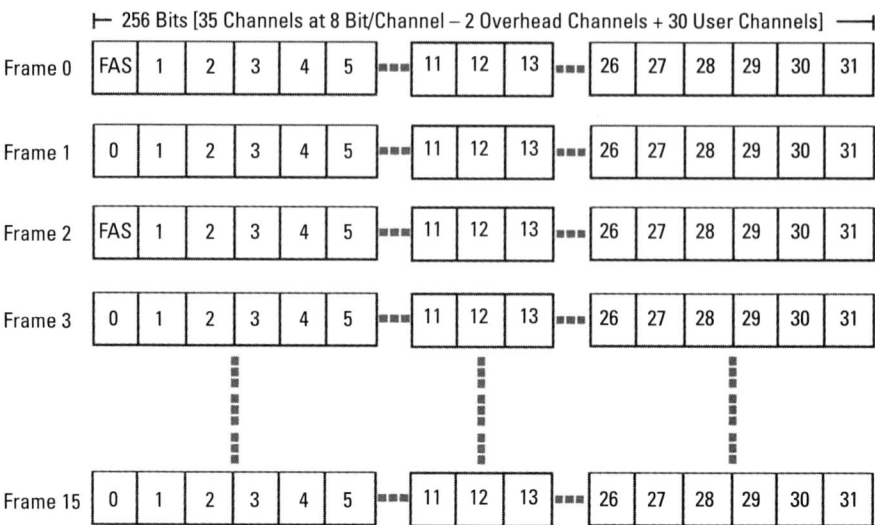

Figure 5.14 16 Frame E1 Multiframe.

Table 5.12
E1 Multiframe CRC Sequence

Submultiframe Number	Frame Number	Bit 1 Description
1	0	C1
1	1	0
1	2	C2
1	3	0
1	4	C3
1	5	1
1	6	C4
1	7	0
2	8	C1
2	9	1
2	10	C2
2	11	1
2	12	C3
2	13	E
2	14	C4
2	15	E

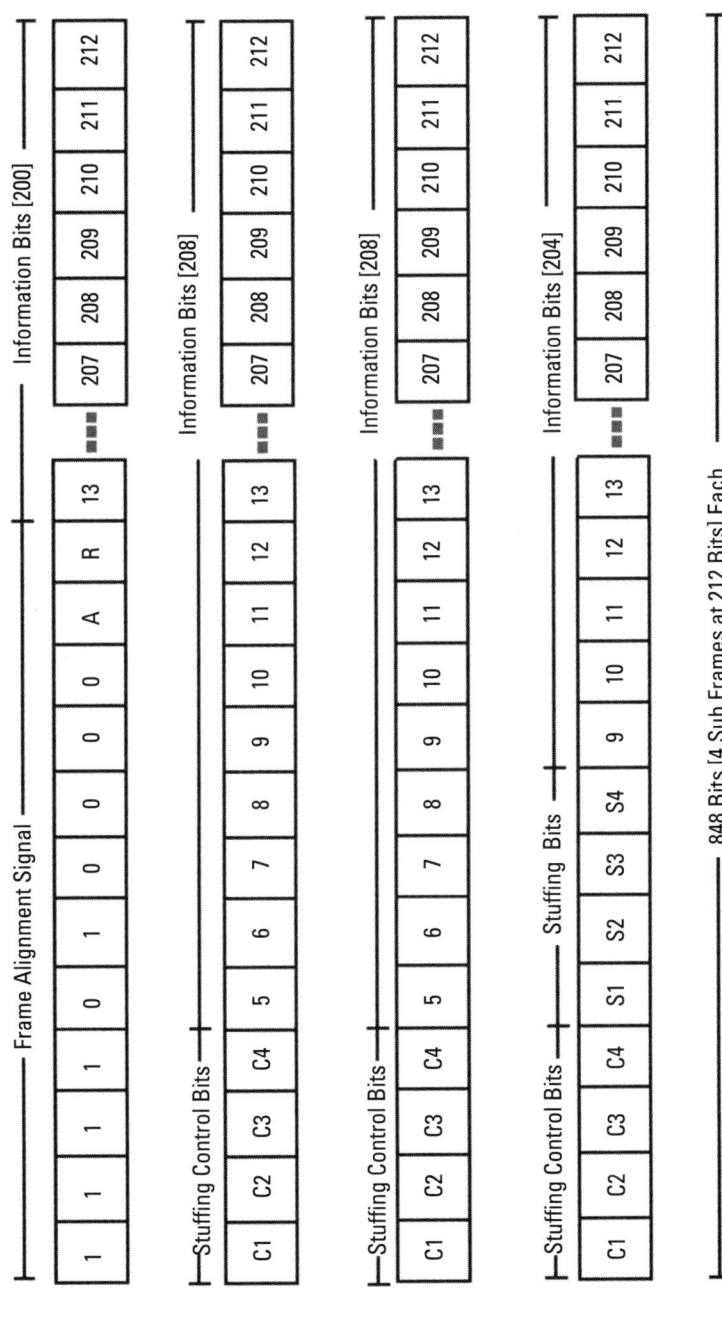

Figure 5.15 PDH E2 frame structure.

are replicated in the second, third, and fourth frames. The receiving equipment polls the values of the incoming C1, C2, C3, and C4 bits and takes the majority value in case of a difference in the bit values. This is done to negate the presence of single-bit errors during transmission.

5.4.3.6 PDH E3 Frame Format

The E3 frame is created by bit multiplexing four E2 frames. The E3 bit rate corresponds to 34.368 Mbps, with a tolerance of + 20 ppm. The E3 frame consists of 1,536 bits, including a 10-bit FAS with a bit pattern of 1111010000. The frame is logically divided into four subframes of 384 bits each. The E3 frame structure is illustrated in Figure 5.16.

With reference to Figure 5.16 the bits C1, C2, C3, and C4 are the stuffing control bits. The three C1 bits control stuffing for the first E2 signal while the C2, C3, and C4 bits control the stuffing for the second, third, and fourth E2 signals, respectively. A value of 1 in the control bit position indicates stuffing while a value of 0 indicates no stuffing. The C1, C2, C3, and C4 bits are replicated in the second, third, and fourth frames. The receiving equipment polls the values of the incoming C1, C2, C3, and C4 bits and takes the majority value in case of a difference in the bit values. This is done to negate the presence of single-bit errors during transmission.

5.5 Ethernet Over PDH

SDH/SONET networks were thought of as a universal transport mechanism that would be able to provide convergent services catering to the demands of the future as well. However SDH/SONET frames were optimized to carry voice traffic only and were not efficient in carrying data or video traffic. Ethernet over PDH (EoPDH) is one of the available techniques that provides Ethernet connectivity over telecommunication networks. EoPDH is a homogeneous methodology for transporting native Ethernet frames over the standardized PDH networks. The technology facilitates the delivery of carrier-grade Ethernet services over the existing transport infrastructure. EoPDH employs frame encapsulation techniques to transport variable bit rate (VBR) Ethernet frames over constant bit rate (CBR) PDH containers over a physical link. The protocols involved include the generic framing procedure (GFP), the concatenation techniques contiguous concatenation (CCAT) and virtual concatenation (VCAT), and the link capacity adjustment scheme (LCAS). GFP is defined as per ITU-T G.7041 recommendation while LCAS is conformant to ITU-T G.7042 recommendation, and VCAT is based on ITU-T G.7043 recommendation.

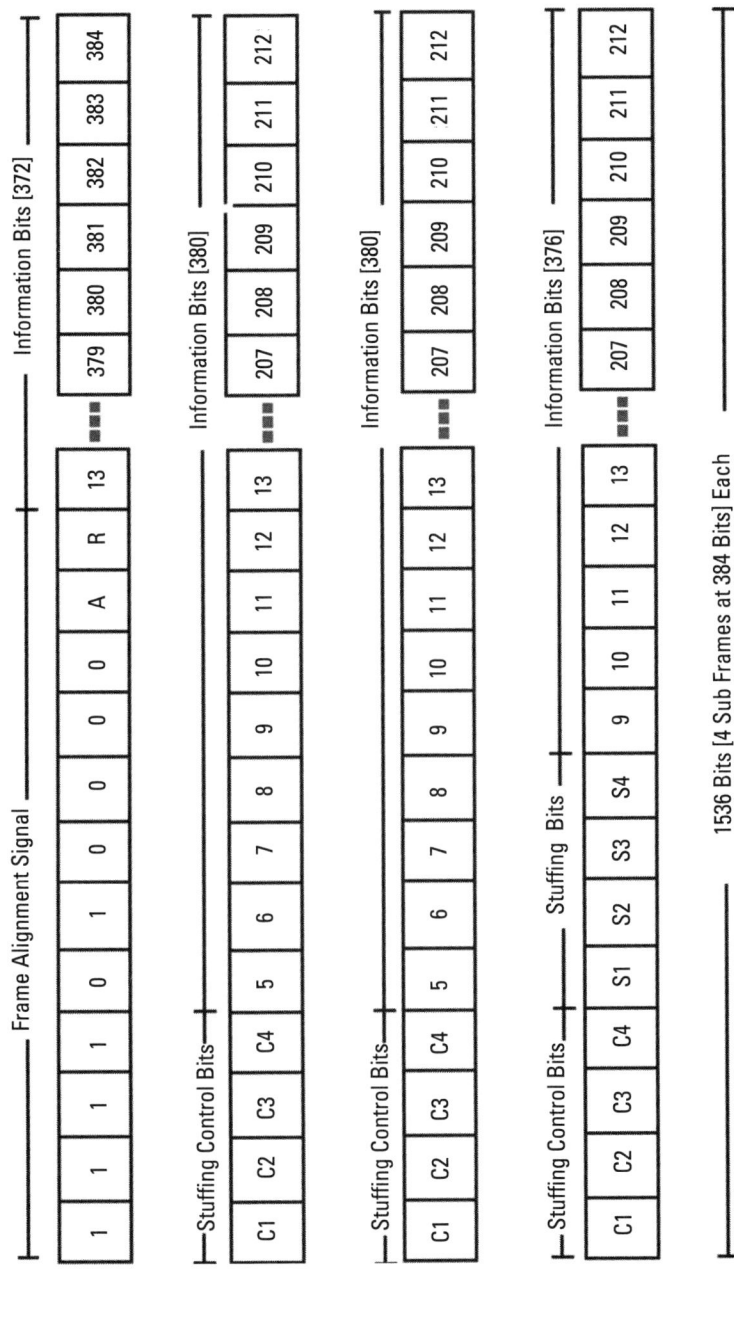

Figure 5.16 PDH E3 frame format.

5.6 Asynchronous Network Limitations

The primary limitation of the asynchronous (NADH/PDH) networks was their lack of flexibility and scalability. These were due to the absence of a common reference clock in the network as well as the use of copper as a medium for transmission. There were few attempts to have PDH over fiber in Europe along with the deployment of a fifth-order multiplexing rate, but these were nonstandardized. The major limitations of these networks are summarized as follows:

1. The early transmission systems were essentially point-to-point systems. NADH employed a bit-interleaving technique for multiplexing that worked well with point-to-point transmission. The lack of a common clock (asynchronous) was also less pronounced in a point-to-point network. These systems worked well as all the inputs were multiplexed at one location and were terminated at the next location.

2. The problem arose when it was necessary to add or drop a lower order rate at some intermediate location. As indicated earlier, in an asynchronous network a signal can be demultiplexed only at the level multiplexed. Due to the lack of a common reference, and the subsequent use of a bit-interleaving technique, the start of individual lower order signals cannot be determined within the higher order signal. This implies that for adding/dropping a lower order signal at an intermediate location, the higher order signal had to be first demultiplexed to bring it on par with the order of the signal to be added/dropped, and again multiplexed as per the hierarchy for onward transmission. Due to this limitation building of multipoint networks was complex and required multiple boxes at a single location—commonly referred to as multiplexer mountains, and hence service provisioning was also complicated.

3. There is no provision on an asynchronous/PDH network to compile and maintain a connection map (outlining the different paths interconnections in the network). The path management or tracking must be done manually.

4. The asynchronous/PDH networks predominantly employed a point-point topology or a star topology. These networks are not suited to the ring topology that provides for path-protection facilities.

5. There are no in-band channels for OAM&P on asynchronous/PDH networks. This implies that a separate out-of-band channel had to be configured for network and performance management.

6. Due to the lack of a common reference clock and the subsequent use of the bit-stuffing technique, and also due to the fact that the transmission was over copper cables, the bandwidth offered by the network was restricted.

To overcome these limitations ANSI proposed a fiber-based network called SONET, and the ITU and ETSI put forth an equivalent SDH.

5.7 SONET: Evolution

The SONET standardization process was initialized by the T1X1 subcommittee of the ANSI-accredited Committee T1 in 1985 with the standardization of the carrier-to-carrier optical interfaces. The standardization process was initiated on multiple fronts that included the problem of interconnection of optical fiber-based transmission terminals of multiple vendors commonly referred to as the mid-fiber meet. In addition, the T1 committee launched a proposal to standardize common optical interface parameters, including wavelength and optical power levels. This proposal was adopted as a draft standard for SMF optical interface specifications. Another proposal on long-term network operations was adopted by the T1M1 subcommittee as a draft standard on fiber-optic systems maintenance

The proposal by Bellcore on fiber plant standardization led to the interconnection of multivendor optical fiber–based terminals as well as the functional elements of the optical network. This included multiplexing equipment from various manufacturers and the introduction of digital cross-connect (DXC) systems. To a large extent, this solved the problem of mid-fiber meet or the interconnection of fibers. The proposal also included a hierarchical structure of digital signals that were multiples of a basic rate. This was achieved by the use of a synchronous bit-interleaving technique that would facilitate easy deployment of the multiplexing systems in the network. This proposal was eventually adopted as a standard and led to the use of the term SONET. Thus the standards developed encompassed the bit rates, hierarchy, optical interface specifications, and OAM interactions. One of the major stumbling blocks in the formulation of the new standards, in the initial stages, was the choice of the base transmission rate. Bellcore had proposed a rate of 50.688 Mbps, while AT&T wanted the base rate to be pegged at 146.432 Mbps. A new rate of 49.920 Mbps was eventually agreed upon.

The SONET standards as finalized by the T1X1 subcommittee were designed for use by the telecommunication networks in the United States. These standards were subsequently examined by the CCITT for global deployment. However, procedural difficulties arose primarily due the differences in the working of the T1 committee and the CCITT. Also, globally, many of the key telecom markets were still highly regulated, and the regulators of these countries were not too concerned about standardization.

Accordingly, a CCITT study group XVIII, initiated after the formalization of the SONET standards, began to explore a new synchronous digital hierarchy along with its associated NNI. CCITT wanted to evolve a common standard to be implemented globally. The standardization process was, however, slow, hampered by the fact the CCITT decision making body does not meet frequently.

The SONET optical base carrier rate was christened optical carrier-first-order, or level 1 (OC -1). As an interim arrangement to facilitate the transition from copper- to fiber-based networks, an electrical equivalent signal structure synchronous transport structure–level 1 (STS-1) was also introduced. The STS-1 could be represented as a rectangle organized into 13 rows and 180 columns. The CEPT SDH structure, on the other hand, was represented as a rectangle of nine rows and 270 columns. Eventually, an NNI of around 150 Mbps was agreed upon with a view of accommodating higher bandwidths in the future. The North American signal of 50 Mbps could be represented as a structure of 13 rows * 60 columns or nine rows * 90 columns. The European base rate of 2 Mbps, which amounted to around 32 bytes of data, could be more efficiently carried in a nine-row format as nine rows * four columns or 36 bytes, while a 13-row format it would have to be structured as 13 rows * three columns or 39 bytes. The North American base signal of 1.5 Mbps or 24 bytes could be efficiently represented in either a nine-row format as a structure of nine rows * three columns or 27 bytes or the 13 row format as a structure of 13 rows * two columns or 26 bytes. This prompted CEPT to insist on a nine-row format for international standardization. The proposal meant a change in the bit rates as well as the structure of the NADH. Fortunately, however, the structure presented an efficient structure for the NADH T3/DS3 signal as compared to the European E3 signal. Ultimately, the CCITT proposed a 150-Mbps NNI along with a byte-interleaving structure for the higher-order SONET signals and accommodation of the 34-Mbps E3 signal based on the nine-row structure. This proposal was finally accepted after an extensive round of deliberations and heralded the birth of the SDH.

5.8 Summary

The field of telecommunications has been service-oriented right from its birth, the invention of the telephone. Telephone technology has progressed in a distinct cycle, starting with the introduction of a new service, reaching maturity and subsequent widespread usage, and culminating with its saturation and the consequent demand for better or new services. Each new service provides a new growth/revenue model until it is replaced with a new service. This repetitive pattern holds well even today with the only difference being the reduced cycle times.

The rapid developments in the field of optical fiber transmission and the associated progress in the field of semiconductor technology, including VLSI and multilayer printed circuit boards, led to the development of complex circuits and standards. These developments fueled demands for improved and increasingly sophisticated services that required large bandwidth, better performance monitoring facilities, and greater network flexibility. This heralded the development of the transport networks, which form the common backbone capable of providing reliable, error-free, high-capacity triple-play networks over increasingly longer distances.

The first attempt at standardizing the functioning of the telecommunication network by the ITU culminated in the development of the NADH as well as the PDH. The primary limitation of these asynchronous networks was their lack of flexibility and scalability. These problems were due to the absence of a common reference clock in the networks as well as the use of copper as a medium for transmission. There were a few attempts to have PDH over fiber in Europe along with the deployment of a fifth-order multiplexing rate, but these were nonstandardized. The early transmission systems were essentially point-to-point systems. The problem arose when it was necessary to add or drop a lower order rate at some intermediate location. A signal can be demultiplexed only at the level multiplexed. Due to the lack of a common reference, and the subsequent use of a bit-interleaving technique, the start of individual lower order signals could not be determined within the higher order signal. This implied that for adding/dropping a lower order signal at an intermediate location, the higher order signal had to be first demultiplexed to bring it on par with the order of the signal to be added/dropped, and again multiplexed as per the hierarchy for onward transmission. Thus, these networks were not suited to function within a ring topology. The problem was further compounded given the fact that there are no in-band channels for OAM&P on these asynchronous networks. The bandwidth offered by the network was restricted due to the lack of a common reference clock. To overcome these limitations, the ANSI developed a fiber-based SONET, and the ITU and the ETSI created the SDH. SDH/SONET networks' bandwidth was limited to 10 Gbps, and the frame structure was not optimized for transporting packet-switched data. As a result, SDH/SONET networks have been replaced by OTNs.

5.9 Review

5.9.1 Review Questions

1. Originally, ITU was known as _____.
    a. International Telegraph Union
    b. International Telecommunication Union
    c. Indian Telecommunication Union
    d. Indian Telegraph Union
2. An E1 is formed by multiplexing _____ 64K channels together.
    a. 30
    b. 32
    c. 24
    d. 34
3. A DS3 is formed by multiplexing _____ 64K channels together.
    a. 30
    b. 64
    c. 128
    d. 672
4. The standard G.704 provides the specifications for electrical interfaces in an asynchronous network:
    a. True
    b. False
5. The channel bandwidth of a PCM system is _____.
    a. 48 Kbps
    b. 64 Kbps
    c. 36 Kbps
    d. 128 Kbps
6. The following statement(s) is/are true with respect to asynchronous networks:
    a. The primary limitation of the asynchronous (NADH/PDH) networks was the lack of flexibility as well as scalability.
    b. In an asynchronous network a signal can be demultiplexed only at the level multiplexed. Due to the lack of a common reference, and the subsequent use of a bit-interleaving technique, the start

of individual lower order signals cannot be determined within the higher order signal.

c. There is no provision on an asynchronous network to compile and maintain a connection map.

d. There are no provisions for in-band channels for OAM&P operations in an asynchronous network.

   i. a only
   ii. a and b
   iii. a, b, and d
   iv. All statements are true

7. The E3 frame is created by bit-multiplexing five E2 frames. The E3 bit rate corresponds to 34.368 Mbps, with a tolerance of + 20 ppm.

   a. True
   b. False

8. The time duration of an E1 frame is 125 μs, and it contains 32 channels with eight bits per channel or a total of 32 * 8 = 256 bits per time slot.

   a. True
   b. False

9. In an E1 frame the voice channels are encoded using A-law in accordance with the ITU-T G.711 specification.

   a. True
   b. False

10. The DS3 signal can only be constructed by multiplexing seven * DS2 signals, which, in turn, are constructed by multiplexing four * DS1 channels together. A DS3 signal cannot be directly be constructed by multiplexing 28 * DS1 signals together.

    a. True
    b. False

5.9.2 Exercises

1. Describe the signaling structure employed in NADH and PDH networks and highlight the significant differences.
2. Describe the procedure for CRC computation in an E3/DS3 frame.
3. What is bit stuffing? Why is it necessary in PDH networks?
4. What are the key recommendations of ITU-G.703?

5.9.3 Research Activities

1. Detail the significant differences between NADH and PADH specifically and the European and American signal transmission conventions in general.
2. Briefly describe the Japanese, Korean, Hong Kong, and Canadian digital hierarchies.
3. List the various steps involved in bit stuffing with reference to an E2 frame.
4. Describe the EoPDH technique and its advantages.
5. Use the URLs mentioned in the Referred Reading section to research and understand the concept of multiframes and CRC in NADH/PDH multiplexing.

5.10 Referred Standards

G.702: Digital Hierarchy Bit Rates

G.704: Synchronous frame structures used at 1.544-Mbps, 6.312-Mbps, 2.048-Mbps, 8.448-Mbps, and 44.736-Mbps hierarchies

G.742: Second-order digital multiplex equipment operating at 8.448 Mbps employing positive justification

G.747: Second-order digital multiplex equipment operating at 6.312 Mbps (multiplexing three tributaries at 2.048 Mbps)

G.751: Digital multiplex equipment operating at the third-order bit rate of 34.368 Mbps and fourth-order bit rate of 139.264 Mbps employing positive justfication

G.752: Characteristics of digital multiplex equipment based on a second-order bit rate of 6.312 Mbps employing positive justification

G.753: Third-order digital multiplex equipment operating at 34.368 Mbps with or without justification

G.755: Digital multiplex equipment operating at 139.264 Mbps (multiplexing three tributaries at 44.736 Mbps)

G.7041: Generic framing procedure

G.7042: Link capacity adjustment scheme

G.7043: Virtual concatenation of plesiochronous digital hierarchy signals

G.804: GFP frame mapping into plesiochronous digital hierarchy

Y.1730: Requirements for OAM functions

Y.1731: OAM functions and mechanisms

5.11 Recommended Reading

5.11.1 Books

Dryburgh, L., and J. Hewett, *Signaling System No. 7 (SS7/C7): Protocol, Architecture, and Services,* 2004.

Bellamy, J. C., *Digital Telephony,* Third Edition, John Wiley & Sons, 2000.

Warier, S., *The ABCs of Fiber Optic Communication,* Norwood, MA: Artech House, 2017.

5.11.2 URLs

http://www.pulsewan.com/data101/sdh_basics.htm.

http://members.cox.net/michael.henderson/Papers/Framing.pdf.

http://www.cpe.ku.ac.th/~nguan/presentations/datacom/carrier.pdf.

http://sst.umt.edu.pk/newsite/courses/Fall2008/TE-375/Plesiochronous%20Digital% 20 Hierarchy.ppt.

http://en.academic.ru/dic.nsf/enwiki/24101.

http://online-pdf-search.co.cc/search-pdf/Plesiochronous+Digital+Hierarchy+(PDH)/.

http://www.itu.int.

http://www.atis.org.

http://www.maxim-ic.com/appnotes.cfm/appnote_number/3849/CMP/WP-39.

http://www.rad.com/RADCnt/MediaServer/18551_Ethernet.pdf.

http://www.telrad.com/pages/products/EoPDH.aspx.

6

Synchronous Optical Networks

6.1 Chapter Objectives

Asynchronous networks (as detailed in Chapter 5) suffered from bandwidth limitations and a lack of exact network time synchronization. Transport networks, on the other hand, offer a common backbone allowing for extremely reliable, error-free, high-capacity transmission that is capable of carrying voice, data, and video information over long distances. The importance of transport networks in the field of modern-day communications cannot be understated. This chapter traces the evolution of the SONET and covers the basics of both SONET and SDH. The chapter also includes the details of key standards governing the architecture and functioning of the modern-day optical networks. Later, Part 4 provides a detailed treatise of the synchronous frame, including the overhead functionalities and pointer processing. An understanding of the different elements of the transport network and their internetworking is an essential precursor to building competencies related to effective network maintenance and troubleshooting. This chapter provides engineers and technicians with an innate understanding of the physical network interconnections, conventions, and terminologies required to effectively maintain and troubleshoot optical fiber networks, along with a framework for understanding TDM-based transport networks. In addition, the chapter details the architecture and functioning of the SDH network to provide readers with a conceptual basis for understanding the OTN networks presented in Chapter 7 and to offer insight into some of the disadvantages in transporting packet data over SDH. Current TDM transport networks are based on OTN technologies.

Key Topics

- An overview of the transport network;
- The need for and benefits and functions of a transport network;
- The evolution of SONETs;
- Transport network architecture;
- The components of transport networks.

6.2 Introduction

The advent of digital transmission systems and the consequent conversion from the legacy analog networks was initiated by North America, which introduced T1 transmission systems in the early 1960s. The subsequent conversion in Europe followed a few years later. The new networks were referred to as asynchronous networks in North America, while their European counterparts were known as PDH networks.

One of the major drawbacks of these networks, other than the absence of a common network reference clock, was the lack of in-band channels for OAM&P purposes. This is significant given that OAM&P costs account for a substantial portion of the overall network cost. In fact, in certain cases (especially for new networks in the initial stages of deployment) the OPEX may equal the CAPEX of a telecommunication service provider. Another important aspect that has to be factored in is the presence of multiple service providers within a geographical area. This was especially true in North America where the end of the state monopoly over telecom operations witnessed the emergence of a number of new service providers both at the intracity as well as intercity levels. This made it difficult for service providers as well as customers to draw service level agreements (SLAs) as it was difficult to pinpoint the part of the subnetwork where a problem existed. Further it was evident that the copper networks, which employed line timing for synchronization, would be unable to provide for the higher bit rates that would be required in the future. The new network architecture would need to employ optical transmission with each NE synchronized to a common reference.

6.3 Transport Network: An Overview

The first optical network architecture was unveiled by the ANSI-accredited Committee T1, which had commenced work on the new architecture in the mid 1980s [1]. The rates and formats for the new optical network were defined in the T1X1 subcommittee. The OAM&P inputs and requirements were prepared by another subcommittee referred to as T1M1. These two subcommittees together developed the object-oriented network element models for SONET. The development of the SONET standards was followed by the development of the SDH standards by CCITT. CCITT was the result of amalgamation of two separate ITU committees, the CCIF and the CCIT and was subsequently renamed the ITU-T [2]. ITU-T was interested in ensuring a common standard for transport networks across the globe. ITU-T had a great number of discussions with ANSI and its European counterpart ETSI to come to an agreement on standard bit rates. This effort proved to be unsuccessful. However, it managed to ensure compatibility between SONET and SDH networks.

The explosive growth in the field of semiconductor technology witnessed the introduction and the proliferation of PCs and the advent of data networks and, more recently, video networks. This however called for redundancy in physical infrastructure, processes, and procedures, adding to unwarranted complexities for end users, requiring them to approach multiple service providers to meet their needs. Thus the need of the hour was convergence or a convergent framework capable of delivering multiple services to a user using a common framework. This need resulted in the emergence of transport networks as opposed to plain transmission networks.

In the abstract sense, a transport network can be regarded as the set of facilities and equipment that ensures error-free, reliable, and high-speed and high-capacity data transport between end customers. As transport networks expand geographically their size and capacity increases exponentially with the resultant need for centralized OAM&P systems. It would be extremely difficult, if not impossible, to run a transport network with any degree of reliability or cost-effectiveness without an effective OAM&P system. The lack of OAM&P capabilities had prevented Ethernet from becoming a viable transport network technology, except for in very small networks.

The development of technology for digital transmission, including the developement of integrated circuits (IC) and digital signal processing (DSP), was the turning point in the evolution of transmission networks. The use of digital signals provided the opportunity for signal regeneration with no significant performance degradation and a better SNR. With the introduction of digital transmission technology, the most appropriate multiplexing technology became TDM, which could be employed both over existing copper cable systems as well as on microwave systems [3].

6.4 Transport Network: Need, Benefits, and Function

The initial telecommunication networks were designed for carrying voice traffic only. The electrical signals (representing speech) were carried over copper wires strung through overhead poles and later through underground copper cables. As the technology progressed, different technologies for delivering voice emerged. These included the CDMA and GSM wireless techniques. There was parallel development in the field of semiconductor technologies resulting in the proliferation of PCs and the subsequent growth of the internet. These developments resulted in the significant growth of data traffic. This was followed by the development of applications like medical imaging, technologies like DTH televisions, and digital TV broadcasting, which required streaming video feeds to be delivered across locations (DOCSIS standards). Each of these services/technologies required independent network infrastructure along with the supporting systems. Furthermore, end users need to individually subscribe to each of these services, resulting in redundancies in hardware and software and the associated processes and procedures.

The transport network provides a common framework or backbone that can deliver multiple services to the end user. It is an optimized network that supports converged services as well as applications. The transport network provides extremely reliable, error-free, high-capacity transmission networks that are capable of carrying voice, data, and video information over long distances. The importance of transport networks in the field of modern-day communications cannot be understated.

A network is the interconnection of equipment, or NEs, using a suitable medium. A network is designed to provide specific services. The basic services, including the voice services, are provided by the portion of the network referred to as the access network, the primary interface with the end customer. The primary function of the transport network is to interconnect the access NEs. The transport network provides a high-speed and high-capacity pathway (capacity and speed are interlinked) for transporting the signals (data, information, and control) between the access network entities. The transport signal does not generate any traffic (other than the control and management information required to operate and manage its entities); rather it carries the traffic generated by the access NEs from one location to another in a reliable and error-free fashion.

The transport network is based on universal standards and hence facilitates the interconnection of carriers irrespective of geography. The use of standards also ensures that networks are not vendor-dependent and hence provides better scalability. Being a common framework, the transport network can have high-bandwidth availability and hence can facilitate transmissions at higher speeds (as bandwidth and speed are directly correlated), as well as dedicated protection schemes. Without transport networks these features would have to be directly incorporated on the access NEs, resulting in higher cost as well as equipment complexity. Figure 6.1 summarizes the key benefits of the transport networks [3].

6.4.1 Evolution of Synchronous Optical Networks

The SONET standardization process was initialized by the T1X1 subcommittee of the ANSI-accredited Committee T1 in 1985 with the standardization of the carrier-to-carrier optical interfaces. The standardization process was initiated on multiple fronts that included the problem of interconnection of the optical fiber–based transmission terminals of multiple vendors commonly referred to as the mid-fiber meet. In addition a proposal to standardize common optical interfaces parameters, including wavelength as well as optical power levels, was taken up by the T1 committee [4]. This proposal was adopted as a draft standard for SMF optical interface specifications. Another proposal on long-term network operations was adopted by the T1M1 subcommittee as a draft standard on fiber-optic systems maintenance [5]

A proposal by Bellcore on the fiber plant standardization led to the interconnection of multivendor optical fiber–based terminals as well as the functional elements of the optical network. To a large extent, this solved the problem of mid-fiber meet or the interconnection of fibers. The proposal also included a hierarchical structure of digital signals, which were multiples of a basic rate. This was achieved by the use of a synchronous bit-interleaving technique that would facilitate easy deployment of the multiplexing systems in the network. This proposal was eventually adopted as a standard and led to the use of the term SONET. The standards developed encompassed the bit rates, the hierarchy, the optical interface specifications, and the OAM interactions. One of the major stumbling blocks in the formulation of the new standards, in the initial stages, was the choice of the base transmission rate. Bellcore had proposed a rate of 50.688 Mbps while AT&T wanted the base rate to be pegged at 146.432 Mbps. A new rate of 49.920 Mbps was eventually agreed upon. The SONET standards as finalized by the T1X1 subcommittee were designed

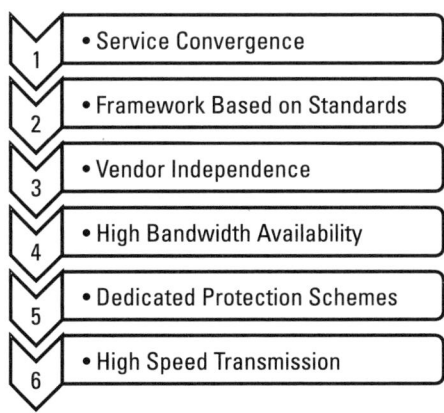

Figure 6.1 Transport networks—key benefits.

for use by the telecommunication networks in the United States. When the standards were examined by the CCITT for global deployment, procedural difficulties arose, primarily due the difference in the working of the T1 committee and the CCITT. Also globally many of the key telecom markets were still highly regulated and hence the regulators of these countries were not too concerned about standardization.

The SONET optical base carrier rate was christened OC -1. As an interim arrangement to facilitate the transition from copper to fiber-based networks an electrical equivalent signal structure STS-1 was also introduced. The STS-1 could be represented as a rectangle organized into 13 rows and 180 columns. However the CEPT SDH structure was represented as a rectangle of nine rows and 270 columns. Eventually an NNI of around 150 Mbps was agreed upon with a view of accommodating higher bandwidths in the future. The North American signal of 50 Mbps could be represented as a structure of 13 rows * 60 columns or nine rows * 90 columns. The European base rate of 2 Mbps, which amounted to around 32 bytes of data, could be more efficiently carried in a nine-row format as a nine rows * four columns or 36 bytes, while in a 13 row format it would have to be structured as 13 rows * three columns or 39 bytes. The North American base signal of 1.5 Mbps or 24 bytes could be efficiently represented in either a nine-row format as a structure if nine rows * three columns or 27 bytes or the 13-row format as a structure of 13 rows * two columns or 26 bytes. This led to the adoption of the nine-row format for international standardization. The proposal meant a change in the bit rates as well as the structure of the NADH. However the structure was efficient for the NADH T3/DS3 signal as compared to the European E3 signal. The CCITT finally proposed a 150-Mbps NNI along with a byte interleaving structure for the higher-order SONET signals and accommodation of the 34-Mbps E3 signal based on the nine-row structure. This proposal was finally accepted after an extensive round of deliberations, heralding the birth of the synchronous digital hierarchy. Thus SONET was used extensively within America while SDH was used as an international gateway.

6.4.2 Optical Transport Networks

The exponential growth in data traffic coupled with the need to rationalize the CAPEX/OPEX forced carriers to look at options that allowed them to leverage existing network infrastructure while catering to bandwidth demands seamlessly. The result was the OTN, which allowed carriers to integrate their SDH/SONET networks in order to deliver a transparent framework that could efficiently handle diverse traffic types, while providing a scalable option to enhance the network throughput 10 times from 10 Gbps (STM-64/OC-192) to 40 Gbps and further to 100 Gbps. OTN provides a transparent hierarchical network that is designed to work with WDM networks providing the functionalities of transport, multiplexing, routing, in-band management, network monitoring with FEC, and traffic protection capabilities. The SDH/SONET networks were replaced by OTN networks. OTN architecture and functioning is covered in detail in Chapter 7.

6.5 Transport Network: Architecture

The transport network is not a single homogeneous network but is divided into three different layers each having a specific function. This architecture simplifies the OAM&P of the network. The architecture of a transport network (abstract view) is illustrated in Figure 6.2. The remaining chapters add detail to this simplified network view.

The architecture of a transport network can be explained by the following analogy. A human being needs to employ external means to travel from one distant location to another. It is beneficial to employ mass travel modes rather than to create a personalized means of travel for each individual. Accordingly there can be different modes of travel with the primary differentiation being the speed and cost. As shown in Figure 6.3 the primary means of travel over a distance is road, rail, or air. These are shared modes of travel that are accessed through an access point. The access point for a rail network is a railway station, while that for road travel could be a bus depot; the airport serves as an access point for air services. The modes of travel correspond to the architecture of a transport network. The access point serves as a point of aggregating or collecting passengers wanting to use a shared transport service and hence corresponds to the collector/aggregation layer. Individuals reach the access points through

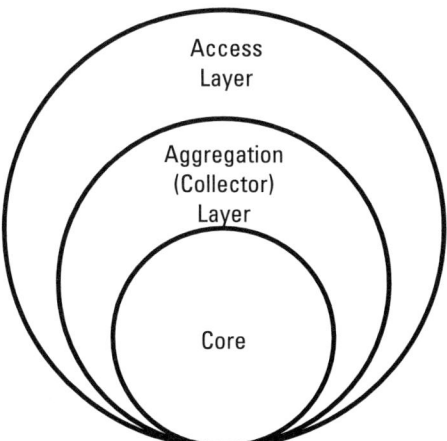

Figure 6.2 Transport network architecture—abstract view.

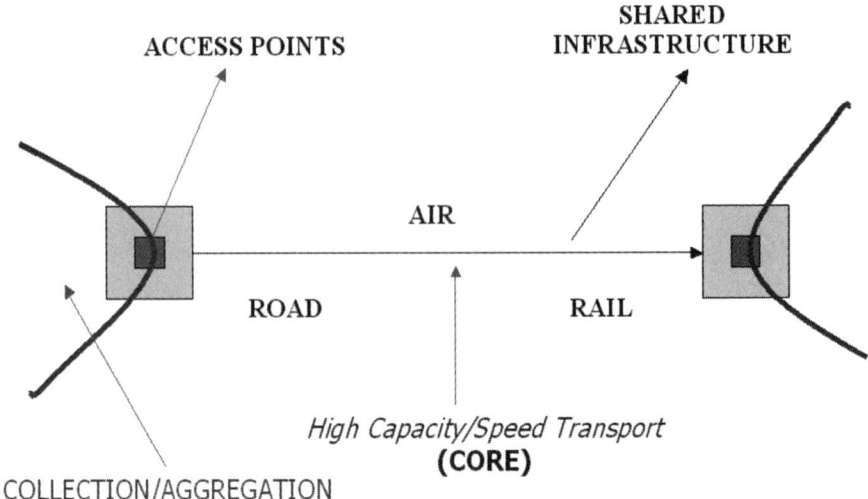

Figure 6.3 Transport network analogy.

a variety of modes of short-distance travel including walking, auto-rickshaws, taxis, suburban buses, and train services or metros. These short-distance modes of travel represent the access layer. The layers are explained in detail as follows [6]:

1. Access layer: The access layer is the fundamental and primary layer in any transport network. The access layer functions as the user interface and connects the access network NEs onto the transport network. The end user for the transport network is the access NE, and a customer is the end user of the access network element. For example, in a wireless CDMA/GSM network the base transceiver stations/basic switching centers (BTSs/BSCs) are connected to the mobile switching centers (MSCs) via the access network. The MSCs themselves are connected to SDH/SONET equipment at the collector/aggregation layer. A service provider thus provides basic voice and data services using this layer. The access layer is connected to the core through the collector/aggregation layer in case long-distance access is needed. Local as well as intracity connectivity is possible directly or via the collector layer. The bandwidth employed at this layer in modern-day networks corresponds to 155 Mbps/622 Mbps.

2. Collection/aggregation layer: The information (in the form of voice/data/video signals) coming in from the various access NEs and from

the access layers is collected and bundled together before being passed to the core layer for long-distance delivery. The primary function of the collector/aggregation layer is bandwidth management. Bandwidth management includes two important functions grooming and consolidation. Grooming (bringing together payloads over multiple paths and separated in the time domain) refers to the transport and the subsequent termination of the signals via protected as well as unprotected paths within the network. Consolidation refers to the filling up or assigning of unused bandwidth on the outgoing transport frames. The incoming frames from the access networks may only be partially full. The collector layer equipment consolidates the traffic from multiple inputs onto its outgoing channels. This concept is analogous to a courier company employee picking up a small parcel from an individual for intercity delivery. The courier company would not arrange to send the parcel individually; rather it takes it to a local, regional, or central collection location (depot) where multiple deliveries are consolidated for onward transmission over a variety of transport mechanisms including road, rail, ship, and air. The bandwidth employed at this layer is on the order of 2.5 Gbps.

3. Core: The core is the backbone of any transport network and hence is the most important of all the three layers. Since the actual information is going to be transported from the source to destination using the core (for long-distance transmission), the data rate or the information-carrying capacity in the core is very high compared to the other two layers. The speed (high throughput and low latency) of the core is also exceptionally high because delays can be extremely critical at this stage especially for data and video transport. This stage employs digital or optical cross-connects for efficient multiplexing and to provide a faster and more efficient service. The currently supported line rate[1] at the core layer is 100 Gbps (for TDM networks). Terabit Ethernet (TbE), or 400G Ethernet, according to standards IEEE P802.3cd–400G, is slated for release in December 2017.

Figure 6.4 summarizes the three layers of a transport network along with their primary functions.

---

1. Note that the line rate or bit rate is per optical wavelength. Further throughput is not equivalent to the bitrate. Throughput calculations do not include the overheads and hence will be lower.

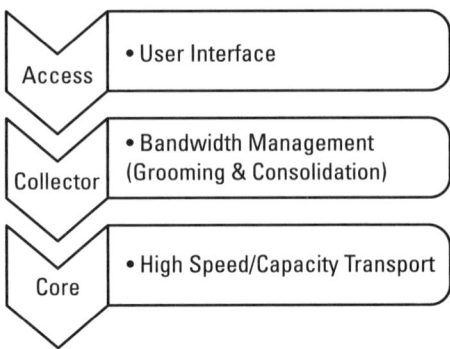

Figure 6.4 Transport network layer functions.

6.6 Transport Network: Components

A network is designed to provide services. The telecommunication network is designed to provide reliable, converged (triple-play) services to end users. A network is an interconnection of equipment using a suitable medium. The method of interconnections is referred to as the topology. Figure 6.5 shows the components of the transport network, and the equipment, media, and topologies employed in transport networks are summarized as follows.

1. Network equipment: The primary function of a transport network is multiplexing; hence the main equipment items deployed in the network are multiplexers. The multiplexers corresponding to the access and the collection layers are referred to as low-capacity (low-cap) multiplexers or low-cap equipment. The multiplexers/cross-connects corresponding to the core layer are referred to as high-capacity (high-cap) equipment of high-density DXCs (HD-DXCs). The basic differences between a multiplexer and a regenerator are explained in the following section. In addition to the multiplexers and cross-connects, equipment is required to counter the effects of attenuation, dispersion, jitter, and wander. Commonly used equipment in the network to counter the effects of attenuation/distortion includes amplifiers and regenerators. This equipment is briefly described as follows. (Please note that the equipment architecture is the same for SDH and SONET.)
   - Multiplexers/demultiplexers: A multiplexer is a device that combines multiple input channels onto a single (or multiple in case of protection) output channel typically using the TDM technique. Multiplexing and demultiplexing are complementary functions,

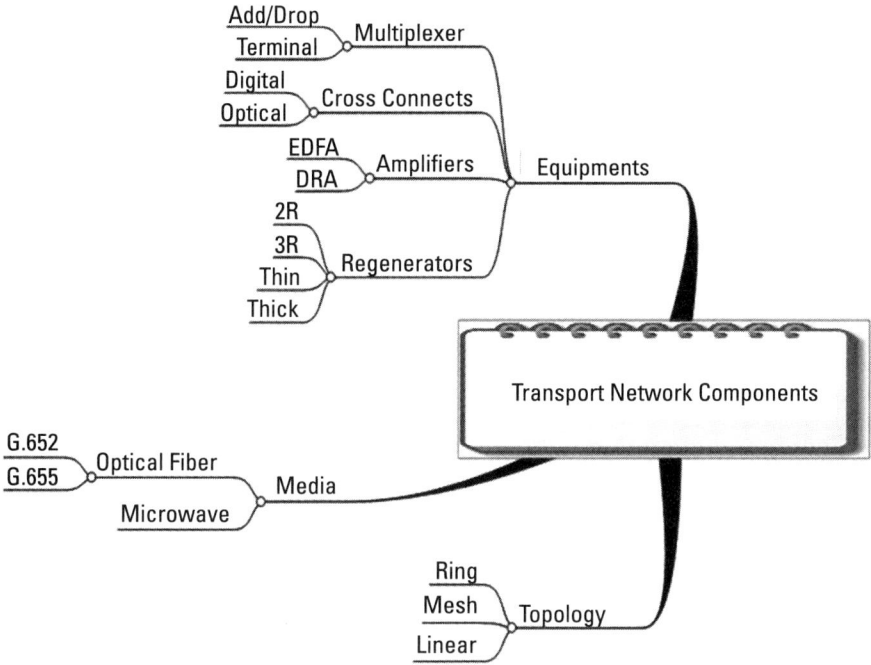

Figure 6.5 Transport network components.

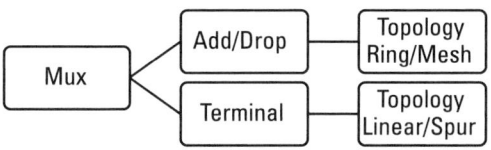

Figure 6.6 Mux classification.

and the device that performs them is referred to as a mux. Figure 6.6 presents the classification of a mux.

Muxes can be classified as terminal multiplexers (TMs) or add/drop multiplexers (ADMs). In modern-day equipment there are no physical (hardware) differences between a terminal and an ADM. The difference lies in the method of interconnection or topology. Figure 6.7 presents a point-to-point configuration. In this configuration, NE A multiplexes the signals coming in from its tributary ports and transmits the resultant SDH frame on its transmit port to NE B every 125 μs. Simultaneously NE B does the same operation and transmits an SDH frame to NE A every 125 μs. On the

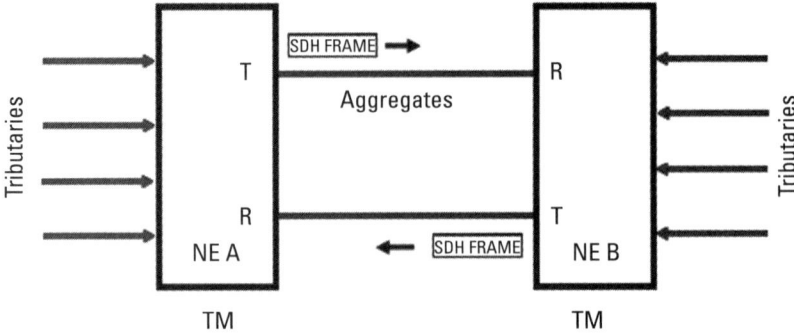

Figure 6.7 Terminal multiplexers in point-to-point configuration.

receipt of the frames NE A and NE B demultiplexer and terminate the signals. Subsequently a new frame is generated and transmitted. Hence, the multiplexers in this configuration are referred to as terminal multiplexers. The concept is analogous to a train that originates at one station and terminates at another. At the intermediate stations passengers can board or disembark the train. Figure 6.8 depicts a linear configuration of muxes. In this example the frame termination happens only at the terminal ends (NE A and C) and not at the intermediate node (NE B).

Figure 6.9 shows a ring configuration. In the case of a ring configuration a frame keeps on passing (there are changes in the overheads) through the NEs every 125 μs in both directions (clockwise and counterclockwise). Hence all the NEs in a ring are ADMs.

Figure 6.10 presents a ring configuration with an additional spur. In this case the NEs that are a part of the ring are ADMs while the frame is terminated at the spur and hence the mux is configured as a TM.

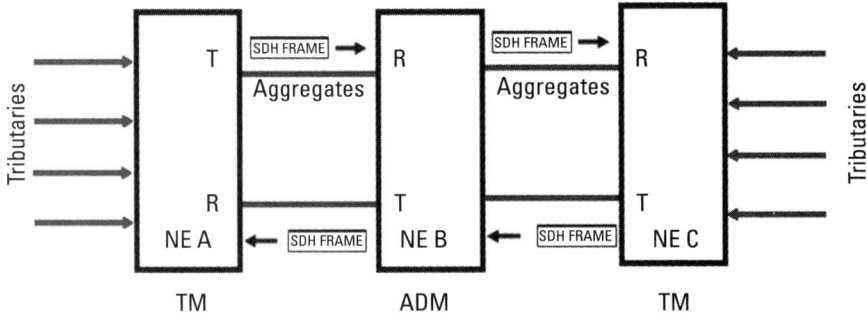

Figure 6.8 ADM and terminal multiplexers in linear configuration.

Synchronous Optical Networks

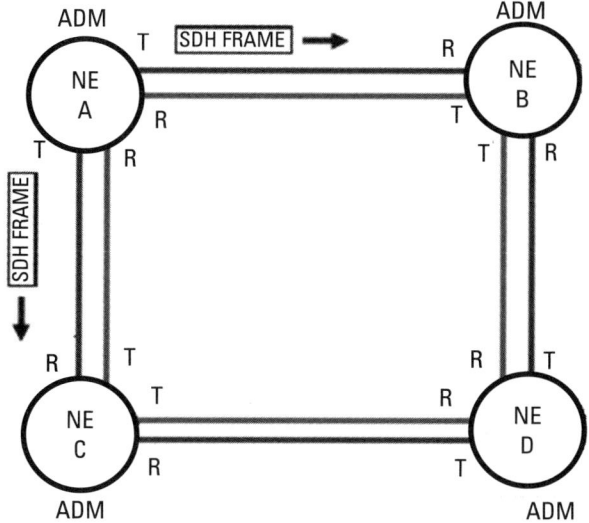

Figure 6.9 ADMs in ring configuration.

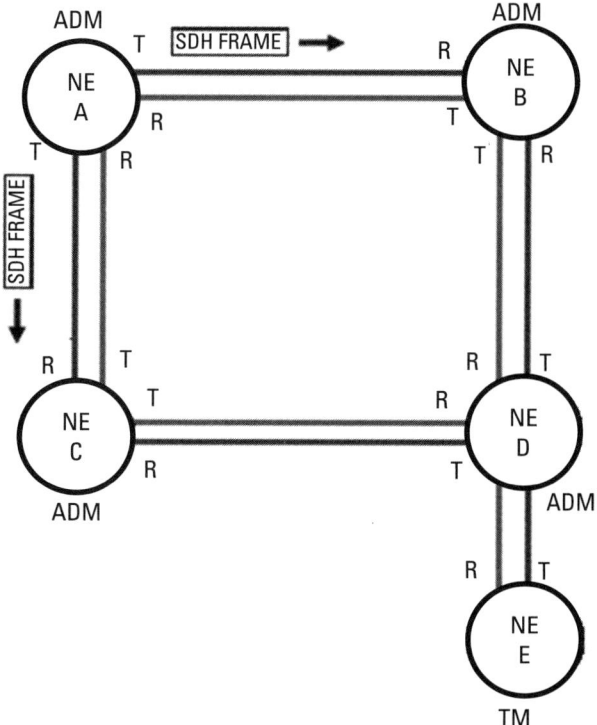

Figure 6.10 ADM and TM in ring configuration.

Figure 6.11 illustrates a mux. A mux takes in a number of input channels referred to as tributaries or tribs and multiplexes them to form an aggregate channel (agg) and vice versa. These linkages are established by making (programming) logical connections on the mux using a variety of physical interfaces and appropriate user interfaces. A connection requires a minimum of two end points. The first end point is referred to as the source end, while the second end point is referred to as the destination end. A connection can be unidirectional (one way only, one direction, transmit to receive) (refer to Figure 6.12) or bidirectional (two-way, transmit and receive) (see Figure 6.13). The connection types are illustrated in Figure 6.14, and the corresponding descriptions of the connection types are provided in Table 6.1.

In the case of a ring topology (described next) a mux will have two transmit/receive paths. They are referred to as the west side and east side or west aggregate and east aggregate or aggregate 1 (agg 1) and aggregate 2 (agg 2) or aggregate A (agg A) and aggregate B (agg B), or in the case of higher-order muxes as line 1 and line 2 or line A and line B.

Depending upon the topology employed a lower-order multiplexer would have one or two aggregate channels. During transmission the mux sends out the same frame on both the paths. This implies that every mux can receive the frame on both of its paths. The traffic flows through the network by means of the logical con-

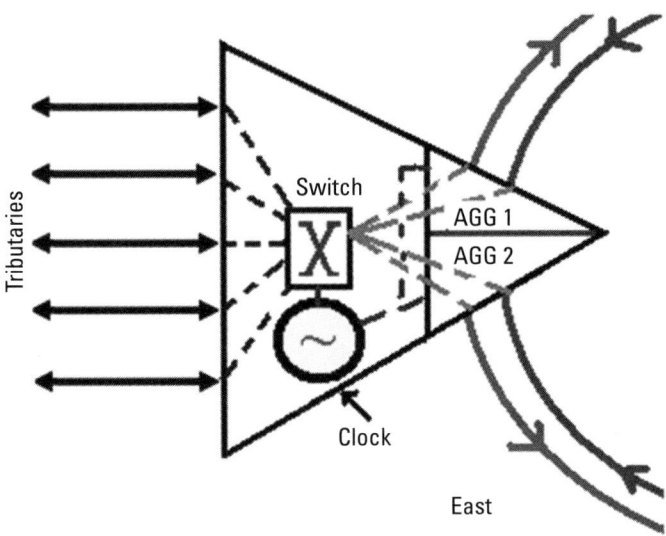

Figure 6.11 Multiplexer/demultiplexer—functional architecture.

Synchronous Optical Networks

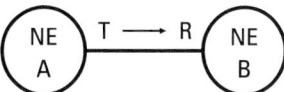

Figure 6.12 Unidirectional connection.

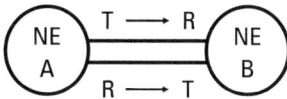

Figure 6.13 Bidirectional connection.

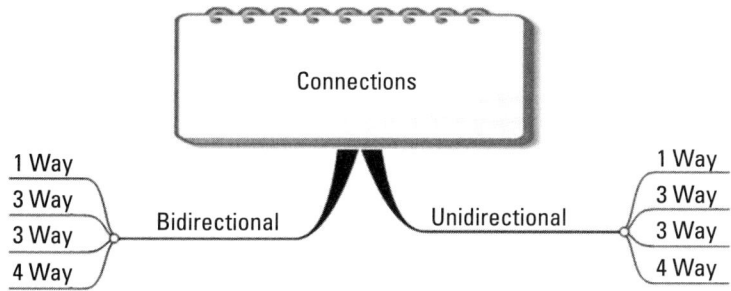

Figure 6.14 Connection types.

Table 6.1
Basic Mux Connection Types: Description

S.N	Connection Type	End Points	Entities	Description
1	Unprotected add/drop	Two-way	1 trib and 1 agg	Information from a tributary is transmitted in one direction of to one aggregate only—used for unprotected networks—linear or spur topology
2	Protected add/drop	Three-way	1 trib and two Agg	Information from a tributary transmitted on both directions or onto both aggregate channels—used for protected networks is based on ring topology
3	Ring interconnection	Four-way	Four tribs or four agg or 2 tribs and 2 agg	Used for interconnection of rings
4	Pass-through	Two-way	Two agg	The information enters the mux through one aggregate and passes back to the network through the other link—used for the connections that are not to be dropped on a mux

nections established on each of the muxes within the network. If a signal or channel (hereto referred to as payload) has to be added/dropped on a mux there would be an add/drop connection provisioned on the muxes (source as well as destination). The remaining connections would be simply forwarded from one aggregate channel to the other. The other major components of a mux include dedicated hardware (or software on some of the lower order muxes) for making the cross-connections. This unit is referred to as a switch, a cross-connect, or a payload manager (in the case of software-based switching). The switch unit includes a time slot assigner (TSA) that assigns an incoming signal to a time slot on the outgoing channels (or vice versa) based on the connections provisioned on the mux. Just individuals require a watch to keep time, a mux also needs a timing source. This source is inbuilt onto the mux and is referred to as a local oscillator [3].

- Cross-connects: A cross-connect is a mux with certain additional capabilities. A cross-connect differs from a mux in the fact that it supports full cross-port connectivity as well as payload interchange. A cross-connect is like a junction box and is used to connect network segments. In a cross-connect signal can be transferred between any combination of ports: aggregate to aggregate, aggregate to tributary, tributary to aggregate, and tributary to tributary. In a mux only aggregate-to-aggregate and aggregate-to-tributary connections are possible. A tributary-to-aggregate connection is especially useful for unidirectional connections as well as for protection purposes. A trib-trib connection facilitates movement of payloads between muxes without blocking the agg channels. This technique is also referred to as hair pinning. Figure 6.15 presents the cross-connect types and their features.

It may be noted that the switching capacity of a cross-connect will be much higher (in multiples) than its transmission rate or the line rate. For example if the line rate is 10 Gbps the cross-connect capacity can be as high as 1,280 Gbps. Another important point to note is that connections are nonblocking. In case of blocking, the provisioning of a connection can bring down other temporary connections temporarily.

The number of connections that can be affected and the duration of the downtime is probabilistic and varies from equipment to equipment. Blocking happens because the switching hardware or software on both are unable to handle the connections up to

Synchronous Optical Networks

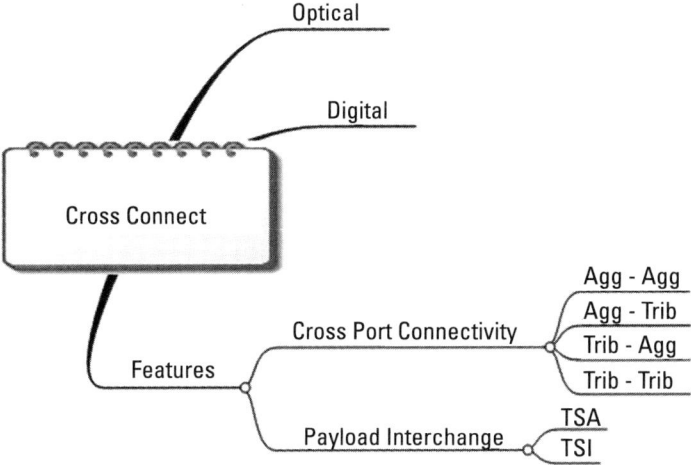

Figure 6.15 Cross-connect types.

the maximum capacity (the switch capacity is kept at least 25% higher than the cross-connect capacity). Figure 6.16 illustrates this concept [3]. The functional block diagram of a cross-connect is as shown in Figure 6.17.

Payload interchange or time slot interchange (TSI) is a similar function performed by dedicated chipsets (or software) on the mux wherein the payload coming in on a particular timeslot on the incoming signal is assigned to another timeslot on the outgoing channel. This is usually the case in case of payload movements from a lower order ring to a higher order ring.

Figure 6.18 summarizes the role of a cross-connect in the transport network. In Figure 6.18, NE D interconnects four rings (multihoming). NE D handles the interconnection traffic switching in addition to its subtending traffic

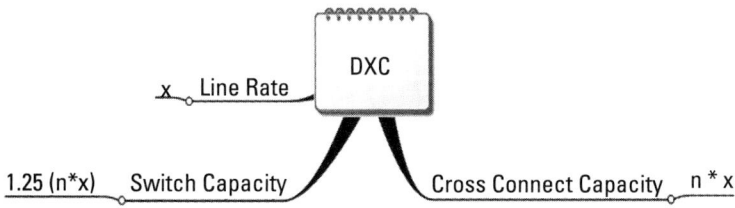

Figure 6.16 Cross-connect—key parameters.

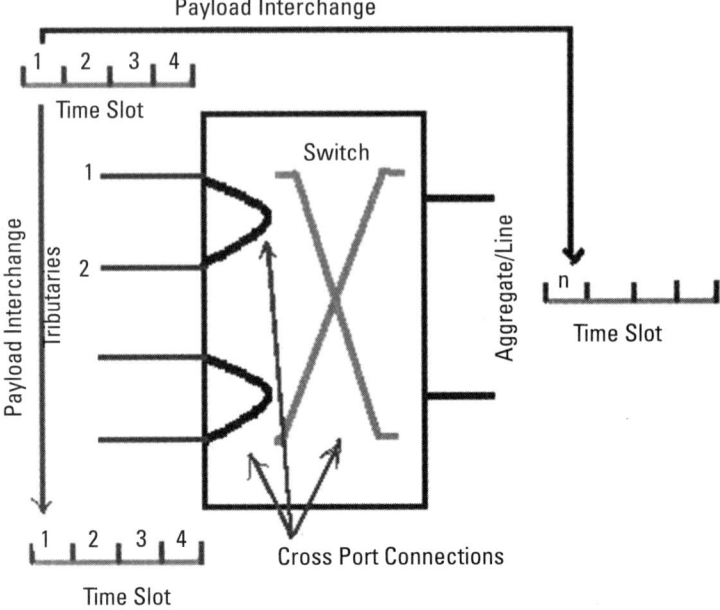

Figure 6.17 Cross-connect—functional blocks.

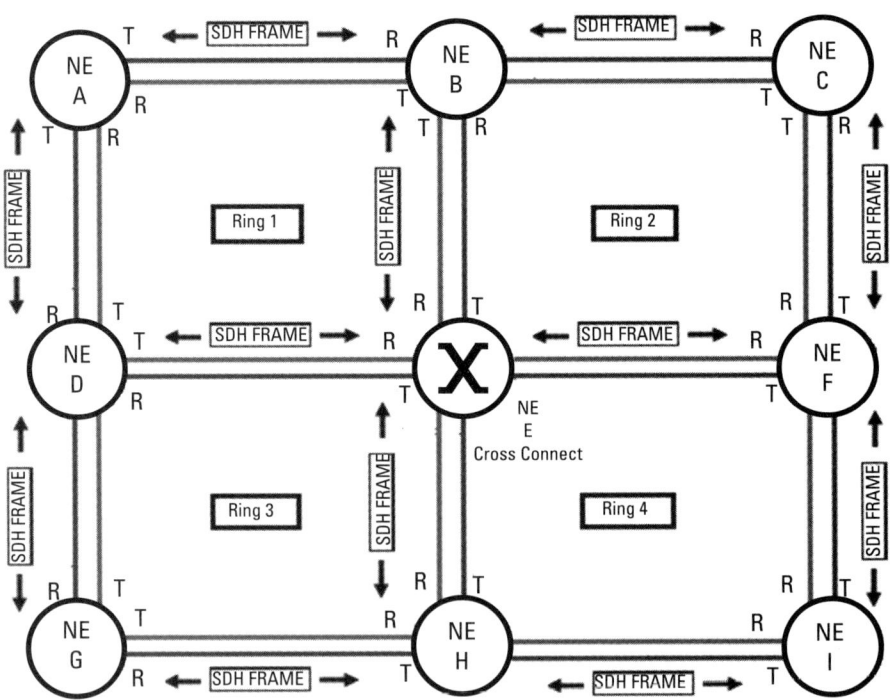

Figure 6.18 Cross-connect—functional blocks.

Example 6.1

In case there was only one train operational between a pair of stations and the train had only one compartment and one berth, the passenger could be uniquely identified by the train number only.

Further let us assume that the trains are only partially filled with a total of only 80 passengers as against the maximum capacity of 400. The trains from different locations arrive at a common junction on their way to the destination. The railways have decided to consolidate the four individual trains into a single train for the onward journey to the destination (collector layer—bandwidth management function). In this scenario the old identification numbers (train number, compartment number, and berth number) cannot be retained for all the passengers. This will lead to some (or all) of the passengers having to be reassigned to different trains, compartments, and berths.

- Amplifiers: An optical signal propagates for a distance before it gets attenuated. The distance depends upon a variety of factors that include the type of signal, type of fiber, number of splices, type of splices, and number of connections. An optical link budget details these parameters. An amplifier is a device that increases the signal strength or imparts a fixed gain to the input signal, thereby increasing the length of the optical link. Amplifiers are explained in more detail in Chapter 7.
- Regenerators: Optical regeneration is used to increase the length of the data path and ensure the stability of critical transmission parameters for long-distance spans (e.g., the length of the fiber and the power of the laser). Modern-day high-speed transmission systems mandate reamplifying, reshaping, and retiming (3R) regeneration, which itself is responsible for bringing the optical signal back to a more readable form (after it has been impaired due to a variety of causes). Regenerators are detailed in Chapter 8.

2. Media: Optical fiber is the only medium that is optimized for transport networks. However microwave links are also used in places where laying of fibers is not possible or extremely difficult.

   As per TeleGeography's global bandwidth research service the demand for international bandwidth grew at a compounded annual rate of 57% between 2007 and 2011. The aggregate capacity requirements have been doubling every two years. In order to support the high bandwidth requirements, transmission rates on the order of exa bits per second (Ebps)—1,024 peta bits per second (Pbps) and 1,024 terabits per second (Tbps)—are required. Optical fiber is the only medium that can support these high bandwidth requirements. In September

2012 Nippon Telegraph and Telephone Corporation (NTT) along with Fujikura Ltd. Hokkaido University and Technical University of Denmark demonstrated ultra-large capacity transmission of 1 petabit (1,000 terabit) per second over a 50-km length using a 12-core (light paths) optical fiber. This speed represents the equivalent to sending 5,000 HDTV videos of two hours in a single second.

The most commonly deployed fiber type is G.652 for SDH/SONET and DWDM network applications. It is also referred to as standard SMF. G.652 is a single-mode optical fiber and cable that has a zero-dispersion wavelength around 1,310 nm. The fiber was originally optimized for use in the 1,310-nm wavelength region but can also be used in the 1,550-nm region. G.652 is characterized by high chromatic dispersion at the 1,500 to 1,625-nm wavelength window, generally used for long-haul and DWDM (C band and L band) transmission. The cost of dispersion compensation, especially at 10 Gbps and above, can be very high for networks using G.652 fiber. However, its low cost makes it the most widely used fiber for access, collector, and metro networks.

G.653 is a single-mode optical DSF with a core-clad index profile tailored to shift the zero-dispersion wavelength from the natural 1,300 nm in silica-glass fibers to the minimum-loss window at 1,550 nm. G.653 fibers have the lowest dispersion profile in the C-band and support signal rates up to 40 Gbps; hence, they are preferred for long-haul networks. However these fibers are not suitable for DWDM networks due to their vulnerability to four-wave mixing effects.

The ITU-T G.654 fiber is optimized for operation in the 1,500 to 1,600-nm region with the low-loss region corresponding to the 1,550-nm band. The G.654 fibers have a larger core area and hence can handle higher power levels. However they suffer from the effects of high chromatic dispersion at 1,550 nm. These fibers have been designed for extended long-haul undersea applications.

The effects due to the nonlinear characteristics of a fiber can be overcome by shifting the zero-dispersion wavelength outside the 1,550-nm operating window. The fibers, referred to as NZDSFs, have a small amount of chromatic dispersion at 1,550 nm, which minimizes nonlinear effects, such as FWM, SPM, and XPM. This is especially suitable for DWDM networks and eliminates or reduces the need for costly dispersion compensation. There are two types of fiber families referred to as NZD+ and NZD–, in which the zero-dispersion value falls before and after the 1,550-nm wavelength, respectively. G.655 is a NZDSF fiber that is optimized for DWDM transmission in the C and L bands[3].

- Microwave links: A microwave link uses a beam of radio waves in the microwave frequency range to transmit information between two fixed locations on the Earth. These links are used for connectivity in terrains that do not facilitate laying of fibers, as well as in urban areas where ROW permissions may not be issued for laying fibers using conventional T&D methodology. Microwave links require LoS connectivity, and their bandwidth is restricted to 155 Mbps only.

  In microwave-based mobile backhaul networks, the PDH microwave is adopted in the access layer, while the SDH microwave is used in the convergence or core layer. Traditional TDM microwave systems are inefficient in using air interface bandwidth. A significant percentage of the current mobile traffic is voice and hence the bandwidth of 155 Mbps is sufficient for transport. With the growth of 3G/4G/5G the current systems will become untenable.

  In order to ensure long-term service development, microwave technology for service access has evolved to IP cores while providing higher capacity to meet network needs. The 4G mobile broadband networks (LTE) support an end-to-end mobile broadband solution by adopting a packet transport network (PTN). Such networks combine IP-based microwave systems with newer optical fiber systems to offer pure packet architecture. The currently available IP-based microwave links support speeds in excess of 500 Mbps.

3. Network topologies: The transport network is the backbone that facilitates the convergence of multiple services over a common framework. A failure in the backbone can have a cascading impact on the services affecting a large number of customers. Hence backbone networks need to have inherent schemes for traffic protection. The only topology that is suitable for a transport network is ring and mesh. Figures 6.9 and 6.18 illustrates the different flavors of the ring and mesh topologies.

   The most commonly used topology in access and collector rings is the two-fiber bidirectional ring. A 2F bidirectional ring provides each network element two transmit paths and two receive paths, commonly referred to as the working and protection paths or east and west connectivity. A network node transmits simultaneously on both these paths. This implies that every node receives the same data from two different paths and can switch in case of failure. The 2F bidirectional ring provides protection from a single physical failure. For protection from multiple network failures a 4F-ring or mesh protection needs to be deployed. The core networks usually employ shared mesh architecture. Figure 6.19 illustrates the commonly used topologies in the transport network.

6.7 Summary

The transport network refers to the common framework or backbone that provides extremely reliable, error-free, high-capacity, long-distance, and multicapacity transmission capabilities. Key topics covered in this chapter are listed as follows.

1. The need and benefits of a transport network;
2. The evolution of the transport technologies and standards;
3. The key components of a transport network, including multiplexers, cross-connects, amplifiers, and regenerators.

Other topics covered by the chapter are summarized as follows.

1. A digital cross-connect is akin to a junction box and helps in interconnecting multiple network segments. It supports cross-port connectivity and payload interchange.

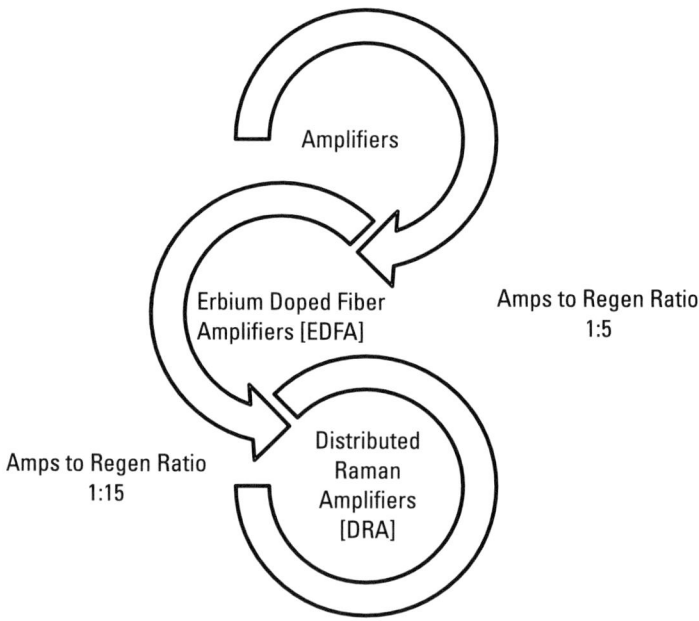

Figure 6.19 Transport network topologies.

2. Traffic is engineered through the network through connections. The most commonly used connections are three-way connections that use the ring architecture to provide protection to traffic.

3. Four-way connections, also referred to as ring-to-ring connections, help to provide a protected interconnection of multiple network segments or rings.

4. A regenerator helps in correcting amplitude and timing distortions from the optical signals. However, this technique requires optical-to-electrical conversion of the input signal followed by regeneration and subsequent conversion back to optical—commonly referred to as OEO. A number of regenerator sections can thus introduce latency in the system.

5. EDFAs are commonly deployed in the network. They provide a cost-effective solution to the attenuation of the optical signals in the C band.

6.8 Review

6.8.1 Review Questions

1. PDH represents the first digital transmission–based asynchronous network to be deployed in America.
    a. True
    b. False

2. SDH network uses in-band communication channels for OAM&P.
    a. True
    b. False

3. The transport network provides a _____ that can deliver multiple services to the end user.
    a. Layer
    b. Common framework/backbone
    c. Utility
    d. Wavelength

4. The transport network is also referred to as a _____ network.
    a. Triple-play
    b. Data

c. Video

d. Voice

5. A digital cross-connect can also be referred to as a_____.

   a. Punch point

   b. Switching matrix

   c. Junction box

   d. Transport medium

6. A transport network consists of _____ layers.

   a. Seven

   b. Five

   c. Four

   d. Three

7. The _____ layer of the transport network is responsible for bandwidth management.

   a. Access

   b. Aggregation

   c. Core

   d. Bandwidth management is not a transport function

8. A regenerator and multiplexer can be used interchangeably on the transport network.

   a. True

   b. False

9. A ring topology does not support the deployment of terminal multiplexers.

   a. True

   b. False

10. An add/drop multiplexer can perform TSI for its tributary and aggregate payloads.

    a. True

    b. False

6.8.2 Exercises

1. Briefly describe the architecture of a transport network.

2. What is meant by a cross-connect? How is it different from a multiplexer?

3. How is traffic engineered through a transport network?
4. List the different types of connections and their usage.
5. Briefly describe amplifiers. Include the different types and their principle of operation.
6. Describe the need and function of a regenerator.

6.8.3 Research Activities

1. Briefly describe the advantages of deploying a transport network.
2. Discuss the challenges faced by ITU-T in bringing about a universal transport standard. Include details on the six crucial issues and a note on the 9R vs. 13R problem.
3. List the key drawbacks of the current transport network architecture. Prepare a broad architecture of the future networks.
4. What is meant by the term OTN? How is it related to the current topic?

6.9 Case Study: SDH Network Architecture

6.9.1 Background

ESW, Inc., is entering the telecom services market in one of the fastest growing economies in the world. The company aims to provide triple-play services to 29 states; 4,000 cities; and over 640,000 villages using a terrestrial fiber-optic network spanning over 200,000 kilometers. The network is expected to cater to over 70 million customers with more than 30 million rural customers requiring only voice services (see Figure 6.20).

6.9.2 Requirements

1. Cost-effective network architecture;
2. Scalability—a network architecture that can be scaled up to meet growing traffic demands;
3. Flexibility—or ability to offer different class of services including support for Ethernet/IP networks;
4. Centralized FCAPS support;
5. Centralized BSS/OSS;
6. Ability to recover from physical network failures— expected to be frequent due to the terrain and infrastructural development activities.

180 Engineering Optical Networks

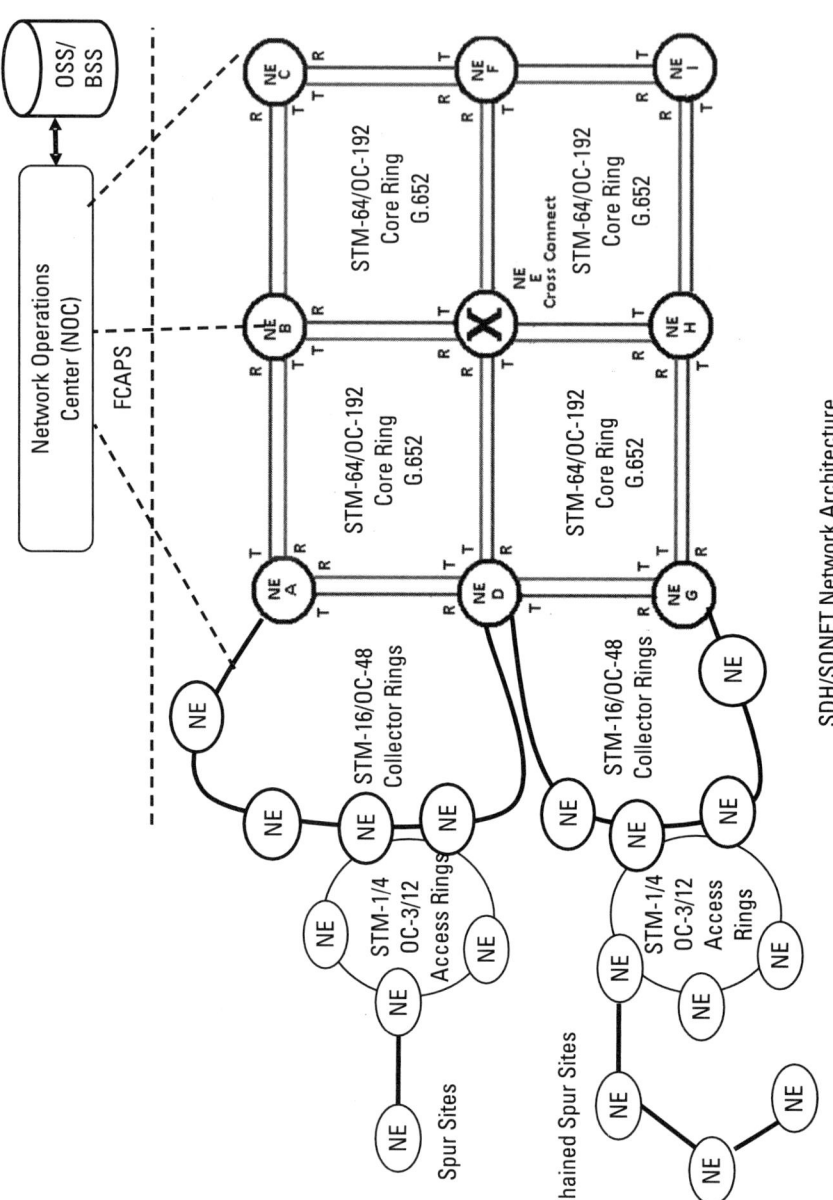

Figure 6.20 Proposed network architecture of ESW Inc.

6.9.3 Key Challenge

Cost-effective scalable and flexible network architecture.

6.9.4 Proposed Solution

A three-tier photonic network based on SDH/SONET technology was proposed, outlined as follows:

1. Tier-1: Intracity links on 10G STM-64 rings;
2. Tier-2: Intercity on 2.5 STM-16-Gbps rings;
3. Tier-3: Access rings of 622 Mbps STM-4/155 Mbps STM-1 rings or spur sites on 155-Mbps STM-1 rings (unprotected).

The major highlights of the proposed solution for ESW is presented in Table 6.2.

As is evident from the background, the ESW network is based on purely SDH/SONET nodes. In order to keep the CAPEX at the minimum while providing for enhanced bandwidth at all the layers and improved network resiliency, the following recommendations were made:

1. Fiber Plant—The choice of fiber plant has a direct impact on network performance and scalability. A green field operator has the freedom to choose a fiber plant that meets its long-erm requirements. As per the case study requirements, the proposed network is based on SDH/

Table 6.2
Key Features of the Proposed Network Architecture for ESW

S.N	Feature	Existing
1	Topology—core	Ring
2	Topology—collector	Ring
3	Topology—access	Ring Unprotected spur
4	Transport standard/technology	SDH/SONET
5	Core bandwidth	10 Gbps
6	Collector bandwidth	2.5 Gbps
7	Access bandwidth	155/622 Mbps
8	10 GigE/1 GigE transport	GFP+VCAT+LCAS
9	Protection from fiber cut	Single (Only two paths per NE)
10	Protection mechanism	SNC-P MSP Port level protection Card level protection

SONET technology and may be required to support WDM infrastructure in the future. Based on the limited information available the choice of fiber plant narrows down to the following:

a. G.652D: A SMF with significantly reduced optical attenuation at a water absorption wavelength of around 1,383 nm. The fiber has low attenuation over a wide wavelength range. It is suitable for DWDM as well as non-DWDM applications and speeds of 100G. It is a cost-effective solution especially for networks with speeds greater than 10G due to the higher span lengths possible and the consequent reduction of regenerators and amplifiers.

b. G.655: Suited for DWDM networks with large network spans. The cost of the fiber is relatively higher than G.652D[2]. In view of the issues of potential damage to the physical network (physical fiber cuts due to infrastructural development activities, especially in the urban areas), it has been proposed that it should be installed in small network segments.

6.9.5 Recommendation

It is recommended that ESW use G.652D cable for its network. Further keeping in mind the need for future expansion, it is recommended that ESW use 96/48/16 core G.652 cables for the core, collector, and access networks respectively. The fiber laying would be by standard T&D in rural areas and HDD in urban areas, and DBC is recommended for small access rings and spur sites in urban areas. Table 6.3 lists the important characteristics of the G.652 fiber.

2. Scalability: The choice of SDH/SONET provides the network scalability up to 10 Gbps (STM-64/OC-192) with direct add/drop capability for a range of tributaries from E1/T1 to STM-64/OC-192.

3. Topology: A physical ring topology, with bidirectional data flow, for all the tiers would facilitate leveraging the protection benefits of the SDH/SONET network.

4. Protection:

   a. Path-level protection:

      i. Subnetwork connection—protected (SNC-P): SNC-P is a protection scheme that works with ring topology. The ring topology provides dual paths in the network. The working traffic flows in one direction while the protection traffic flows in the

---

2. Varies between cable manufacturers; in many cases the difference may not be significant.

Table 6.3
Characteristics of G.652.D (Zero Water Peak Fiber)[1] and G.655 NZDSF Fibers

S.N	Parameter	G.652D	G.655
1	Attenuation		
	at 1,310 nm	≤ 0.334 dB/km	≤ 0.22 dB/km
	at 1,550 nm	≤ 0.194 dB/km	≤ 0.19 dB/km
2	Zero dispersion wavelength	1302 ~ 1,322 nm	1,520 nm
3	Chromatic dispersion		
	at 1285 ~ 1,330 nm	≤ 3.5 ps/(nm · km)	—
	at 1,550 nm	≤ 18 ps/(nm · km)	≤ 6 ps/(nm · km)
	at 1,625 nm	≤ 22 ps/(nm · km)	≤ 11 ps/(nm · km)
4	PMD (uncabled fiber)	≤ 0.15 ps/√km	≤ 0.1 ps/√km

1. Please note that the table mentions generalized values. Actual values are dependent upon specific fiber types by different manufacturers. Kindly refer to specific data sheets for actual information.

opposite direction. The protection mechanism employs tail-end switching wherein the receiving node initiates the protection switching mechanism, based on defects and/or alarms raised along the path.

ii. Dual-ring interconnect (DRI): The DRI mechanism provides an additional layer of path protection between interconnected SNC-P rings. This mechanism provides for dropping and continuing traffic at interconnecting rings with a view of eliminating single points of failure. The DRI mechanism can be provisioned on two or four nodes.

b. Section-level protection:

i. Multiplexed section protection (MSP): MSP is a linear protection scheme wherein a dedicated (1:1) or shared (1: N) protection path is provisioned for each working path. The protection switch is triggered by the raising of section level alarms. This protection mechanism can also be configured between two different subnetworks.

ii. Multiplexed section protection shared protection ring (MS-SPRING): A two-fiber MS-SPRING can be configured wherein the fiber bandwidth is divided among working and protection paths on both the physical fibers. (For example on an STM-16 ring, 8 * VC-4 is reserved for working traffic while the remaining 8 * VC-4 is reserved for protection traffic). The protection switch is triggered by the K1, K2, K3 overhead bytes. On detecting a fiber cut the traffic toward the nodes ahead of the break is switched to the protection bandwidth on the second fiber, which travels in the direction opposite to that of the primary traffic.

    c. Card/port level protection: The card and port level protection schemes are equipment-/vendor-specific and involve redundant hardware in hot standby mode for protection against failure.

5. Network management: The ITU-T defined telecommunications management network (TMN) as a network management framework that facilitates interconnectivity over heterogeneous telecommunication networks. The TMN presents a layered organization that facilitates centralized network management and integration of BSS/OSS systems. The TMN logical layers include the following:

    a. Business management: Includes the functions related to business including trend analysis and billing.

    b. Service management: Definition and management of network services;

    c. Network management: FCAPS support;

    d. Element management: OAM& P functions.

6.10 Referred Standards

G.652: Characteristics of a single-mode optical fiber cable

G.653: Characteristics of a dispersion shifted single-mode optical fiber cable

G.654: Characteristics of a cut-off shifted single-mode optical fiber cable

G.655: Characteristics of a non-zero dispersion shifted single-mode optical fiber cable

G.703: Physical/electrical characteristics of PDH interfaces

G.707: Network Node interface for SDH

G.783: Characteristics of SDH

G.784: SDH management

G.803: Architecture of networks based on SDH

G.813: Timing characteristics of SDH equipment slave clocks (SECs)

G.957: Optical interfaces for equipment and systems relating to the synchronous digital hierarchy

GR-253-CORE (Telcordia generic requirements standard): synchronous optical network (SONET) transport systems: common generic criteria

GR-499-CORE, Transport systems generic requirements (TSGR): common requirements

IEEE P802.3bs 200 Gb/s and 400 Gb/s—Proposed standards for 200Gbps/400Gbps Ethernet

T1.105.02-2001: Synchronous Optical Network (SONET)—Payload Mappings

T1.105.06-2002: Synchronous Optical Network (SONET): Physical Layer Specifications

6.11 Recommended Reading

6.11.1 Books

Senior, J. M., *Optical Fiber Communications*, Second Edition, Prentice Hall Series in Optoelectronics, Prentice Hall, 1992.

Bahaa E. A., M. Saleh, and C. Teich, *Fundamentals of Photonics,* Wiley-Interscience, 1991.

Dutton, H. J. R., *Understanding Optical Communications,* Prentice Hall Series in Networking, Prentice Hall, 1998.

Kartalpoulos, S. V., *Introduction to DWDM Technology,* SPIE and IEEE Press, 1999.

Senior, J., Optical Fiber *Communication: Principles and Practice,* Prentice Hall, 2008.

Keiser, G., *Optical Fiber Communication,* Tata McGraw Hill, 2008.

6.11.2 URLs

http://ecmweb.com/training/electrical_basics/electric_basics_fiber_optics_7/.

http://ecmweb.com/mag/electric_basics_fiber_optics_4/.

http://www.arcelect.com/.

http://www.nfpa.org.

http://www.ul.com/telecom.

http://www.ciscopress.com.

http://www.hubersuhner.com/.

http://www.avap.ch/.

http://www.telebyteusa.com/.

http://search.techrepublic.com.com/search/fiber-optics.html.

http://www.fiberopticproducts.com.

http://www.ask.com/questions-about/Fiber-Optics.

http://www.optiwave.com/.

http://www.fibersolutionsonline.com/.

References

[1] American National Standards Institute, [n.d.]. Retrieved May 1, 2010, [web document]—Source URL < http://www.ansi.org>.

[2] International Telecommunication Union [n.d.]. Retrieved May 1, 2010, [web document]—Source URL < http://www.itu.int/>.

[3] Warier, S., *The ABCs of Fiber Optic Communication,* Norwood, MA: Artech House, 2017.

[4] ITU-T [n.d.]. Retrieved May 1, 2010, [web Document]—Source URL < http://www.itu.int/en/ITU-T/Pages/default.aspx>.

[5] American National Standards Institute [n.d.]. Retrieved May 1, 2010, [web document] Source URL < http://www.ansi.org>.

[6] European Telecommunications Standards Institute [n.d.]. Retrieved May 1, 2010, [web document]. Source URL < http://www.etsi.org>.

7

Optical Transport Network

7.1 Chapter Objectives

The exponential growth in the number of users accessing triple-play services coupled with the explosive progression in the number and type of services has placed high demands on the transport layer, in terms of reliable high-speed throughput, resulting in a complex architecture that provides for reliable transport of fixed granularity bandwidths. The emergence of 4G networks, and the progression to 5G technologies, in the not so distant future will necessitate an exponential increase in the bandwidth capabilities of the core network.

SDH- and SONET-based core networks were not really optimized to carry Ethernet traffic, which makes them an inefficient transport mechanism without the use of concatenation techniques. The transport of Ethernet traffic necessitates the cumbersome process of mapping/demapping at the originating/termination nodes and the use of generic framing protocols and concatenation techniques. The inefficiencies in transporting non-SDH/PDH payloads coupled with the throughput limitations has led to the obsolescence of SDH networks.

This chapter details the architecture of OTNs along with a detailed discussion of their inherent advantages over legacy TDM technologies. There has been significant investment in TDM-based core over the past several years, and service providers are naturally reluctant to transition to packet-based core. Network operators with significant investments in SDH/SONET technologies have migrated to OTNs with a view to preserve their investments while enhancing and the bandwidth management capabilities of their existing networks. Green field operators have displayed a marked tendency to deploy photonic packet-switched technologies. Each generation of the transport networks has built on the disadvantages of the previous generation and has exhibited a marked improvement in terms of flexibility, scalability, bandwidth, and OAM&P capabilities. The chapter concludes with detailed coverage of the needs, functionalities, and architecture of an optical control plane along with a discussion of current standards.

Key Topics

- OTN Network architecture, interfaces and bit rates;
- The linkages between SDH/SONET, OTN, IP/MPLS, and DWDM networks;
- The business imperatives for transitioning to OTN;
- The OTN network organization;
- Optical control plane functionality.

7.2 Introduction to TDM Core Networks

The traditional TDM-based core network consisted of an SDH/SONET transport layer over which the service layer was deployed. The service layer included traditional TDM services as well IP-based services. In the past, TDM-based core networks centered on SDH/SONET technologies were the preferred choice of carriers. However, the development of DWDM networks provided a scalable option for SDH/SONET-based networks to meet the perceived bandwidth requirements in the near future. Developing an SDH/SONET network with an overlaid DWDM network was simpler and considered more cost-effective than IP networks over DWDM. The higher cost of transponders required to interface IP-based networks (with non-DWDM frequencies) with WDM systems was a deterrent, however. The use of SDH/SONET systems added processing overheads related to transporting Ethernet frames over STM/OC frames, but, nevertheless, they met the industry bandwidth requirements at the time.

The exponential growth in data traffic coupled with the need to rationalize the CAPEX/OPEX forced carriers to look at options that allowed them to leverage existing network infrastructure while catering to bandwidth demands seamlessly. The OTN offered carriers enough promise to integrate their SDH/SONET networks by delivering a transparent framework that efficiently handles diverse traffic types, while providing a scalable option to enhance the network throughput 10 times from 10 Gbps (STM-64/OC-192) to 40 Gbps and further to 100 Gbps. OTN provides a transparent hierarchical network that is designed to work with WDM networks providing the functionalities of transport, multiplexing, routing, in-band management, network monitoring with FEC, and traffic protection capabilities.

The rapid developments in the packet core network technologies provided carriers with an alternative for developing scalable and multidimensional networks that could cater to the bandwidth requirements in the long term [1]. The development of new physical interfaces like WDM-PHY, on the router, has facilitated easy integration of IP networks with DWDM networks. This has been complemented by the emergence of reconfigurable optical add/drop multiplexing (ROADM) technologies that provide unparalleled flexibility in designing optical networks.

These developments over the past decade have rendered SDH/SONET-based grooming services obsolete due to the enhanced bandwidth, flexibility, transparency, and effective transport for diverse traffic types and have facilitated the acceptance of IP-based core network architectures by carriers.

7.3 Business Imperatives

The legacy asynchronous transmission networks provided low-bandwidth transport capabilities with virtually nonexistent network management capabilities and protection features. SDH/SONET evolved to overcome the inefficiencies of and to efficiently multiplex, PDH/NADH signals. This was achieved by defining transport containers capable of accepting PDH/NADH bit rates and frame structures and aggregating them to 10 Gbps. This technique was suited for handling prevalent client bandwidths but hindered the scalability of these networks in efficiently carry transporting larger payloads.

SDH/SONET network elements were interconnected in a ring topology through the deployment of optical fiber plants. This interface functioned as the photonic and physical layers of the OSI stack (layers 0 and 1). The developments in the field of WDM technologies and the subsequent deployment of DWDM networks, with a view toward harnessing the bandwidth capabilities of optical fiber, witnessed the emergence of another option for the transport of high-bandwidth signals.

The initial approach to increasing the bandwidth carrying capabilities of an optical fiber was to increase the signal transmission rates. This signal transmission rates were, however, restricted to 10 Gbps using SDH/SONET technologies. The advent of WDM provided a means to transmit multiple rays of light over the optical fiber, thereby enhancing its bandwidth-carrying capabilities manifold. The primary advantage of WDM systems is the transmission of each client signal in its native format. This helps in overcoming the administrative/processing bottleneck involved in mapping/demapping the client signals to the underlying transport layer frame types, SDH/SONET. The obvious drawback, however, is the lack of end-to-end OAM&P capabilities for the client signals. For SDH/SONET networks, there was also the need to enhance their FEC capability, especially on long-haul routes, where the SNR decreases. Moreover, SDH/SONET systems were not designed to efficiently transport Ethernet frames. Figure 7.1 illustrates the basic architecture of a photonic network.

Example 7.1—Transport of Gigabit Ethernet (GigE) Data Through an SDH Network

A client needs multiple GigE (links) to be provisioned between its central office and regional offices. The service provider provisions the links using its SDH/SONET network. However, the minimum switching granularity of an SDH/SONET network is VC-12/VC-11 (2 Mbps/1.5 Mbps) while the maximum is VC-4 (STM-1/OC-1). The provisioning of a GigE pipe through an SDH network would require concatenation of VC-4 containers. This is done by logically grouping a minimum of VC-4 containers to form a contiguous pipe as outlined:

- Required capacity of pipe = 1,000 Mbps;
- Number of containers required to be provisioned on SDH/SONET network = 8;
- Bandwidth provisioned = 155.520[1] * 8 = 1,244.16 Mbps;
- This implies that the bandwidth utilization is 75%.[2,3]

Example 7.2—VCAT

Example 7.1 assumes that a CCAT technique is used to join the payloads. CCAT mandates that all the VC-4s being concatenated should be adjacent in the time domain. The use of VCAT can improve the bandwidth utilization as only seven containers (VC-47c) would be required for provisioning a GigE pipe as outlined:

- Required capacity of pipe = 1,000 Mbps;
- Number of containers required to be provisioned on SDH/SONET network = 7;
- Bandwidth provisioned = 155.520 * 7 = 1,088.64 Mbps;

---

1. STM-1 container size is 9 rows * 270 columns = 2,430 bytes. This translates to a throughput of 2,430 * 64kbps = 155.520 Mbps. Each byte represents a 64-Kbps PCM channel. This includes 81 bytes of overheads. The net payload size is 2,349 bytes, which translates to throughput of 2,349 * 64 Kbps = 150.336 Mbps.
2. 244.16/1,000=24.16% wasted bandwidth.
3. 80% bandwidth utilization excluding the overheads.

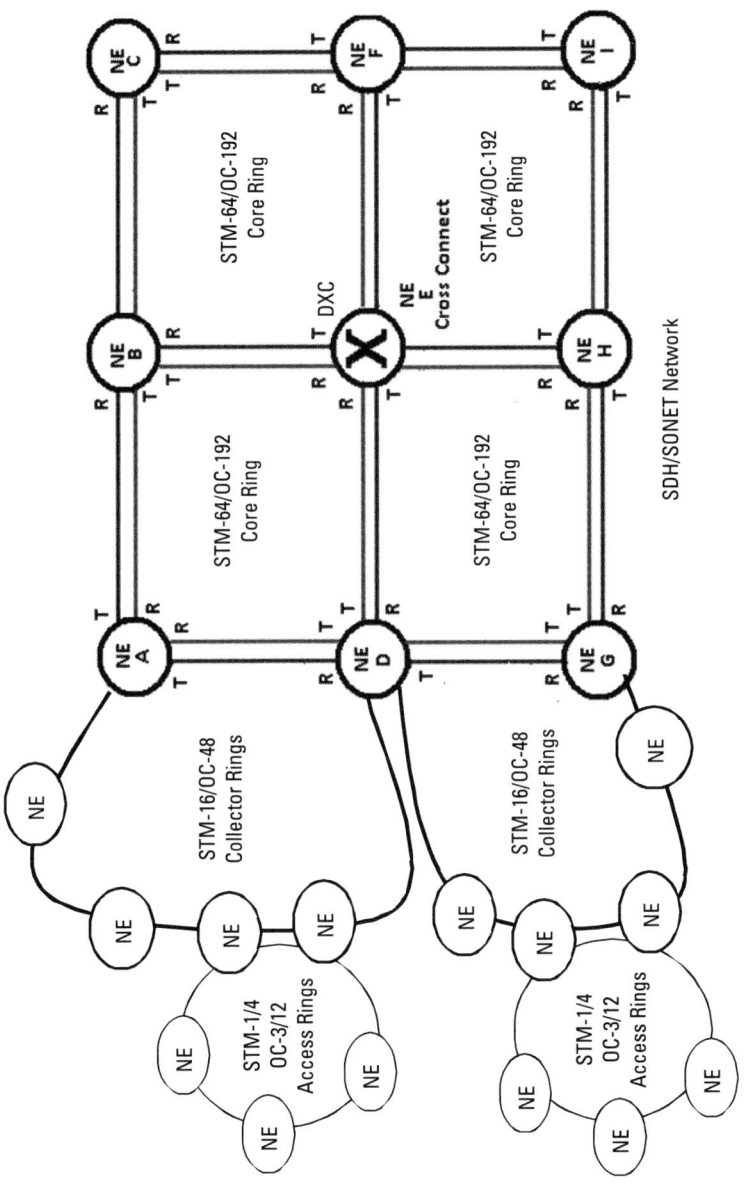

Figure 7.1 Legacy photonic network architecture.

- This implies that the bandwidth utilization is 91%.

However, the use of VCAT necessitates the use of a control plane to manage the bandwidth management and reservation on a link. The functionality of the control plane is discussed in a subsequent section.

The development on the WDM front implied that service providers would need to deploy and manage two distinctly separate transport networks to ensure flexibility and operational agility and to achieve optimum return-on-investment (ROI) on the sizable CAPEX required to deploy photonic networks. SDH/SONET technologies can transparently transport PDH signals but require adaptation/partial termination for data signals and/or multiplexing lower-order SDH/SONET signals. This is an issue when transporting client signals over multiple service provider networks, since the management communication channel over the overhead bytes cannot transparently pass through different service providers' networks. The alternative would be the deployment of multiple client-specific management systems that would add to the network complexity. The other option was to explore the possibility of adding new overhead channels that would support the OAM&P functions, on the WDM network, while providing transparent transport of the client signals.

7.4 Network Organization

OTN evolved to overcome the disadvantages of legacy SDH/SONET networks. OTN networks are designed to transparently, reliably, and efficiently transport higher bandwidth client signals, with OAM&P functionality directly over a fiber network or using established WDM infrastructure. The OTN architecture is based on the ITU-T G.872 standard and is designed to leverage the strengths of the WDM network thereby facilitating seamless interconnection with legacy SDH/SONET networks and IP networks.

7.4.1 Standards

The OTN architecture is based on several ITU-T standards. Table 7.1 lists the ITU-T standards related to OTN.

7.4.2 Network Elements

An OTN consists of NEs, or ONEs, linked by an optical fiber network in a ring and/or mesh topology. The elements support the following functions:

1. Multiplexing/demultiplexing;

Table 7.1
OTN Standards

ITU-T Standards	Description
G.709	Interfaces for the optical transport network
G.798	Characteristics of optical transport network hierarchy equipment functional blocks
G.870	Terms and definitions for OTNs
G.872	Architecture for the OTN
G.873.1	OTN: linear protection
G.873.2	ODU_k shared ring protection (SRP)
G.878	OTN architecture
G.959.1	OTN physical layer interfaces

2. Switching;
3. Transport of client payloads;
4. Supervisory capabilities;
5. End-to-end (E2E) network management;
6. Protection switching;
7. Signal regeneration.

OTN requires a separate set of NEs to realize a fully flexible and scalable network architecture. However, since the basic network standards and interfaces are based on SDH/SONET standards, existing SDH/SONET NEs can be used to support OTN functionality using PIUs with OTN functionality. The primary element of an OTN network is an ADM. The ADM supports bandwidth ranging from 2.5 to 100 Gbps along with mapping, multiplexing, cross-connect, framing, and multiprotocol dynamic bit rate switching functionalities. The ADM consists of PIUs that support multiple protocols (i.e., SDH, SONET, IP, ATM, FR, and SAN), switching units, muxponders and/or optical transponders[4] (OTRs), and redundant power units and time synchronization modules. Figure 7.2[5] shows a schematic diagram of a ONE [2].

7.4.3 OTN Transport Hierarchy

This section provides a simplified view of the OTN network [3]. The purpose is to present the birds-eye view of the network so that the reader can appreciate the network structure without being bogged down by intricacies.

---

4. May include transponders for providing encryption functions.
5. The illustration is for representative purposes and does not conform to the actual layout of an equipment manufacturer. The diagram also does not indicate any traffic protection slots that may be included on the equipment for card level protection.

The optical channel is implemented by means of a framed signal with overheads supporting the management functionality, as specified in G.709. OTN system performance is enhanced through the use of FEC, employing the frame overhead bytes. The mapping of the client signals onto the optical channel is facilitated by two layers referred to as the optical data unit (ODU) and optical transport unit (OTU). Table 7.2 lists and briefly describes the four layers. The concept of photonic and digital layers is illustrated in Figure 7.3.

This results in the introduction of a network with two digital layers, the ODU and OTU. The photonic layers, along with their functionality and associated overhead bytes are outlined in Table 7.2 [4].

As indicated in Table 7.2, OCh maps to the optical domain while the OPU_k, ODU_k, and OTU_k exist in the electrical domain [3]. Additional layers are defined in the optical domain for future use. The underlying client signal (for example, SDH/SONET) is encapsulated (is mapped/demapped at the terminal points without any modification to the payload during transit) within the OPU_k with rate justification provided as needed. The OPU_k layer is thus analogous to the path section in SDH/SONET systems. The ODU_k is analogous to the line overhead within the SDH/SONET frames, while the OTU_k is akin to the section overhead in the SDH/SONET frames. The FEC is contained within the OTU_k. Figure 7.4 illustrates the basic OTN (OTU_1) frame structure.

Note 7.1

The OTN frame structure and transmission is different than the SDH/SONET frame transmission. The OTN frame consists of four rows of 4,080 bytes that are transmitted left to right, top to bottom.

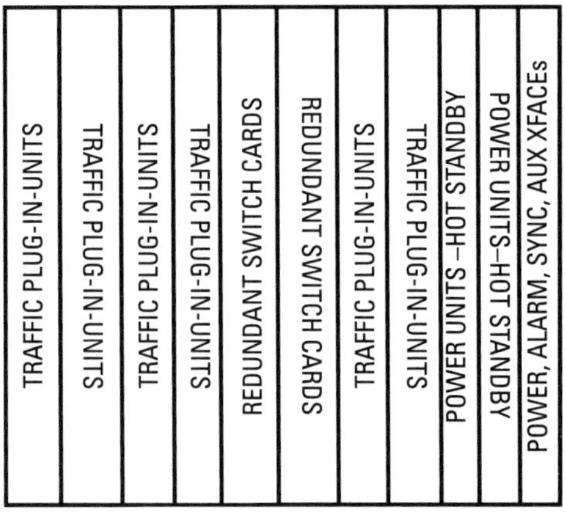

Figure 7.2 ONE.

Table 7.2
OTN Hierarchy

Layer	Domain	Functionality	Equivalence in SDH/SONET
Optical channel—O_{ch}	Optical	Optical channel (end-to-end optical path)	—
OTU_k*	Electrical	Represents physical optical port, performs performance monitoring, FEC	Section
ODU_k	Electrical	Path-level monitoring, alarm indication signals and automatic protection switching (APS)	Line
OPU_k	Electrical	Client signal encapsulation, rate justification	Path

*The value of k ranges from 1 to 3.

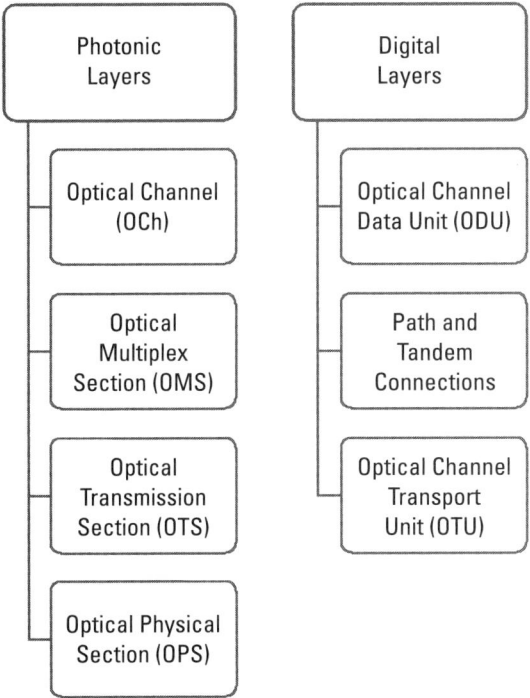

Figure 7.3 OTN layers.

Example 7.3—OTN Transport Mechanism

OPU_k can transport an entire SDH/SONET higher-order payload or multiple lower order payloads. This implies that four STM-16/OC-48 payloads can

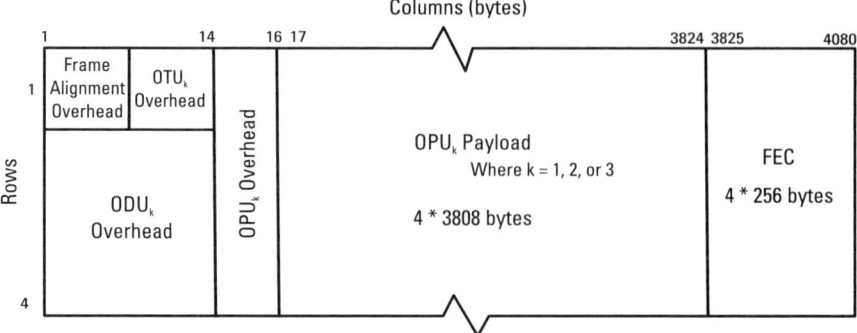

Figure 7.4 OTN frame structure.

be mapped onto an OTU_2 structure. The transport of these payloads is bit-transparent, which ensures that the overheads are not modified and that the integrity of the individual payloads is maintained. In addition the transport is timing-transparent, which implies that the input timing of the payloads is preserved (asynchronous mapping mode) during the demapping ($ODU_2 \rightarrow ODU_1$). In addition the OTN transport mechanism is delay-transparent. This implies that the timing relationship of the four STM-16/OC-48 signals that were multiplexed onto an ODU_2 is preserved. For example if four STM-16/OC-48 signals are mapped into ODU_1 and then multiplexed into an ODU_2, their timing relationship is preserved until the demapping operation at the far end (destination). Table 7.3 outlines the OTN bit rates as defined by G.709 and application type. The corresponding ODU structure, which represents the server layer for the client signals, is outlined in Table 7.4. A graphical representation of the hierarchy is presented in Figure 7.5.

Table 7.3
OTN Line Rates

Container	Data Rates (Gbps)	Application
OTU-1	2.66	Transport of STM-16/OC-48 signals
OTU-2	10.70	Transport of STM-64/OC-92 or WAN PHY (10-GigE) signals
OTU-2e	11.09	Transport of full-line rate LAN PHY signals from IP/Ethernet switches/routers with 10-Gbps interfaces
OTU-2f	11.32	Transports a 10-fiber channel
OTU-3	43.01	40-Gbps signal (i.e., four multiplexed channels of STM-64/OC-192)
OTU-3e	44.58	40-Gbps (4*OTU2e)
OTU-4	112	100-Gbps Ethernet

Table 7.4
ODU Information Structure

ODU Type	Data Rate (Gbps)	Application
ODU-1	2.499	Transport of SDH/SONET, Ethernet packets
ODU-2	10.037	Transport of ODU-1, SDH/SONET, Ethernet packets or WAN PHY
ODU-2e	10.399	Transport of 10 GBE or fiber channel
ODU-3	40.319	Transport of ODU-1/ODU-2, SDH/SONET, 40 GBE
ODU-3e	41.785	Transport of up to four ODU2e signals
ODU-4	104.794	Transport of ODU-1/ODU-2, SDH/SONET, 100 GBE
ODUflex (CBR)	239/238 * Client bit rate	Transport of common bit rate signals (fiber channel, CPRI)
ODUflex (GFP)	Any configured bit rate (multiple of 1.25 Gbps)	Transport of Ethernet, MPLS, IP packets using generic framing procedure

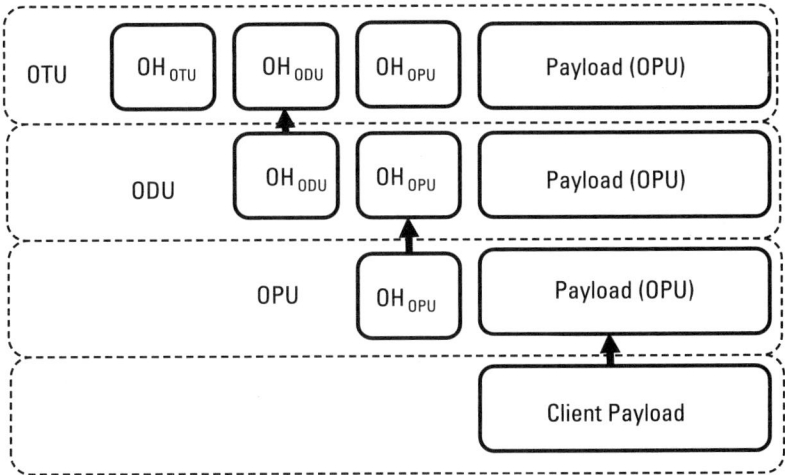

Figure 7.5 OTN hierarchy.

7.4.4 Network Architecture

OTN is based on the ITU G.872 "Architecture for the Optical Transport Network (OTN)" standard. The layered network is comprised of the following:

1. Optical channel (OCh);
2. Optical multiplex section (OMS);
3. Optical transmission section (OTS).

OTN is referred to as digital wrapper technology [2]. A digital wrapper is a method for encapsulating existing data frames irrespective of the native protocol. This digital wrapper is flexible in terms of frame size and allows multiple frames to be wrapped into a single entity. The wrapper concept provides efficient management of client streams with significantly less effort while providing management functionalities on a per wavelength basis for WDM systems.

OTN supports transport of a range of protocols including SDH/SONET, IP, ATM, frame relay, and storage area networks (SANs) in their native formats without the cumbersome payload mapping/demapping process associated with SDH/SONET networks. OTN supports the following network architectures:

1. Point-to-point;
2. Ring;
3. Mesh.

Figure 7.6 presents a bird's-eye view of an OTN network and its interrelationship with legacy TDM, WDM, and IP networks:

Example 7.4—Support for Native Format on OTN

The bit rate of a 10 GigE LAN PHY interface (from a switch or router) is 10.3 Gbps, while that of an STM-64/OC-192 interface is 9.953 Gbps. It is thus evident that payloads from the 10 GigE interface cannot be directly mapped to that of the STM-64/OC-192 interface (in native format). In contrast the OTN frame is elastic and can accommodate diverse traffic types in their native format. The OTU-2e (11.09-Gbps) can accommodate signals from a 10 GigE LAN PHY while the standard STM-64/OC-192 frames can be accommodated withOTU-2.

OTN consists of the following optical entities, as defined by ITU-T recommendation G.872:

1. OCh: This layer network provides the end-to-end networking functionality for transparent transport of diverse client signals, encapsulated in the G.709 frame structure. Its primary functions include the following:
   - Optical signal transmission;
   - Ensuring integrity and maintenance of optical signals;
2. Optical multiplex section (OMS). The optical section layer multiplexes wavelengths within the OTN. The main functions include the following:
   - Networking functionality for multiwavelength optical signals;
   - Ensuring integrity and maintenance of optical signals.

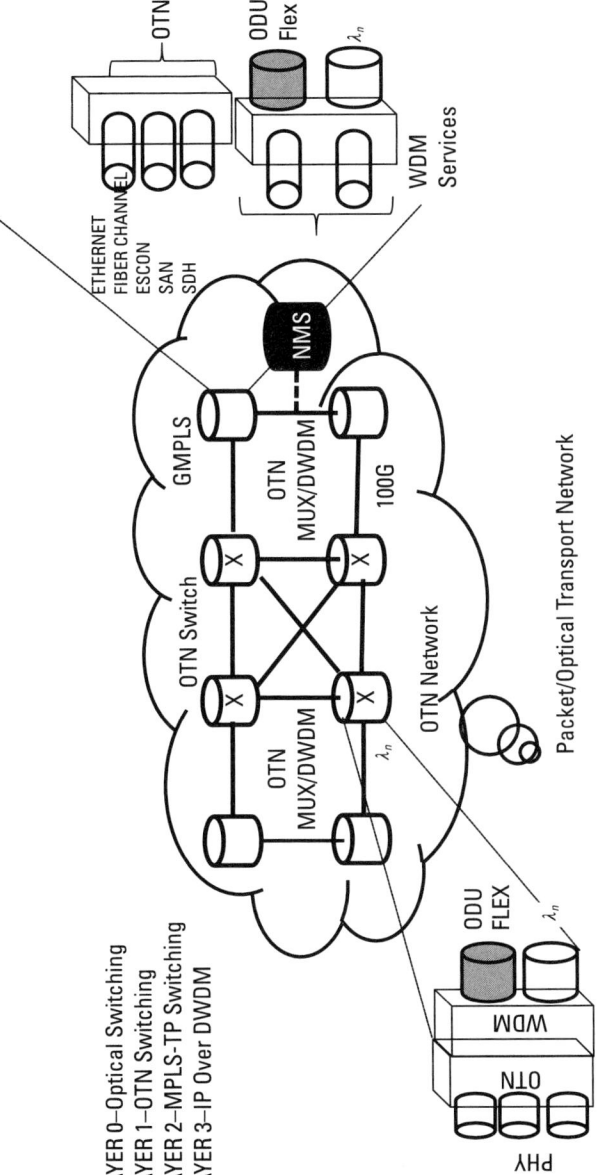

Figure 7.6 Next-generation photonic network architecture.

3. OTS: The OTS refers to sections between ONEs including terminal equipment (TE), amplifiers, and regenerators. The primary function of the OTS includes the following:
   - E2E networking of optical channels with encapsulation of client signals;
   - Flexible routing and OAM functions.

Figure 7.7 illustrates the architecture of an OTN.

7.4.5 Physical Interfaces

ITU-T G.959.1 defines the physical layer interdomain (IrDI) interface for OTNs. These interfaces (Figure 7.8) are based on the presumption of the use of WDM infrastructure. They can, however, be used appropriately as intradomain interfaces (IaDI) as well as on non-OTNs. The resources of a telecom operator or a service provided are organized into domains referred to as administrative domains. E2E management functionality is provided within the administrative domain. The IrDIs may be unidirectional, point-to-point, or single and multichannel depending on the operator requirements. This facilitates compatible interfaces to span multiple administrative domains and support intraoffice, short-haul and long-haul applications (without line amplifiers). Multichannel IrDI supports WDM equipment including amplifiers [5].

To summarize:

1. IrDI refers to the physical interface between two administrative domains. In other words IrDI is the interface between carriers (operators or vendors) with regeneration (3R) at the terminal ends of the interface.

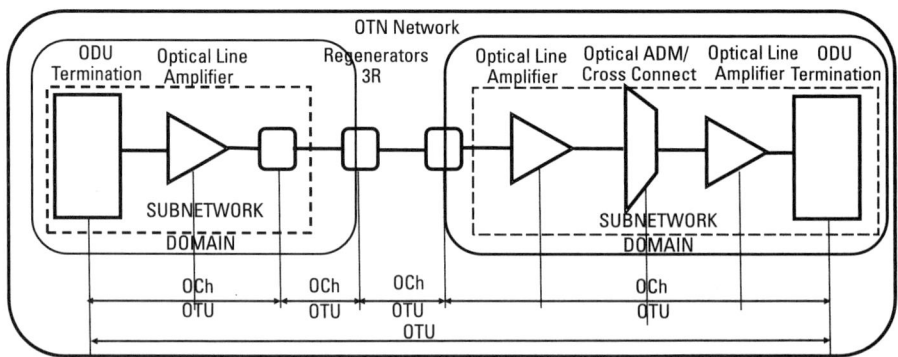

Figure 7.7 Architecture of an OTN.

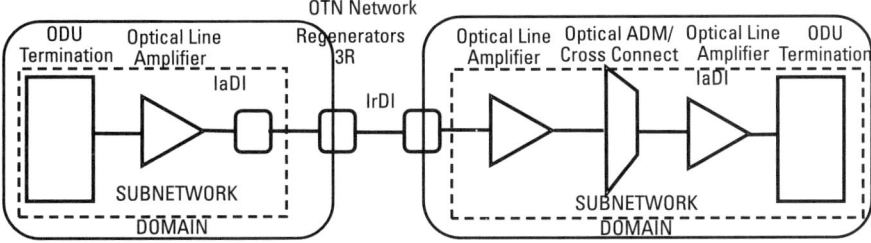

Figure 7.8 OTN interfaces.

2. The IaDI is the physical interface that lies within an administrative domain or the interface within carriers (operators or vendors).

7.4.6 Logical Interfaces

The OTN as specified in ITU-T G.872 defines two interface types:

1. Single optical transport unit interfaces:
   a. Excluding optical layer overhead;
   b. Including optical layer overhead (SOTUm).
2. Multi-optical transport unit interfaces:
   a. Excluding optical layer overhead;
   b. Including optical layer overhead (SOTUm).

The interfaces that do not include optical layer overheads perform 3R processing at each end of the link. The interfaces include management information and may be associated with one or more application identifiers. The application identifier could either include a standardized application code or a vendor-specific identifier. The application identifiers are used to interconnect ONE from different vendors. Equipment from the same vendor would have the same vendor-specific identifiers

7.4.7 Connection Management

The optical network offers fixed bandwidth connections between two end points. [6]. There are three primary types of connections:

1. Unidirectional point-to-point connections;
2. Unidirectional point-to-multipoint connections;
3. Bidirectional point-to-point connections.

An asymmetric connection is considered a combination of two unidirectional connections and/or a special case of bidirectional connection.

7.4.8 Automatic Protection Switching

The ONEs are capable of detecting specific triggers that are indicative of path/network issues as specified in ITU-T recommendations G.806 and G.798. OTN supports a linear protection switching architecture in which the protection switching can be triggered at both the ends of a protected section or subnetwork connection. One end of the link is referred to as the head end while the other end is referred to as the tail end[6]. The head end performs the bridging function wherein the data being transmitted is copied on to the working and the protection paths while the tail end performs the protection switching from the working to the protection paths. There are two types of protection architectures supported:

1. 1+1 architecture—A dedicated protection entity for every working entity.
2. 1:n architecture—One shared protection entity for n working entities.

The protection switching can be unidirectional and/or bidirectional. In case of unidirectional switching only the affected link (transmit or receive) switch onto the protection path. In the case of bidirectional switching both the transmit and receive links switch onto the protection path irrespective of a failure condition on one or both the links. An APS channel using the overhead bytes is used to coordinate the protection switching functions[7].

Further the protection switching can be revertive or nonrevertive. In revertive operation, traffic is restored to the working entities after a switch reason has cleared. This scenario is applicable in the case of a 1+n protection scheme where a protection channel(s) is shared between multiple working channels. In contrast in a nonrevertive operation, normal traffic is allowed to remain on the protection entity even after a switch reason has cleared. This scenario is applicable in case of 1+1 protection scheme.

7.5 Optical Control Plane

SDH/SONET networks initially consisted of two operational planes:

---

6. The end initiating the protection switching is referred to as the head end while the switching end is referred to as the tail end.
7. Except for 1+1 unidirectional protection scheme

1. User/transport plane—User data that is transmitted through PIU, switching matrices (fabrics), equipment backplane, and onto the fiber plant (and vice versa in the receive direction);
2. Management/OAM&P plane—Management traffic that flows from the NE, EMS, NMS, and/or OSS.

Optical control plane (Figure 7.9) is a combination of hardware and software that automates the connection management and/or restoration management within photonic networks. The concept of the control plane has been extended from the SDH/SONET networks to OTN and WDM networks. The control plane is positioned in between the user and the transport planes and provides ONEs with additional intelligence including the overall network architecture and resources. This additional intelligence facilitates dynamic provisioning of user service requirements, switching in case of failures, and restoration from failures. The control plane deployment is more effective in networks with a meshed core but can also be used on a ring network (interconnection of multiples rings) [7].

The control plane automates the call[8]/connection management functions according it with the ability to automatically compute optimal client connection paths prior to the establishment of connections. Further the use of statistical techniques (algorithms) improves the operational efficiency and network resource utilization. The control plane also plays an important role in protection

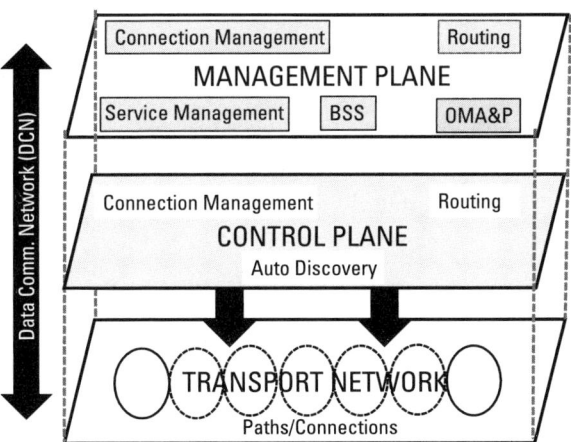

Figure 7.9 Optical control plane.

---

8. In control plane terminology a path or connection is referred to as a call.

switching by automatically computing diverse protection paths, reserving bandwidth required for protection switching and automatic route computation for path restoration.

Example 7.5—Control Plane Protection

A customer network may comprise of legacy SDH/SONET rings subtending on to a meshed OTN core. In case of a path failure (due to a fiber cut on an SDH/SONET ring), the first response would be provided by the SDH/SONET 50-ms APS mechanism. A ring network can provide protection from only a single physical failure. In case of another failure on the same ring[9] the APS would not be able to initiate any protection switching. The control plane now steps in to do the following:

1. Check for alternate path (either dynamically or based on hard-coded[10] information);
2. Perform path reservation;
3. Switch traffic onto the reserved path;
4. Monitor the failed path for change in status;
5. Perform automatic restoration once the original path has been restored.

7.5.1 Standards

Needs/Objectives

- Set of guidelines leading to the design of automated optical networks with full interoperability among vendors (multivendor);
- Specifications for the minimum set of features to be supported by devices conforming to common standards.

The initial ITU-T recommendation G.807/Y.1302 specified the requirements for the control plane of automatically switched transport networks (ASTNs) for automating connection management across a transport network. This recommendation was withdrawn and merged with G.8080/Y.1304, which specifies the architecture for automatically switched optical network (ASON). The IETF has extended the use of MPLS to include TDM-based networks like SDH/SONET and OTN. The IETF has been responsible for the development

---

9. Prior to the restoration of the earlier failure.
10. Switching time would be less than that for dynamic switching.

of protocols based on the ITU-T framework and/or industry demands. In addition another nonprofit industry forum, the Optical Internetworking Forum[11] (OIF) has been instrumental in developing a framework for the application of IETF control plane protocols to the ITU-T control plane architecture with a view of facilitating multivendor interoperability. Figure 7.10 summarizes this relationship. Table 7.5 summarizes the key optical control plane standards.

Figure 7.11 illustrates the interrelationship between the various ITU-T ASON standards.

7.5.2 Logical Interfaces

There are three standard logical interfaces for ASON [8]:

1. User-network interface (UNI): UNI is a bidirectional interface used for signaling between the control plane of the service initiator and provider.

2. Internal network-network interface (I-NNI): I-NNI is a bidirectional interface used for signaling between trusted inter- or intradomain control plane entities.

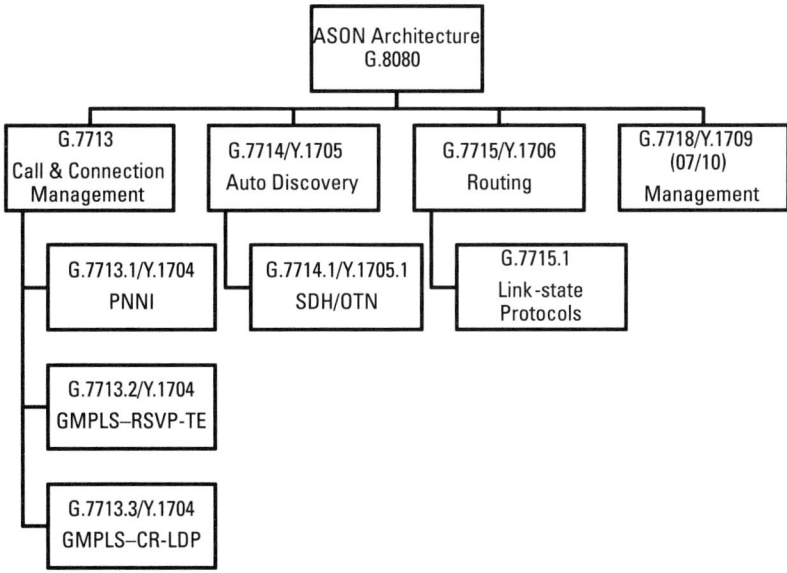

Figure 7.10 Control plane standards.

---

11. An industry forum that promotes multivendor interoperable networking solutions.

Table 7.5
Optical Control Plane Standards

Standard	Body	Description
G.8081/Y.1353 (02/12)	ITU-T	Terms and definitions for ASON
G.8080/Y.1304 (02/12)	ITU-T	ASON architecture
G.7713	ITU-T	Distributed call and connection management (DCM)
G.7713.1/Y.1704	ITU-T	Distributed call and connection management—PNNI implementation
G.7713.2/Y.1704	ITU-T	Distributed call and connection management—GMPLS RSVP-TE implementation
G.7713.3/Y.1704	ITU-T	Distributed call and connection management—GMPLS CR-LDP implementation
G.7714/Y.1705	ITU-T	Generalized automatic discovery techniques
G.7714.1/Y.1705.1	ITU-T	Protocol for automatic discovery in SDH and OTN networks
G.7715/Y.1706	ITU-T	Architecture and requirements for routing in automatically switched optical networks
G.7715.1	ITU-T	ASON routing architecture and requirements for link state protocols
G.7716/Y.1707 (01/10)	ITU-T	Architecture of control plane operations
G.7717/Y.1708	ITU-T	Connection admission control
G.7718/Y.1709 (07/10)	ITU-T	Framework for ASON management
RFC 3945	IETF	Generalized multiprotocol label switching (GMPLS) architecture
RFC 6373	IETF	MPLS transport profile (MPLS-TP) control plane framework
RFC 3471	IETF	Generalized multiprotocol label switching (GMPLS) signaling functional description
RFC 3473	IETF	Generalized multiprotocol label switching (GMPLS) signaling resource reservation protocol-traffic engineering (RSVP-TE) extensions
RFC 4202	IETF	Routing extensions in support of generalized multiprotocol label switching
RFC 4206	IETF	Label switched paths (LSPs) hierarchy with generalized multiprotocol label switching (GMPLS) traffic engineering (TE)
OIF-ENNI-RSVP-02.2	OIF	Recovery amendment to E-NNI 2.0—RSVP-TE signaling (February 2014)
OIF-ENNI-OSPF-02.2	OIF	Recovery amendment to E-NNI 2.0 OSPFv2-based routing (February 2014)
OIF-ENNI-REC-AM-01.0	OIF	Recovery amendment to E-NNI 2.0 common part (February 2014)
OIF-PCE-IA-01.0	OIF	Path computation element (June 2013)
OIF-OTN-TCM-01.0	OIF	Guidelines for application of OTN TCM white paper (June 2013)
OIF-ENNI-ML-AM-01.0	OIF	Multilayer amendment to E-NNI 2.0—common part (April 2013)
OIF-ENNI-RSVP-02.1	OIF	Multilayer amendment to E-NNI 2.0—RSVP-TE signaling (April 2013)

Table 7.5 (continued)

Standard	Body	Description
OIF-UNI-02.0-R2-RSVP	OIF	User network interface (UNI) 2.0 signaling specification release 2—RSVP extensions for user network interface (UNI) 2.0 signaling release 2 (January 2013)
OIF-SEP-03.2	OIF	Security extension for UNI and E-NNI 2.1 (October 2012)
OIF-SLG-01.3	OIF	OIF control plane logging and auditing with Syslog (October 2012)
OIF-SMI-03.1	OIF	Security for management interfaces to network elements 2.0 (October 2012)
OIF-E2E-SEC-01.0	OIF	End-to-end transport of UNI client authentication, integrity, and data plane security support information (May 2012)
OIF-CP-MGMT-01.0	OIF	Management plane (OSS) support for control plane networks white paper (2012)
OIF-RSVP-PVT-EXT-01.0	OIF	OIF application of vendor private extensions in RSVP implementation agreement (October 2011)
OIF-ENNI-OSPF-02.0	OIF	External network-network interface (E-NNI) OSPF-based routing—2.0 (intra-carrier) implementation agreement (July 2011)
OIF-CWG-CPR-01.0	OIF	OIF carrier working group guideline document: Control plane requirements for multidomain optical transport networks (July 2010)
OIF-E-NNI-Sig-02.0—E-NNI	OIF	Signaling specification (April 2009)
OIF-G-Sig-IW-01.0	OIF	OIF guideline document: signaling protocol interworking of ASON/GMPLS network domains (June 2008)
OIF-UNI-02.0	OIF	Common—user network interface (UNI) 2.0 signaling specification: common part (February 2008)
OIF-CDR-01.0	OIF	Call detail records for OIF UNI 1.0 billing (April 2002)

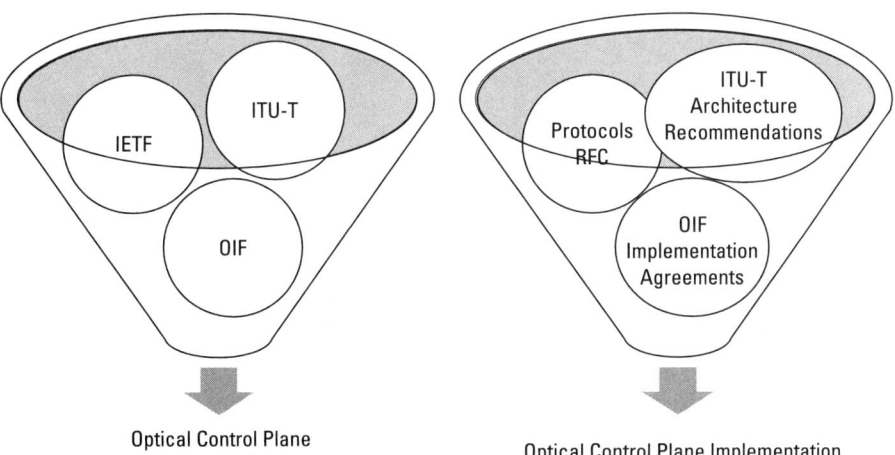

Figure 7.11 ASON standards.

3. External network-network interface (E-NNI): E-NNI is also a bidirectional interdomain control plane signaling interface.

The interfaces facilitate the exchange of the following [8]:
1. Authentication and connection admission control (UNI, E-NNI);
2. Connection service messages (UNI, E-NNI, I-NNI);
3. End-point name and address (UNI);
4. Network resource control information (I-NNI);
5. Reachability information (E-NNI);
6. Topology information (I-NNI).

7.5.3 Control Plane Architecture

The control plane, situated between the management and the transport planes, consists of a set of interconnected entities that communicate through signaling channels and facilitate automated call and connection management including automated resource discovery and routing [9]. The key entities are listed in Table 7.6.

Note 7.2

The DA, TAP, and LRM together facilitate the control plane discovery mechanism.

Note 7.3

The DA operates within the transport plane and provides name separation between the transport and the control planes, in conjunction with TAP.

Table 7.6
Control Plane Functional Entities

Entities	Function
Call controller (CC)	Call management
Connection controller (CallC)	Connection management
Discovery agent (DA)	Name separation between control and transport planes
Link resource manager (LRMA)	Subnetwork[1] point (SNP) resource management
Link resource manager Z(LRMZ)	Provide topology information
Network call controller (NcallC)	Intradomain call control
Routing controller (RC)	Intradomain routing
Termination and adaptation performer (TAP)	Provides relationship between the control and transport plane names of a resource

[1] SNP refers to transport plane.

7.5.4 Control Plane Functions

The key functionalities of the control plane include the following [8]:

1. Automatic discovery;
2. Call management;
3. Connection management;
4. Routing/switching;
5. Protection switching and restoration.

The discovery, call management and protection switching, and restoration functions of the control plane are described as follows.

1. Automatic discovery: Automatic discovery adds a layer of intelligence to the underlying NE while eliminating the need for explicit configuration. The control plane incorporates the following discovery mechanisms:
    a. Neighbors:
        - Physical media adjacency;
        - Layer adjacency;
        - Logical (control layer) adjacency.
    b. Resource;
    c. Service:
        - The first step is to verify the physical connectivity between two ports between which a call/connection is to be established.
        - The next step is to verify the logical connectivity (for each layer) between the call/connection end points.
        - The associations between the logical link end points are established by verifying the layer adjacencies. These associations are used to set up network topology maps that support the routing functionalities, identification of link end-points required for connection management
        - The control adjacencies between the control entities of neighboring transport plane NEs are subsequently set up.
2. Call and connection management: A call is a service provisioned end-to-end within the optical network [9]. The concept is similar to connections in SDH/SONET networks. A connection is the underlying transport entity that supports the call. The concept is similar to the path in SDH/SONET networks. A call may be supported by a single or multiple connections. The call and connection control mechanisms

are separately handled within the control plane. The call control function involves call setup, maintenance, and admission control. Connections are dynamically set up and released as per the initiated call sessions. Call control is applicable at ingress ports (UNI) and network boundaries (E-NNI). There can be three types of connections specified by the control plane [9]:

a. Permanent;
b. Soft-permanent;
c. Switched.

These three connections are described as follows.

- A permanent connection also referred to as a provisioned connection, can be either manually configured or set up by the control plane. There is no further intervention or action required (automatic routing and/or signalling) to support this connection.
- A soft permanent connection refers to a temporary or dynamic or switched connection between two permanent end points (generally at the edge of the network/domain). This type of connection does not require a user-network signaling interface (UNI).
- A switched connection is an on-demand connection between specified end-points using the control plane routing/switching and requires a UNI.

7.5.5 Protection Switching and Restoration

The control plane defined by the ASON standards drastically improves network reliability by providing the ability to automatically detect and recover from network failures—physical as well as logical. The control plane supports predefined or hard-coded protection resources for a call-automatic, dynamic or switched calls, and hybrid mechanisms. In the event of a failure the control plane performs the protection switch, monitors the call/connection, and performs automatic restoration when the original working resources has recovered from its failure condition. The transport plane supports a 50-ms protection switching scheme. The control plane–enabled protection switching mechanism typically takes longer (depending on the type of connection)—tens of milliseconds while the restoration function may take hundreds of milliseconds to seconds. The control plane enables a range of QoS parameters to be defined on the network. A network fault is initially handled by the transport plane in conjunction with the management plane. In case of multiple failures and/or the inability of the transport plane to respond, the request is forwarded to the control plane for action.

7.6 FEC

Data transmission through telecommunication links are subject to errors that occur due to unreliable and/or noisy communication channels. The receiver needs to have a mechanism to understand that the information being received over the channel is errored and to request for retransmission. FEC, also referred to as channel coding, is a technique based on the pioneering work of American mathematician Richard Hamming [10] that uses an error-correcting code (ECC) to redundantly encode (reducing transmission bandwidth) the transmitted data. This redundancy facilitates limited error detection, and possibly correction, without the need for retransmission. The different type of codes are illustrated in Figure 7.12.

SDH/SONET systems uses section overhead (SOH) bytes to transmit FEC information, referred to as in-band FEC. However, the number of bytes

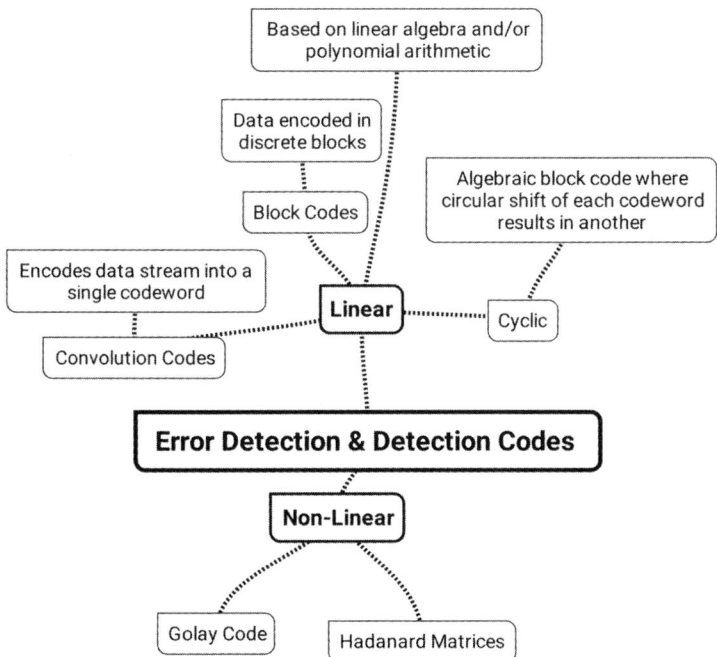

Figure 7.12 Types of channel codes.

reserved/used for FEC is limited, thereby reducing its effectiveness. One of the major advantages of OTN is the use of FEC, which is effective in dispersion-limited and OSNR[12] limited systems.

Example 7.6—OTU Data Rates

The OTN OTU1 frame consists of 4,080 bytes (Figure 7.4).

Therefore, $n = 4,080/16 = 255$[13].

The OPU_1 consists of 3,808 payload bytes and an additional 16 bytes of overhead.

Therefore $k = 3,824/16 = 239$

The ODU_1 data rate is 2.488 Gbps (OPU_1 payload rate) * 239/238 {3808/16 = 238}= 2.499Gbps

In a similar fashion the OTU_1 data rate = 2.499 * 255/238 = 2.667 Gbps.

7.6.1 Reed-Solomon Codes

ITU-T G.709 specifies 16-byte interleaved codecs using Reed-Solomon (RS) codes for OTUk transmission. A nonbinary block code RS (255, 239) is used for FEC processing.

A RS code is specified as RS (n, k), which implies that parity bytes are added to k bytes of data (being transmitted) to form a code of n bytes. Using this technique a decoder, at the receiving end, can correct a maximum of t within the code word of n words, where 2t=n-k.

Example 7.7—RS Coding Gain

RS (255,239) implies that 16 bytes (255-239) are used or in other words that 16 parity bytes are added to 239 information bytes. The code can correct eight byte errors:

$$2t = n - k$$
$$2t - 255 - 239 = 16$$
$$t = 8$$

This translates to a coding gain of approximately 6 dB.

The additional bandwidth consumed is: 16/239 = 6.69% or approximately 7%.

---

12. OSNR is the measure of the ratio of signal power to the noise power on an optical link.
13. The data would be transmitted in 16 frames

Example 7.8—RS Code Interleaving

RS codes consider errors on a symbol basis. Example 7.7 shows that RS (255,239) code can correct eight symbols in a codeword ($t = 8$ bytes). A byte/symbol is considered to be errored if one or all bits are errored. The preceding statement also implies that the code can detect a maximum eight symbols with bit errors.

It is easier to detect a symbol when all bits are errored as compared to when there is only a single-bit error. FEC is therefore suited to correcting burst errors. The efficiency of the RS codes can be improved by using interleaving techniques. The process of interleaving spreads the effect of burst errors among multiple code words.

Example 7.9—Advantages of Interleaving

Assume that there are 64 code words being transmitted in a group. The use of RS code provides the ability to correct eight bit errors.

- In the best-case scenario there would be one bit error in each symbol, which implies that a maximum of eight bit errors can be corrected in the entire group.
- In the worst-case scenario each of the symbols may have all bits errored. This implies that a maximum of $8 * 8 = 64$ bit errors can be corrected in the entire group.
- The interleaving of code words in the group will facilitate the correction of 512 symbol errors, irrespective of the fact that these may be in one long burst.

7.6.2 Bose-Chaudhuri-Hocquenghem Codes

Bose-Chaudhuri-Hocquenghem Codes are multiple error detection and correction cyclic codes, based on Hamming codes. The code words are formed dividing the information bits received, represented by a polynomial by another polynomial (generator)[14]. The primary advantages of using BCH codes include the following:

1. Control over the number of correctable symbol errors;
2. Simplified decoding[15].

---

14. All code words are multiples of the generator polynomial.
15. Using an algebraic method referred to as syndrome decoding.

7.6.3 Low-Density Parity Check (LDPC) Block Codes

LDPC block codes (third-generation codes) were invented by Rober Gallager but not used until the late 1990s. LDPCs are codes generated from sparse bipartite graphs[16]. Figure 7.13 illustrates an example of LDPC code. The n left nodes are referred to as message nodes and the r right nodes are known as check nodes. The graph provides a linear block code of length n and (n-r) dimensions as illustrated in Figure 7.13. The code words are formed from the vectors in such a manner that the sum of the neighboring positon among nodes is zero. LDPC codes represent one of the best choices for high-throughput, real-time transmissions (like reality technologies) for the following reasons:

1. Modulation of the coded signal, enabling it to withstand channel distortions;
2. Capacity-approaching codes (making efficient use of the channel bandwidth);
3. Efficient decoding by using low-latency parallel iterative algorithms;
4. Higher coding gain than RS codes (20–40%, translating to greater than 10 dB).

7.6.4 Advantages of FEC

The transmission quality over a fiber optic link is dependent on several parameters, some of which are inherent to the transmission system being used, while

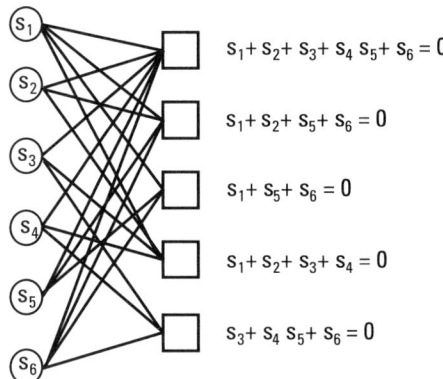

Figure 7.13 LDPC codes.

---

16. In graph theory mathematical relations between objects (pairwise) is represented by graphs. A bipartite graph consists of two independent sets (disjoint) in which no two vertices (within the same set) are adjacent. A graph with the number of edges close to the maximum possible is referred to as dense graph while a graph with a few edges is referred to as a sparse graph.

the others can be controlled through proper planning and deployment. The optical power budget or link budget (LB) determines the span of an optical link. The LB takes into account the attenuation over the fiber, connectors, and other effects and helps the network planner design optimal spans. A larger LB helps in maintaining the link BER with a reduced number of NEs like amplifiers and regenerators. However a higher link budget generally involves higher CAPEX due to the cost of high-quality optical components.

The OSNR requirements become rigorous for line rates greater than 10G due the use of advanced modulation techniques. There is a direct correlation between OSNR and performance of optical links (low OSNR and high BER—especially on DWDM links). FEC provides for higher coding gain, at higher rates, by encoding optical signals with error detection and correction bytes (typically 7% additional overheads), which can be used by the receiver to detect and correct errors.

FEC provides a cost-effective option to enhance the link budget while maintaining a low BER. The significant benefits include the following:

1. Enhanced link performance;
2. Enhanced span length;
3. Reduction in number of amplifiers and/or regenerators;
4. Enhanced transmission quality (due to error-detection and -correction techniques).

FEC is effective in OSNR-limited systems as well as dispersion-limited (CMD) systems. It is, however, less effective in countering the effects of PMD.

Example 7.10—FEC Coding Gain

The FEC scheme defined by G.709 can result in up to a 6.2-dB improvement in SNR [3].

The above statement implies that an optical can have a decreased transmit power of 6.2 dB and still maintain the link BER in comparison with a link that does not employ FEC[17]. The gain in the LB provided by the LB can be also used to enhance the span length and/or increase channels in DWDM systems[18]. Additionally the enhanced span lengths result in a reduction of regenerators on the network, yielding cost optimization and reduced system latency. Further the

---

17. This statement is true based on the assumption that optical impairments like chromatic dispersion (CD) and polarization mode dispersion (PMD) are minimal.
18. The type of amplifier deployed on a DWDM link may limit the number of channels due to the restrictions on their output power.

gain can also be used to increase the number of grooming elements, OADMs and optical cross-connects (OXCs) in the optical network.

The only disadvantage of using the FEC scheme is that bandwidth needs to be assigned over the link for transmitting coding information, reducing the information-carrying capacity of the link.

Hard-Decision FEC

In conventional network equipment, the receiver makes a decision on the status of the received bit (0 or 1) based on the comparison of the received signal power to a threshold value.

Example 7.11—Hard FEC

The receiver compares the voltage of the incoming pulses with a threshold value.

- Signal transmit voltage = 5V;
- Threshold value = 3.2V;
- If the value of the incoming pulse is greater than 3.2V it represents 1 bit, and if the value is less than 3.2V the incoming pulse represents a 0 bit.

Soft-Decision FEC

In cases of soft-decision FEC, the decoder uses additional bits (overheads) to make the decision on the digital representation of the incoming signal along with an additional parameter referred to as the confidence interval, which provides information on the range or factor by which the input signal differs from the threshold levels. Soft FEC provides a coding gain of over 20–40% over systems using conventional FEC.

7.7 Tandem Connection Monitoring (TCM)

The use of multicarrier networks necessitates a method to track connections with a view of identifying, localizing, and recovering from faulty network conditions. Tandem connection monitoring (TCM) provides networks with the capability of monitoring calls/connections that pass through one or more operators' networks.

OTN provides user-defined path-layer monitoring at multiple (user-defined) end points using TCM bytes of six different levels provided within the OTN overhead. The bytes are labeled TCM1 to TCM6 and allow carriers to individually define their path layers. Figure 7.14 illustrates the concept of TCM.

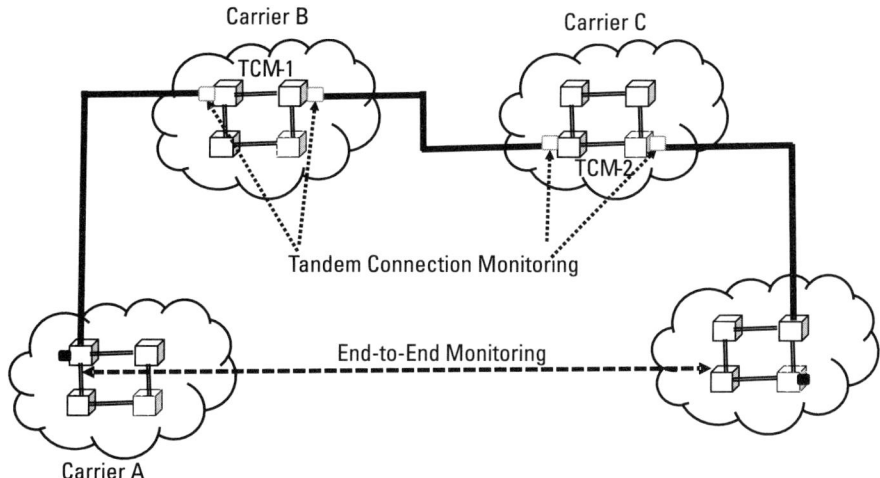

Figure 7.14 TCM—operational principle.

7.8 Switching Architecture

SDH networks and SONETs provide two layers of switching—lower- and higher-order. The lower-order switching was used for multiplexing the PDH/NADH signals. With the advent of newer services like IP and Ethernet, there emerged a gap between the switching rates and the line rates. In order to bridge the gap, the concatenation function was introduced. The transport of IP/Ethernet frames over an SDH/SONET transport structure resulted in inefficient bandwidth usage. The problem was compounded manifold for line rates of 40 Gbps and above. It is evident that switching rates have to be synchronized to the line rates. On the other hand, DWDM networks provide the capability to employ wavelength switching. However, the switching can only be done at the bit rate supported by the wavelength, and grooming facilities would have to be deployed for subwavelength switching.

OTN networks allow operators to switch at any rates greater than 2.5 Gbps, independent of the wavelength being used. The switching rates can mirror the line rates, providing networks with the scalability and flexibility of meeting service and client requirements dynamically. In the early days of OTN a common technique was to use muxponders between point-to-point connections, in order to maximize the utilization of the link (fiber). As the line rates increased from 10 Gbps to 40 Gbps to 100 Gbps and in the near future to 400 Gbps and 1,000 Gbps, the manually patched muxponders became a liability in grooming services between and across wavelengths, besides adding to the CAPEX and OPEX. OTN provides a common electrical switching fabric that supports virtually any type of client signal and represents the initial transformation to the virtual network platform.

7.9 Key Features

The ability to transparently and efficiently transport diverse traffic types is one of the key advantages of OTN. OTN standards were updated in 2012, and several key enhancements were added including the following:

1. 100 GigE support (OTU4→ODU4→OPU4): Facilitates multiplexing of lower-order rates and mapping of the multiplexed output directly using GMP. This eliminates the use of traditional STM-16/OC-48 interfaces thereby enhancing effectiveness.
2. Generic mapping procedure (GMP): GMP asynchronous mapping procedure (AMP) supports original standardized rates, which are multiples of the base OPU1 payloads. GMP AMP provides support for direct mapping of client signals onto any OPU, in a single step, simplifying management tasks while facilitating efficient use of the 100-Gbps rates.
3. ODUflex for CBR and GFP: ODUflex permits higher flexibility at the lower rates where a CBR signal could be mapped directly to a corresponding OTN level. For example, FC-400 or FC-800. ODUflex permits any multiple of 1.25 Gbps (lowest possible rate) to be directly mapped to a relative OPU2/3/4 rate. ODUflex is used for payloads (time- and rate-sensitive) that do not fit onto the defined mapping structure.
4. Support for 1.25 Gbps tributaries: OTN supports internal switching at 1.25Gbps, which is referred to as ODU0/OPU0. This support provides a mechanism for mapping client signals of GigE, FC-100, and STM-4/OC12 and below to be mapped to higher-order rates.
5. Multistage multiplexing: The initial G.709 standard defines a single stage of multiplexing. However, the implementation of ODU/OPU0 and ODUflex rates necessitated the use multistage mapping that provides flexible mapping options. This permits the extension of OTN solutions to end users while providing the ability to map lower rate client signals and handle traffic switching within the network at different OPU levels. The operator benefits from the management simplic-

ity (requirement for multiple management systems) which more than compensates for the effort in implementing multistage mapping

Other important advantages of OTN are described as follows.

- A major benefit of OTN is FEC. As opposed to the in-band FEC in SDH/SONET systems, OTN uses a RS 16 byte-interleaved FEC that uses 4x256 bytes of check information per ODU frame. It is proven that FEC is effective in OSNR as well as dispersion-limited systems. FEC can compensate for reduced OSNR in case lower power levels are used on a link, to counter the nonlinear effects at higher power levels. The FEC for OTN can result in an up to 6.2-dB improvement in SNR [3]. This gain can translate into a higher span length with the same BER. However, FEC is less effective in countering the effects of PMD.
- TCM is used by operators to check the quality of their signals routed over other operators networks using the 'N' bytes in the overhead (in the case of SDH/SONET systems). However, in the case of SDH/SONET networks the TCM is capable of a supporting single layer only, limiting its usage. OTN overcomes this drawback by including support for six layers (six tandem connections); this proves invaluable in tracing the quality of signals between operator networks.
- The transport of client signals in OTN is bit-, timing-, and delay-transparent.
- In SDH/SONET systems the NE switching matrix supported only two levels—lower-order and higher-order. This method, along with the existing support for CCAT and the introduction of VCAT through the next-generation SDH standards was appropriate for the supported line rates of STM-64 (10 Gbps) and IP/Ethernet services. The higher line rates for OTN, as well as the need to support diverse traffic types, necessitate unrestricted switching capabilities. The enhanced switching matrix of OTN equipment helps operators provide services at various bit rates, independent of the wavelength bit rates, using the multiplexing and inverse multiplexing (creation of a larger pipe from smaller fragments) features.
- OBE can be deployed[19] in metro networks to provide core aggregation and switching services. It can also be used for integration with WDM networks through the use of transponders.

---

19. OTN is not efficient in transporting signals < 1GBE.

Example 7.12—Multiplexing of Lower Order Signals

1. 4 * STM-16/OC-48 signals are mapped onto an OTU-2. This process does not require updating of the STM/OC overheads.

2. 4 * STM-16/OC-48 signals are mapped onto an ODU-1 and further multiplexed onto an ODU-2. The multiplexing operations do not alter the timing of the input signals till they are demultiplexed to ODU-1.

7.10 Summary

There are several options that form the basis of the next-generation optical networks. These are OTN, IP over DWDM, IP/MPLS over DWDM, and packet optical transport service. OTN is the next-generation industry standard protocol that facilitates integration of legacy TDM networks and the IP-based packet-switched networks over a WDM system. A major feature of OTN is FEC, which uses 4×256 bytes of check information per ODU frame. The IP over DWDM approach creates a single agile network that integrates a dynamically reconfigurable optical transport layer with tunable optical interfaces. The approach minimizes the OEO conversion and creates a multiservice provisionable platform (MSPP) that provides bandwidth management capabilities at the wavelength level through the use of photonic switching. The key features of an OTN network include the following:

1. The ability to transparently and efficiently transport diverse traffic types is one of the key advantages of OTN. The support for 100 GigE facilitates multiplexing of lower-order rates and mapping the multiplexed output directly using GMP. This eliminates the use of traditional lower-order interfaces STM-16/OC-48 interfaces thereby enhancing effectiveness.

2. GMP AMP supports original standardized rates that are multiples of the base payloads while supporting direct mapping of client signals onto any lower-order payloads, in a single step, simplifying management tasks while facilitating efficient use of the 100-Gbps rates.

3. ODUflex permits higher flexibility at the lower rates where a CBR signal could be mapped directly to a corresponding OTN level. ODUflex is used for payloads (time- and rate-sensitive) that do not fit onto the defined mapping structure.

4. The operator benefits from the management simplicity (requirement for multiple management systems), which more than compensates for the effort in implementing multistage mapping.

5. A major benefit of OTN is FEC. As opposed to the in-band FEC in SDH/SONET systems, OTN uses a RS 16-byte-interleaved FEC that uses 4×256 bytes of check information per ODU frame. It is proven that FEC is effective in OSNR as well as dispersion-limited systems.
6. The transport of client signals in OTN is bit, timing, and delay transparent.

7.11 Review

7.11.1 Review Questions

1. The ITU-T recommendation G.872 defines the frame structure and format of OTN.
   a. True
   b. False
2. The protection schemes in an OTN are defined by the following standard:
   a. G.709
   b. G.873.1
   c. G.798
   d. G.872
3. The E2E optical path in an OTN is specified by the _____ layer.
   a. ODU
   b. OTU
   c. Och
   d. OPU
4. A 10-GBE signal from a router interface can transported over an OTN in its native format.
   a. True
   b. False
5. LAN-PHY 100 GigE is one of the two interfaces that supported the traditional method of transport of IP data over SDH/SONET networks.
   a. True
   b. False
6. The architecture of OTNs is specified in the following ITU-T recommendation:

a. G.870

   b. G.871

   c. G.872

   d. G.8080

7. Four STM-64 signals can be mapped on to an OUT-3.

   a. True

   b. False

8. The smallest OTN container defined is:

   a. OTU-0 (1.25 Gbps)

   b. OTU-0 (2.5 Gbps)

   c. OTU-1 (2.5 Gbps)

   d. OTU-0 (622 Gbps)

9. OTN switches operate at the OTU layer.

   a. True

   b. False

10. OTN uses a RS (255,238) block code for FEC.

    a. True

    b. False

7.11.2 Exercises

1. Compare the features of OTN with IP over DWDM networks.

2. What is meant by the term MSPP? Explain with reference to relevant equipment.

3. Briefly describe the optical control plane. Include three major advantages of having a control plane environment.

4. Describe two advantages of deploying OTN in the network as compared to IP over DWDM networks.

5. Explain in detail the OTN switching hierarchy.

7.11.3 Research Activities

1. Prepare a case for deploying OTN in the networks of the future. Substantiate your case with economic viability and technology constraints.

2. Prepare a use case based on TCM.

7.12 Case Study: OTN Deployment

7.12.1 Background

ESW, Inc., is a leading telecom service provider (tenth-largest by revenues) in South Asia. The company had over 170,000 km of optical fiber network spanning a geographical area of over a million square kilometers. The backbone networks were based on TDM SDH technology in a three tier architecture, as illustrated in Figure 7.15:

1. Tier—1:
   a. Intracity links on 10G STM-64 rings.
2. Tier—2:
   a. Intercity on 2.5 STM-16-Gbps rings.
3. Tier—3:
   a. Access rings of 622 Mbps STM-4/155 Mbps STM-1 rings;
   b. Spur sites on 155-Mbps STM-1 rings (unprotected).

The fiber consisted of 96-/48-/16-core G.652 cables for the core, collector, and access networks, respectively.

7.12.2 Challenges

1. The maximum capacity of the core network, stagnating at 10 Gbps, is presenting a major traffic bottleneck.
2. Most of the metro locations in the network require additional bandwidth to cater to customer demands.
3. There has been a significant increase (nearly 60%) in customers requiring 10 GigE/1 GigE pipes between major metros.
4. In addition, growing urbanization is fueling major infrastructure projects that are resulting in around 70% fiber cuts in urban areas (up to 100 cuts/day). This is severely impacting network uptime and consequently noncompliance of SLA with major customers.
5. Figure 7.15 illustrates the current network architecture. The currently deployed fiber network consists of 96-core G.652 cables in the core network, 48-core G.652 cable in the collector (metro), and 16-/8-core cables in the access network.

224   Engineering Optical Networks

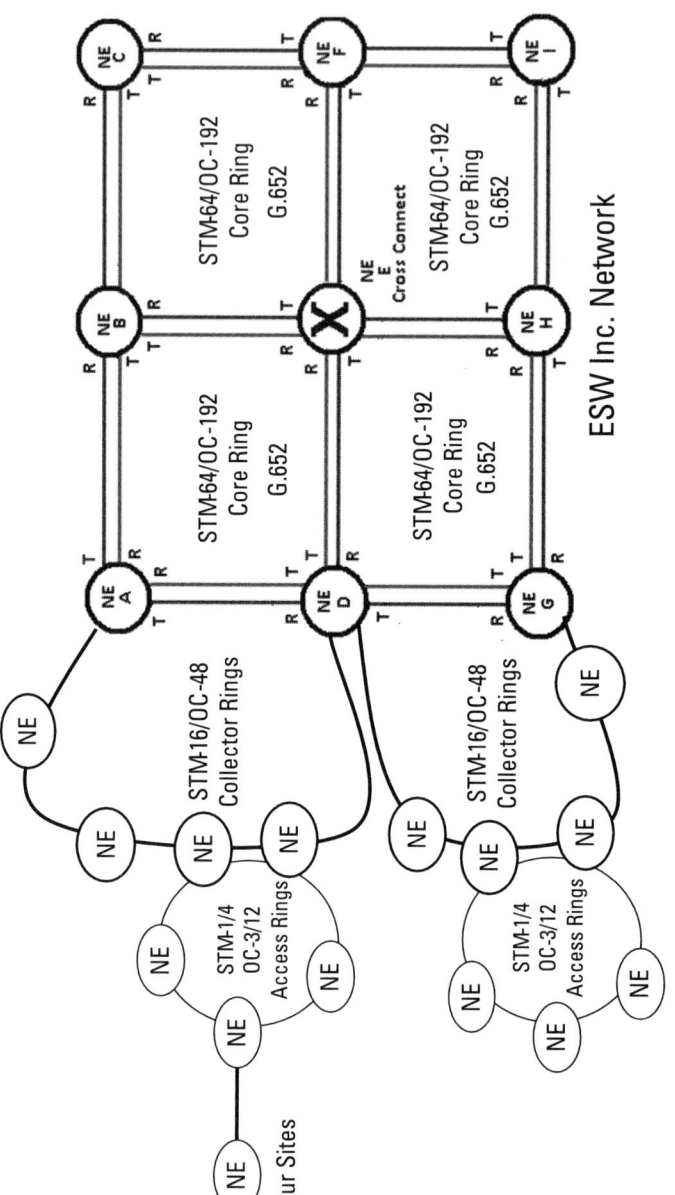

Figure 7.15 Current network architecture of ESW Inc.

7.12.3 Requirements

ESW wants a solution that is cost-effective and that leverages existing infrastructure, while providing additional bandwidth and improved network uptime.

7.12.4 Proposed Solution

The major highlights of the proposed solution for ESW are presented in Table 7.7 and illustrated in Figure 7.16.

As is evident from the background, the ESW network is based on purely SDH/SONET nodes. In order to keep the CAPEX at a minimum while providing for enhanced bandwidth at all the layers and improved network resiliency, the following changes were recommended:

1. No change in fiber plants—The proposed architecture does not envisage any change in the fiber plants.
2. Use of OTN in the core network—The advantages of a photonic network based on the OTN standard has already been detailed in this chapter. The use of OTN improves bandwidth availability at the core significantly by upgrading the 10G rings to 100G capacity. This would take care of the bottlenecks in the intercity links.
3. Mesh topology at the core—In order to tackle the issue of multiple fiber cuts, implementing a shared mesh topology between the ONE in the core has been proposed. The proposal calls for 1:3 shared pro-

Table 7.7
Proposed Network Architecture for ESW—Highlights

S.N	Features	Existing	New
1	Topology—core	Ring	Mesh
2	Topology—collector	Ring	Ring
3	Topology—access	Ring Unprotected spur	Ring Protected spur
4	Transport standard/technology	SDH/SONET	OTN (core) OTN + SDH (collector/access)
5	Core bandwidth	10 Gbps	100 Gbps
6	Collector bandwidth	2.5 Gbps	10 Gbps
7	Access bandwidth	155/622 Mbps	155/622/2.5 Gbps
8	10 GigE/1 GigE transport	No native support GFP+VCAT+LCAS	Native support 10 GigE/1 GigE interfaces
9	Protection from fiber cut	Single (only 2 paths per NE)	Multiple
10	Protection mechanism	SNC-P MSP	Path protection
11	Control plane enabled	No	Yes

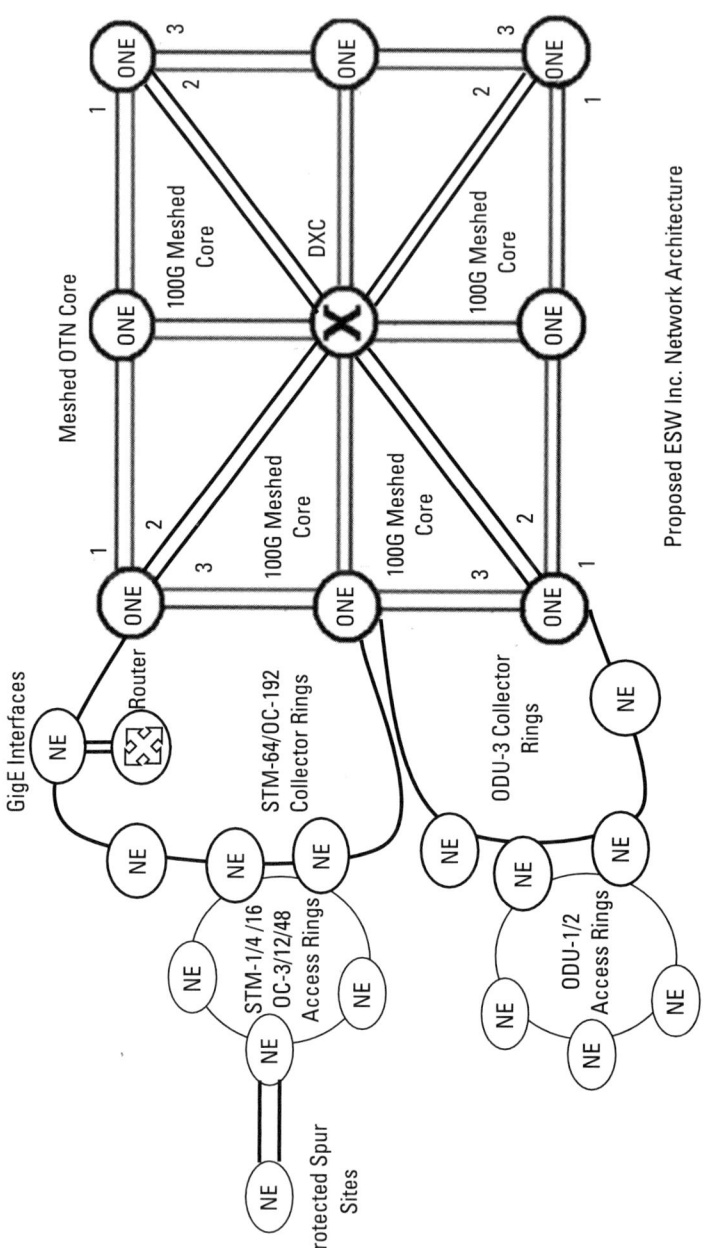

Figure 7.16 Proposed network architecture of ESW Inc.

tection wherein every working path will have a minimum of three protection paths. Each of the ONEs has a minimum of three paths connecting them with neighboring NEs. This provides a cost-effective solution mitigating the issues related to multiple fiber cuts.

4. Protected spur—The spur sites can be extended protection (on a case-by-case basis) by extending another pair of fibers from the primary takeoff site. It should be noted that the pair of fibers needs to be route-diverse to ensure full protection.

5. Control plane—The full benefits of using OTN and a physical topology can be realized by provisioning a control plane with a hybrid protection mechanism. This ensures that dedicated protection paths can be defined for high-priority traffic/control channels resulting in lower switching times while providing automatic tertiary level protection for other traffic paths.

6. Use of 10 GigE/1 GigE physical interfaces—The deployment of OTN facilitates native support to Ethernet/IP traffic and direct interconnection of routing elements using physical interfaces. This eliminates the need for GFP protocol, LCAS, and VCAT techniques. The ability to support native IP traffic caters to the requirement for provisioning additional 10 GigE/1GigE pipes.

7. Use of existing hardware—The SDH/SONET NEs in the erstwhile elements can be retained after rehoming to the new ONEs. In case the earlier SDH/SONET equipment was based on MSPPs, they can be used to provide OTN and GigE interfaces by equipping with the requisite PIUs[20].

7.12.5 Recommendations

In the future ESW may consider upgrading its fiber plant to either G.652D or G.655.

G.652D is a SMF with significantly reduced optical attenuation at the water-absorption wavelength of around 1,383 nm. The fiber has low attenuation over a wide wavelength range. It is suitable for DWDM as well as non-DWDM applications and speeds of 100 Gbps. It is a cost-effective solution especially for networks with speeds greater than 10 Gbps due to the higher span lengths possible and the consequent reduction of regenerators and amplifiers.

G.655 cables are suited for DWDM networks with large network spans and have better values for CD and PMD.

---

20. Vendor-specific.

Table 7.8 lists the characteristics of G.652 and G.655 fiber.

7.13 Referred Standards

G.7713: Distributed call and connection management (DCM)

G.7713.1/Y.1704: ITU-T Distributed call and connection management—PNNI implementation

G.7713.2/Y.1704: ITU-T Distributed call and connection management—GMPLS RSVP-TE implementation

G.7713.3/Y.1704: ITU-T Distributed call and connection management—GMPLS CR-LDP implementation

G.7714.1/Y.1705.1: ITU-T Protocol for automatic discovery in SDH and OTN networks

G.7714/Y.170: Generalized automatic discovery techniques

G.7715.1: ASON routing architecture and requirements for link state protocols

G.7715/Y.1706: Architecture and requirements for routing in automatically switched optical networks

G.7716/Y.1707 (01/10): Architecture of control plane operations

G.7717/Y.1708: Connection admission control

G.7718/Y.1709 (07/10): Framework for ASON management

G.798: Characteristics of optical transport network hierarchy equipment functional blocks

Table 7.8
Characteristics of G.652.D (Zero Water Peak Fiber)* and G.655 NZ-DSF Fiber

S.N	Parameter	G.652D	G.655
1	Attenuation		
	at 1,310 nm	≤ 0.334 dB/km	≤ 0.22 dB/km
	at 1,550 nm	≤ 0.194 dB/km	≤ 0.19 dB/km
2	Zero dispersion wavelength	1,302 ~ 1,322 nm	1,520 nm
3	Chromatic dispersion		
	at 1,285 ~ 1,330 nm	≤ 3.5 ps/(nm · km)	-
	at 1,550 nm	≤ 18 ps/(nm · km)	≤ 6 ps/(nm · km)
	at 1,625 nm	≤ 22 ps/(nm · km)	≤ 11 ps/(nm · km)
4	PMD (uncabled fiber)	≤ 0.15 ps/√km	≤ 0.1 ps/√km

* Please note that Table 7.8 lists generalized values. Actual values are dependent upon specific fiber types by different manufacturers. Kindly refer to specific data sheets for actual information.

G.806: Characteristics of transport equipment—Description methodology and generic functionality

G.8080/Y.1304 (02/12): ASON architecture

G.8081/Y.1353 (02/12): Terms and definitions for ASON

G.873.1: Optical transport network (OTN): Linear protection

OIF-CDR-01.0: Call Detail Records for OIF UNI 1.0 Billing (April 2002)

OIF-CP-MGMT-01.0: Management Plane (OSS) Support for Control Plane Networks White Paper (2012)

OIF-CWG-CPR-01.0: IF Carrier Working Group Guideline Document: Control Plane Requirements for Multi-Domain Optical Transport Networks (July 2010)

OIF-E2E-SEC-01.0: End-to-End Transport of UNI Client Authentication, Integrity, and Data Plane Security Support Information (May 2012)

OIF-ENNI-ML-AM-01.0: Multilayer Amendment to E-NNI 2.0—Common Part (April 2013)

OIF-ENNI-OSPF-02.0-OIF External Network-Network Interface (E-NNI) OSPF-based Routing: 2.0 (Intra-Carrier) Implementation Agreement (July 2011)

OIF-ENNI-OSPF-02.1: Multilayer Amendment to E-NNI 2.0 OSPFv2-based Routing (April 2013)

OIF-ENNI-OSPF-02.2: Recovery Amendment to E-NNI 2.0 OSPFv2-based Routing (February 2014)

OIF-ENNI-REC-AM-01.0: Recovery Amendment to E-NNI 2.0 Common Part (February 2014)

OIF-ENNI-RSVP-02.1: Multilayer Amendment to E-NNI 2.0—RSVP-TE Signalling (April 2013)

OIF-ENNI-RSVP-02.2: Recovery Amendment to E-NNI 2.0—RSVP-TE Signalling (February 2014)

OIF-E-NNI-Sig-02.0: E-NNI—Signalling Specification (April 2009)

OIF-G-Sig-IW-01.0: OIF Guideline Document: Signaling Protocol Interworking of ASON/GMPLS Network Domains (June 2008)

OIF-OTN-TCM-01.0: Guidelines for Application of OTN TCM White Paper (June 2013)

OIF-PCE-IA-01.0: Path Computation Element (June 2013)

OIF-RSVP-PVT-EXT-01.0: Application of Vendor Private Extensions in RSVP Implementation Agreement (October 2011)

OIF-SEP-03.2: Security Extension for UNI and E-NNI 2.1 (October 2012)

OIF-SLG-01.3: Control Plane Logging and Auditing with Syslog (October 2012)

OIF-SMI-03.1: Security for Management Interfaces to Network Elements 2.0 (October 2012)

OIF-UNI-02.0-Common: User Network Interface (UNI) 2.0 Signalling Specification: Common Part (February 2008)

OIF-UNI-02.0-R2-RSVP-OTNv3: RSVP Extensions for User Network Interface (UNI) 2.0 Signaling Release 2 (January 2013)

RFC 3471: Generalized Multi-Protocol Label Switching (GMPLS) Signaling Functional Description

RFC 3473: Generalized Multi-Protocol Label Switching (GMPLS) Signaling Resource Reservation Protocol-Traffic Engineering (RSVP-TE) Extensions

RFC 3945: Generalized Multi-Protocol Label Switching (GMPLS) Architecture

RFC 4202: Routing Extensions in Support of Generalized Multi-Protocol Label Switching

RFC 4206: Label Switched Paths (LSP) Hierarchy with Generalized Multi-Protocol Label Switching (GMPLS) Traffic Engineering (TE)

RFC 6373: MPLS Transport Profile (MPLS-TP) Control Plane Framework

7.14 Recommended Reading

7.14.1 Books

Keiser, G., *Optical Fiber Communication* (Fourth Edition), Tata Mc Graw Hill, 2011.

Jun, Z., and T. H. Mouftah, *Optical WDM Networks: Concepts and Design Principles*, Canada: Wiley-Interscience, 2004.

De Luc, G., *MPLS Fundamentals* (First Edition), Cisco Press, 2007.

Senior, J. M., *Optical Fiber Communications—Principle and Practice* (Second Edition), Pearson Education, 2006.

Elanti, M., et al., *Next Generation Transport Networks—Data, Management, and Control Plane Technologies*, Springer, 2005.

Kartalopoulos, S., *Introduction to DWDM Technology*, Piscataway, NJ: IEEE Press, 2000.

7.14.2 URLs

http://cisco.com/en/US/products/ps5763/products_white_paper0900aecd80395e03.shtml.

https://tools.ietf.org/id/draft-doolan-tdp-spec-00.txt.

https://www.itu.int/ITU-T/studygroups/com15/otn/OTNtutorial.pdf.

https://www.terena.org/activities/ngn-ws/ws2/pennell-ipodwdm.pdf.

http://www.cse.wustl.edu/~jain/cis788-99/ftp/ip_dwdm.pdf.

http://www.swdm.org/.

https://web.archive.org/web/20120312045120/http://squiz.informatm.com/__data/assets/pdf_file/0007/194623/ODU0_ODUflex_White_Paper_2010_02_15_v1_web.pdf.

http://www.lightwaveonline.com/topics/otn.htm.

7.14.3 Journals

Bai, N., et al., "Multimode Fiber Amplifier with Tunable Modal Gain Using a Reconfigurable Multimode Pump," *Opt. Express* 19 (17), 16601 (2011).

Essiambre, R.-J., et al. "Capacity Limits of Optical Fiber Networks," *Lightwave Technology Journal*, 28, No. 4, 2010, pp. 662–701.

Jung, Y., et al., "First Demonstration and Detailed Characterization of a Multimode Amplifier for Space Division Multiplexed Transmission Systems," *Opt. Express* 19 (26), B952 (2011).

Love, J. D., and N. Riesen, "Mode-Selective Couplers for Few-Mode Optical Fiber Networks," *Opt. Lett.* 37 (19), 3990 (2012).

Ramachandran, S., "Dispersion-Tailored Few-Mode Fibers: A Versatile Platform for In-Fiber Photonic Devices," *J. Lightwave Technol.* 23 (11), 3426 (2005).

Randel, S., et al., "6× 56-Gb/s Mode-Division Multiplexed Transmission over 33-km Few-Mode Fiber Enabled by 6× 6 MIMO Equalization," *Optics Express*, 19.17, 2011, pp. 16697–16707.

Richardson, D. J., J. M. Fini, and L. E. Nelson, "Space-Division Multiplexing in Optical Fibers," *Nature Photonics,* 7.5, 2013, pp. 354–362.

Ryf, R., et al., "Mode-Division Multiplexing over 96 km of a Few-Mode Fiber Using Coherent 6 × 6 MIMO processing," *J. Lightwave Technol.* 30 (4), 521 (2012).

Yaman, F., et al., "Long-Distance Transmission in Few-Mode Fibers," *Opt. Expre*ß*s* 18 (12), 13250 (2010).

References

[1] Kompella, K., and P. Belotti, "Transport Networks At A Crossroads: The Roles of MPLS and OTN in Multilayer Networks," *Optical Fiber Communications/National Fiber Optic Engineers Conference,* Session OTuG, Anaheim, CA, March 6, 2011.

[2] ITU-T Recommendation G.872, Architecture for the Optical Transport Network (OTN), January 2017.

[3] ITU. OTN Tutorial. From ITU: https://www.itu.int/ITU-studygroups/com15/otn/OTNtutorial.pdf, 2015.

[4] ITU-T Recommendation G.709: Interfaces for the Optical Transport Network, June 2016.

[5] ITU-T G.872 (10/01), Architecture for the Optical Transport Network (OTN).

[6] ITU-T Recommendation G.959.1: Optical Transport Network Physical Layer Interfaces, April 2016.

[7] Xue, Y. (ed.), "Optical Network Service Requirements," Internet-Draft, Dec. 2002.

[8] ITU-T Recommendation G.8080/Y.1304, Architecture for the Automatically Switched Optical Networks, March 2012.

[9] ITU-T Recommendation G.7713/Y.1704, Distributed Call and Connection Management (DCM), March 2009.

[10] Hamming, R. W. (April 1950). "Error Detecting and Error Correcting Codes" (PDF). *Bell System Tech. J.* USA: AT&T. 29 (2): 147–160. doi:10.1002/j.1538-7305.1950.tb00463.x. Retrieved 4 December 2012.

8
DWDM

8.1 Chapter Objectives

The demand for higher bandwidth continues to increase exponentially, leading to the development of new techniques and technologies for efficient transport of information. The proliferation of technological advances and the deregulation of the telecom sector have led to higher teledensities globally, with a consequent increase in data as well as voice traffic. There has been an explosive demand for bandwidth to carry data traffic, primarily IP, which has increased by over 300% in the past three years as compared to a little over 20% increase in voice traffic over the corresponding period. Leading service providers across the globe have had to double their backbone capacities in the last couple of years to meet this excessive demand. In fact researchers are already predicting a requirement in the order of peta/hexabits per second in the core networks by 2020. The emergence of cloud computing has led to an increase in the number of huge data centers and a consequent increase in bandwidth requirement for SANs. This increased demand for higher bandwidth, especially IP traffic, is putting a huge strain on the existing SDH/SONET-based TDM networks that are not optimized for carrying data traffic and that cannot directly support native interfaces. Further the cost of laying fiber is still very high globally and forms a significant chunk of the CAPEX required to set up a telecommunication network. These issues have led to the development of WDM techniques. WDM facilitates the transmission of multiple rays of light onto a single fiber or in technical terms allows the multiplexing of multiple λ's onto a single fiber. The bandwidth of the core network can thus be increased manifold without any change in the fiber plant. WDM thus provides a cost-effective method to scale up network bandwidth without significant changes in the existing infrastructure. This chapter traces the evolution of WDM systems and presents a detailed architecture of the modern DWDM systems and their integration with traditional transport systems—SDH/SONET as well as IP networks. The chapter provides readers with the conceptual basis of WDM, CWDM, and DWDM systems along with their applications and configurations and an overview of the performance analysis as well as testing process of DWDM systems. The limitations of the existing TDM-based transport networks have heralded the deployment and growth of IP-based core or backbone systems. Telecom operators with a strong investment in SDH systems have upgraded to OTN-based core with mesh architectures and integrated DWDM systems. Greenfield operators prefer an IP-based core with integrated DWDM systems. The objective of this chapter is to provide readers, network engineers, and technicians with the basic theoretical foundations and practical aspects required to understand the layout of DWDM-based systems. A detailed treatise of the myriad components of a DWDM system is beyond the scope of this book.

Key Topics
- WDM;
- WDM standards;
- WDM transmission;
- Optical transmission challenges;
- WDM network components, functioning, and interconnections.

8.2 Introduction

In order to achieve optimum media bandwidth utilization TDM was employed for voice traffic over a network. This technique is analogous to a single highway interconnecting two metros and multiple feeder or service routes connected to the highway. A timeslot is reserved for each feeder route to get a car onto the highway. Thus cars from multiple feeders can come onto the highway and travel through to the destination without any collisions, provided proper time synchronization is maintained. At the destination multiple feeders will take the traffic away from the highway. This technique is obviously better than having each feeder route extended all the way between the two locations. However, there is a limit on the maximum bandwidth that can be obtained by using this technique.

The current SDH/SONET standards provide for a maximum bandwidth of 10 Gbps corresponding to STM-64 and OC-192. The deployment of OTN-based systems has provided a means to scale up the network capacity to 100

Gbps (OTU4). In fact as this book is being written, 400-Gbps interfaces are on the verge of being commercially deployed on long- and ultra-long-haul networks. The deployment of 400-Gbps interfaces seemed a remote possibility until a few years ago, as they were not commercially viable due to their extremely high development cost and footprint. Rapid developments in the field of semiconductor fabrication technologies as well as optical transmission technologies have made the development and deployment of high-density cross-connects with a range of high-capacity interfaces a reality. In fact there has been significant progress on the development of transmission systems up to 1,000 Gbps. In the above scenario a requirement for higher bandwidth will mandate the installation of additional fibers in the network. The laying of fibers is still a very costly proposition across the world and accounts for a significant portion of the network building cost. This is in addition to the significant efforts required to obtain permissions from the various local, state, and central statutory bodies. WDM is a technique that facilitates the transmission of multiple rays of light or multiple lambdas through a single fiber, thereby bringing about an exponential increase in the bandwidth utilization of the fiber plant while doing away with the costs and the hassles of laying extra fibers.

WDM facilitates the integration of legacy TDM-based circuit-switched networks with the next-generation carrier class Ethernet/IP networks offering a scalable low-cost transport mechanism that can support virtually any type of application. The availability of various types of protection schemes ensures rapid recovery from failure. WDM supports a variety of native interfaces including SDH/SONET, Ethernet, fast Ethernet, gigabit networks (GigE), 10 GigE, FICON, ESCON, FC, OTN, ATM, and FR. WDM provides transparent bit rate agnostic interfaces and forms the basis of the modern-day carrier networks. Initially WDM was used in conjunction with SDH/SONET networks but it is now increasingly being used with OTN and Ethernet/IP networks. In fact IP over DWDM is becoming the deployment of choice globally. The shift to a fully optical transparent network is gaining rapid momentum. The primary drawback of the existing TDM-based SDH/SONET networks is the limit to the maximum bandwidth that can be supported and the latency due to the deployment of regenerators and digital cross-connects, which requires the use of O-E-E-O (OEO) conversion. Thus the processing speed of the equipment in the network also plays an important part in determining the speed/capacity of the network. WDM-based networks on the other hand are fully optical networks and hence do not suffer from the disadvantages of the legacy TDM-based networks.

In the interim the deployment of OTN technologies has provided a scalable model for the existing SDH/SONET networks. The emergence of metro networks has placed an additional burden on the core networks. The two primary parameters that decide the use of technologies in the metro networks are bandwidth scalability and the ability to provide multiservice capabilities. The other additional important parameters that are considered are the physical footprint and the OPEX. The challenge for network planners is to provide a flexible and scalable architecture for the core that supports the existing legacy applications as well as next-generation services of the future.

8.3 What is WDM?

WDM is a technique by which multiple rays of light with different wavelengths (λ) are combined and transmitted in the same direction over the same fiber. The concept is as illustrated in Figure 8.1. The different rays of light can carry information from different sources and hence lead to a very high bandwidth utilization of the fiber plant. The use of WDM eliminates the need for laying multiple fibers for capacity increase resulting in major CAPEX and OPEX savings.

WDM systems can be classified, depending on the number of channels employed and the spacing between them, as follows and as outlined in Table 8.1:

1. Narrowband WDM (NWDM);
2. Wideband WDM (WWDM);
3. Coarse WDM (CWDM);
4. Dense WDM (DWDM).

The WDM technique initially supported the multiplexing of only two λ's. The two wavelengths used were 1,533 nm and 1,577 nm and/or 1,310 nm (zero dispersion) and 1,550 nm, which corresponded with the low loss band. It may be noted that the WDM systems that used 1,553 and 1,557 nm required a higher degree of sophistication and hence were more expensive than WDM

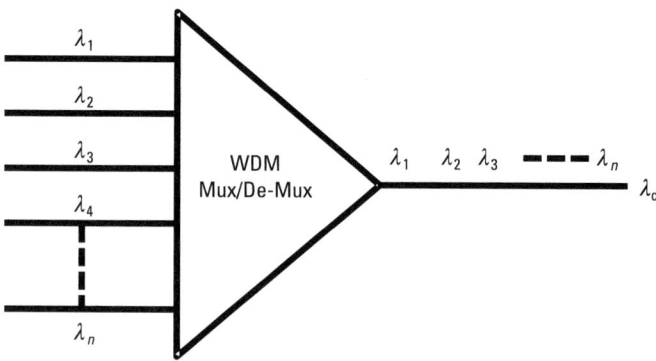

Figure 8.1 WDM.

Table 8.1
WDM Classification

S.N	WDM Type	Abbreviation	Channels	Wavelength (λ) Nm
1	Narrowband	NWDM	2	1,553 and 1,557
2	Wideband	WWDM	2	1,310 and 1,528 to 1,560
3	Coarse	CWDM	18	1,271 to 1,611
4	Dense	DWDM	88 (C-Band) 142 (L-band)	1,530 to 1,565 (C-band) 1,565 to 1,625 (L-band)

systems that used the 1,310- and 1,550-nm wavelengths. CWDM uses a set of frequencies that lie between 1,271 and 1,625 nm with 20-nm spacing in between the two wavelengths. It may be noted that the actual frequencies used by manufacturers of lasers for CWDM applications may be slightly different. The separation between the adjacent wavelengths would however have to be 20 nm. Another important point to be noted is that CWDM channels are equidistant in the wavelength domain as opposed to the DWDM channels, which are equidistant in the frequency domain. DWDM systems initially provided for a channel separation of 100 GHz. This would translate to a nominal separation of around 0.8 nm between adjacent wavelengths. ITU-T allows for closer spacing also by dividing the 100-GHz grid by a factor of 2. Accordingly grids of 50 GHz, 25 GHz, and 12.5 GHz are also supported. The spacing for these grids would be 0.4 nm, 0.2 nm, and 0.1 nm, respectively. The commercially available systems usually provide for a total of 80 λ's, which are split between the C- and the L-bands (40 λ's in each band). However the need for higher bandwidth has led to the development of DWDM systems that provide 88 λ's across the entire C-band spectrum. The evolution of WDM systems is depicted in Figure 8.2.

8.4 Standardization

Initially there were no standards governing the use of the WDM technique. Subsequently ITU-T and ANSI collaborated to roll out a set of standards that would facilitate the deployment of WDM systems in the metro as well as the long-haul networks. The work was executed through the ITU-T study group (SG) 15. Its recommendations initially provided for the use of wavelengths that fell into the 1,530 to 1,565 range. The recommendations were rolled out in a phased manner. The first phase provided for recommendations governing the functioning of point-to-point WDM systems. The second phase extended the capabilities of OADMs and OXCs. Phase 3 includes the optical layer survivability. The ITU-T G.671 recommendation identifies the key transmission parameters for the components that are a part of the WDM systems (WWDM, CWDM, and DWDM) including OADM, filters (fixed as well as tunable),

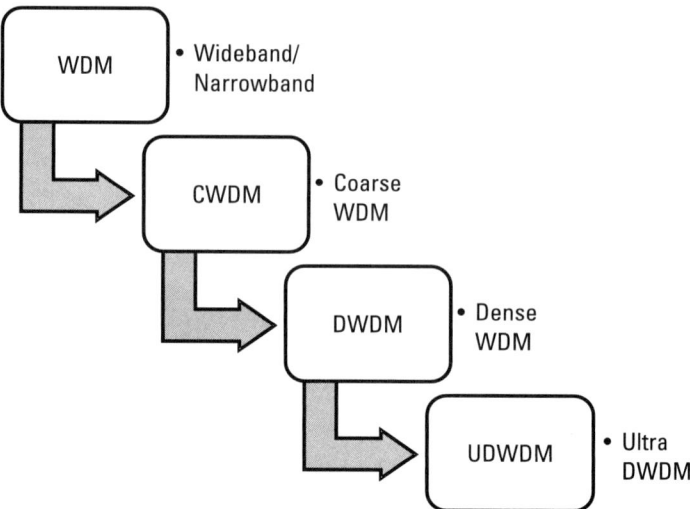

Figure 8.2 WDM evolution.

attenuators, branching elements, connectors, dispersion compensators, splices, and switches. A list of the components of the WDM network along with their functional details is covered in subsequent sections. The interfaces to WDM system elements would be governed based on their deployment. In case of terrestrial long-haul networks the interfaces are governed by ITU-T recommendation G.957, and for single as well as multichannel systems with amplifiers the interfaces are governed by ITU-T recommendations G.691, G.692, and G.959.1. The access network interfaces are governed by ITU-T recommendation G.982. The G.671 recommendation however does not include mechanical specifications as well as installation aspects of the WDM equipment, nor does it provide for methods to test the various parameters of these equipment. The ITU-T recommendation G.694.1 defines the frequency grid for DWDM systems while the ITU-T recommendation G.694.2 defines the frequency grid for CWDM systems. There are 18 channels in five bands with spacing of 20 nm. The bands used are O, E, S, C, and L. The details of these bands are presented in Section 8.4.

8.5 WDM Fundamentals

One of the major factors that limit the range as well as capacity of optical transmission over a fiber is dispersion. The fundamentals of light transmission over an optical fiber as well as the different types of optical fibers are covered in detail in Chapters 1 and 2. The attenuation profile of an optical fiber is presented in Figure 8.3.

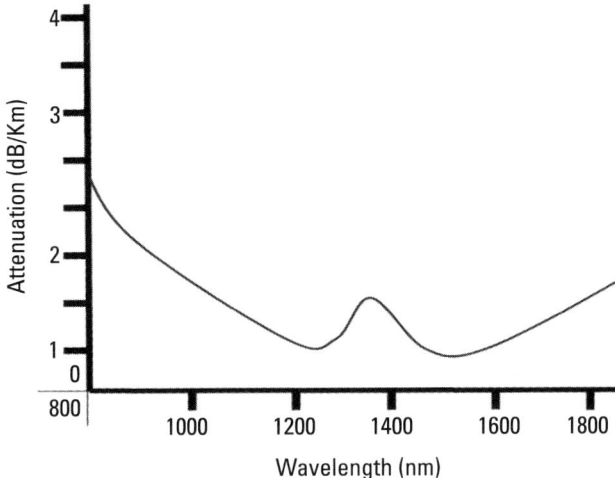

Figure 8.3 Attenuation characteristics of an optical fiber.

The most commonly employed source of light is a laser, which initially was available for operation in two wavelength regions namely the 1,310-nm and 1,550-nm regions. These regions are commonly referred to as the second and the third optical transmission windows. The effects of dispersion are zero at the 1,310-nm window whereas the losses due to the nonlinearities of the optical fiber are the least at the 1,550-nm window. The emergence of the WDM technique witnessed the use of the 1,500 to 1,650 nm band for optical transmission along with the deployment of optical amplifiers to counter the effects of attenuation. The commonly deployed amplifiers are the EDFAs, which provide signal amplification across a range of wavelengths around 1,550 nm commonly referred to as the EDFA window. The common optical transmission windows are as depicted in Figure 8.4 [1].

The rapid advancements in the manufacturing of optical fibers have made it possible to accommodate optical transmissions within a broader spectrum. This has resulted in the availability of a number of new transmission bands. The details of the optical transmission bands are provided in Figure 8.5 [2].

The CWDM systems can operate in the O, E, S, C, and L bands. The span length, using conventional fibers, is highly limited in the O and E bands due to the high signal attenuation. The S band is suited for single-wavelength (fixed-wavelength) transmission in addition to CWDM usage. The O band is designated as such since it was the band used for long-haul transmission originally. The emergence of WDM initiated the use of the C and the L bands for optical transmission. The C band is the most widely used band for WDM/DWDM system deployment largely due to the availability of low-cost amplifier technologies, which provided for a higher span length. The different types of

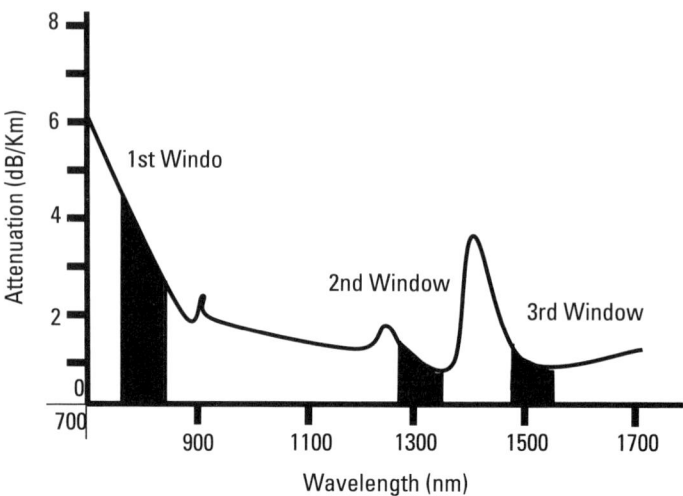

Figure 8.4 Optical transmission windows.

amplifiers employed in commercial networks are detailed late in this chapter. The U band is affected by high transmission losses that would require the use of advanced and costly laser sources for effective usage and thus is not used for any commercial communication currently. As mentioned above, the ITU-T G.694.1 recommendation defines DWDM transmission channels on the C and L bands using a channel spacing of 12.5 GHz, 25 GHz, 50 GHz, and 100 GHz with the reference frequency as 193.1 THz. Accordingly the nominal central frequencies for DWDM can be calculated by using the formula 193.1 + n × m, where n can be a positive or a negative integer including 0, and m corresponds to 0.0125, 0.025, 0.05, and 0.1 depending upon the channel spacing, which can be 12.5 GHz, 25 GHz, 50 GHz, and 100 GHz, respectively. Table 8.2 lists the nominal central frequencies for the different channel spacings.

The total bandwidth available on a medium depends upon several factors including the passband and SNR and is defined by Shannon's theorem. According to Shannon's theorem [3]

$$C = B \log_2 (1 + S/N)$$

where C is the channel information capacity (bits per second) and B is the available bandwidth [passband (in hertz)] over which the SNR (in decibels) is experienced.

The bandwidth of the respective transmission bands can be derived from the above formula by using a value of 60 dB for the SNR. Table 8.3 [4] presents the theoretical bandwidth of the common transmission bands. It is important to note that an increase in the number of channels used in an optical

DWDM

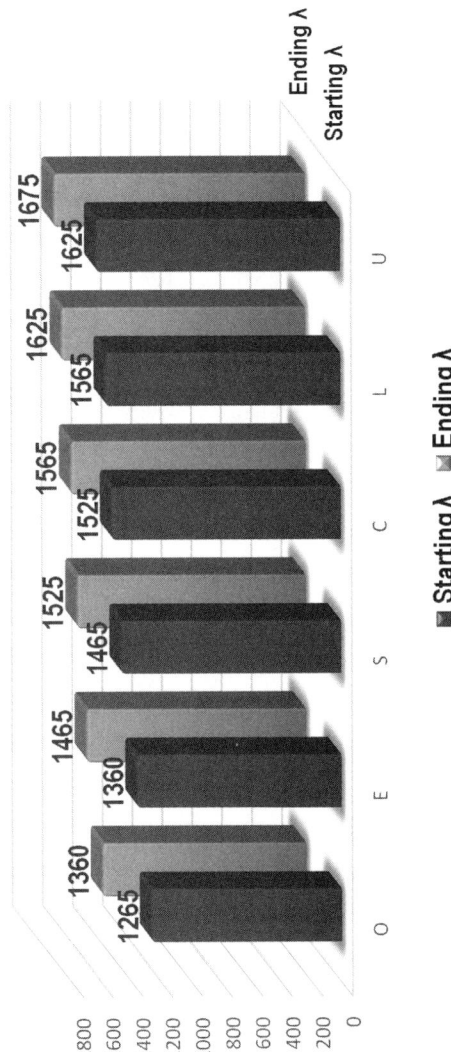

Figure 8.5 Optical transmission bands.

Table 8.2
DWDM Central Frequencies

Channel Spacing	Central Frequencies (THz)
12.5 GHz	193.1 + n × 0.0125
25 GHz	193.1 + n × 0.025
50 GHz	193.1 + n × 0.05
100 GHz	193.1 + n × 0.1

Table 8.3
Channel Bandwidth of Common Optical Transmission Bands

Tx Band	Lower Limit Freq (THz)	λ (nm)	Upper Limit Freq (THz)	λ (nm)	Passband BW (THz)	SNR (dB)	Max Channel BW (Tbps)
O band	220.436	1,360	237.931	1,260	17.495	60	348.70
E band	205.337	1,460	220.436	1,360	15.098	60	300.03
S band	195.943	1,530	205.337	1,460	9.395	60	187.25
C band	191.561	1,565	195.943	1,530	4.382	60	87.34
L band	184.488	1,625	191.561	1,565	7.073	60	140.98
U band	178.981	1,675	184.488	1,625	5.507	60	109.77

transmission band reduces the maximum bandwidth of each channel in addition to increasing the cost of the equipment deployed. Hence most of the commercial offerings tend to strike a balance between both these factors. The bandwidth of the channels can be increased by using advanced modulation techniques. The wavelength limits of the transmission band can be obtained by using the formula:

$$f = c/\lambda$$

where f is the frequency and c refers to the speed of light (in a vacuum) and is taken as 2.99792458 * 108 m/s as defined in the ITU-T G.694.1 recommendation.

The maximum channel bandwidth of the C band, with reference to Table 8.3, is 87 Tbps. The commercially available DWDM multiplexers support 40 λ or 80 λ.

Example 8.1

The maximum line rate using SDH technology is STM-64 or 10 Gbps. Thus the maximum bandwidth supported in the core is:

- 10 Gbps * 40 λ = 400 Gbps or 0.4 Tbps (for a 40-λ system);

- 10 Gbps * 80 λ = 800 Gbps or 0.8 Tbps (for a 80-λ system).

This represents a capacity utilization of less than 1% of the maximum transmission capacity available on the C band.

The ITU-T study group 15 in conjunction with the Optical Interworking Forum (OIF) brought about the development of 100-Gbps interfaces based on G.709 OTN.

Example 8.2

An OTN provides line rates of 100 Gbps. This would augment the backbone capacity to:

- 100 Gbps * 80λ = 8,000 Gbps or 8 Tbps

This represents a capacity utilization of 9% of the maximum transmission capacity available on the C band.

From the above examples it is evident that the current technologies facilitate transmission of only a fraction of the maximum transmission capacity of the C band—87 Gbps (refer to Table 8.3). The deployment of 400G interfaces would provide a major increase in the bandwidth capabilities of the core network.

- 400 Gbps * 80λ = 32,000 Gbps or 32 Tbps

This represents a bandwidth utilization of 36% of the available fiber capacity. It can be observed from the discussion that current technologies are able to only use a fraction of the available bandwidth for optical transmission.

An important fact to be noted is that the passband of the C band is lowest among the bands suitable for telecommunication purposes. However this band is widely used for DWDM deployment owing to its low attenuation characteristics and the availability of EDFAs. Using the formula mentioned in the above section we can list the wavelengths for the DWDM grid. Table 8.4 lists the frequencies as well as the corresponding wavelengths for the C band. The same procedure can be adopted to arrive at similar tables corresponding to the other bands. Further, using the formulas defined in the previous section along with the passband for each of the optical transmission bands (Tables 8.2, 8.3, and 8.5) and the ITU-T 694.1 defined spacing norms, the total number of lambdas in each of the optical transmission bands can be worked out. Figure 8.6 lists the total number of lambdas that can be supported across the optical transmission bands.

Table 8.4
ITU-T 694.1 DWDM Wave Grid

Number	Wavelength (nm)	
	100-GHz Spacing	50-GHz Spacing
1	1,528.77	1,528.77
		1,529.11
2	1,529.55	1,529.55
		1,529.99
3	1,530.33	1,530.33
		1,530.72
4	1,531.12	1,531.12
		1,531.51
5	1,531.90	1,531.90
		1,532.29
6	1,532.68	1,532.68
		1,533.07
7	1,533.47	1,533.47
		1,533.86
8	1,534.25	1,534.25
		1,534.64
9	1,535.04	1,535.04
		1,535.43
10	1,535.82	1,535.82
		1,536.22
11	1,536.61	1,536.61
		1,537.00
12	1,537.40	1,537.40
		1,537.79
13	1,538.19	1,538.19
		1,538.58
14	1,538.98	1,538.98
		1,539.37
15	1,539.77	1,539.77
		1,540.16
16	1,540.56	1,540.56
		1,540.95
17	1,541.35	1,541.35
		1,541.75
18	1,542.14	1,542.14
		1,542.54
19	1,542.94	1,542.94
		1,543.33

Table 8.4 (continued)

Number	Wavelength (nm)	
	100-GHz Spacing	50-GHz Spacing
21	1,544.53	1,544.53
		1,544.92
22	1,545.32	1,545.32
		1,545.72
23	1,546.12	1,546.12
		1,546.52
24	1,546.92	1,546.92
		1,547.32
25	1,547.72	1,547.72
		1,548.11
26	1,548.51	1,548.51
		1,548.91
27	1,549.32	1,549.32
		1,549.72
28	1,550.12	1,550.12
		1,550.52
29	1,550.92	1,550.92
		1,551.32
30	1,551.72	1,551.72
		1,552.12
31	1,552.52	1,552.52
		1,552.93
32	1,553.33	1,553.33
		1,553.73
33	1,554.13	1,554.13
		1,554.54
34	1,554.94	1,554.94
		1,555.34
35	1,555.75	1,555.75
		1,556.15
36	1,556.55	1,556.55
		1,556.96
37	1,557.36	1,557.36
		1,557.77
38	1,558.17	1,558.17
		1,558.58

Table 8.4 (continued)

Number	Wavelength (nm) 100-GHz Spacing	50-GHz Spacing
40	1,559.79	1,559.79
		1,560.20
41	1,560.61	1,560.61
		1,561.01
42	1,561.42	1,561.42
		1,561.83
43	1,562.23	1,562.23
		1,562.64
44	1,563.05	1,563.05
		1,563.45
45	1,563.86	1,563.86
		1,564.27
46	1,564.68	1,564.68
		1,565.09

Table 8.5
Guard Band in WDM Systems

WWDM	CWDM	DWDM 100 GHz	50 GHz	25 GHz	12.5 GHz
100 nm	20 nm	0.8 nm	0.4 nm	0.2 nm	0.1 nm

Table 8.4 lists the ITU-T G.694.1 DWDM wavelengths corresponding to the C band for 100-GHz and 50-GHz spacing. As mentioned above there can be a total of 44 λ's corresponding to the C band 100-GHz spacing and 88 λ's corresponding to C band 50-GHz spacing. The values of the wavelength are computed by using the formulae mentioned in Table 8.2.

8.6 Evolution of WDM Networks to Meet Future Challenges: Flexible WDM Grids

In the future, as the need for more bandwidth increases exponentially, one could look to the 25-GHz spacing for increasing the number of wavelengths. This implies a reduced spacing between the successive wavelengths and consequentially the use of higher quality optics for successful operation—hence a higher cost. This is especially true if it is necessary to adhere to the ITU-T standard band ranges.

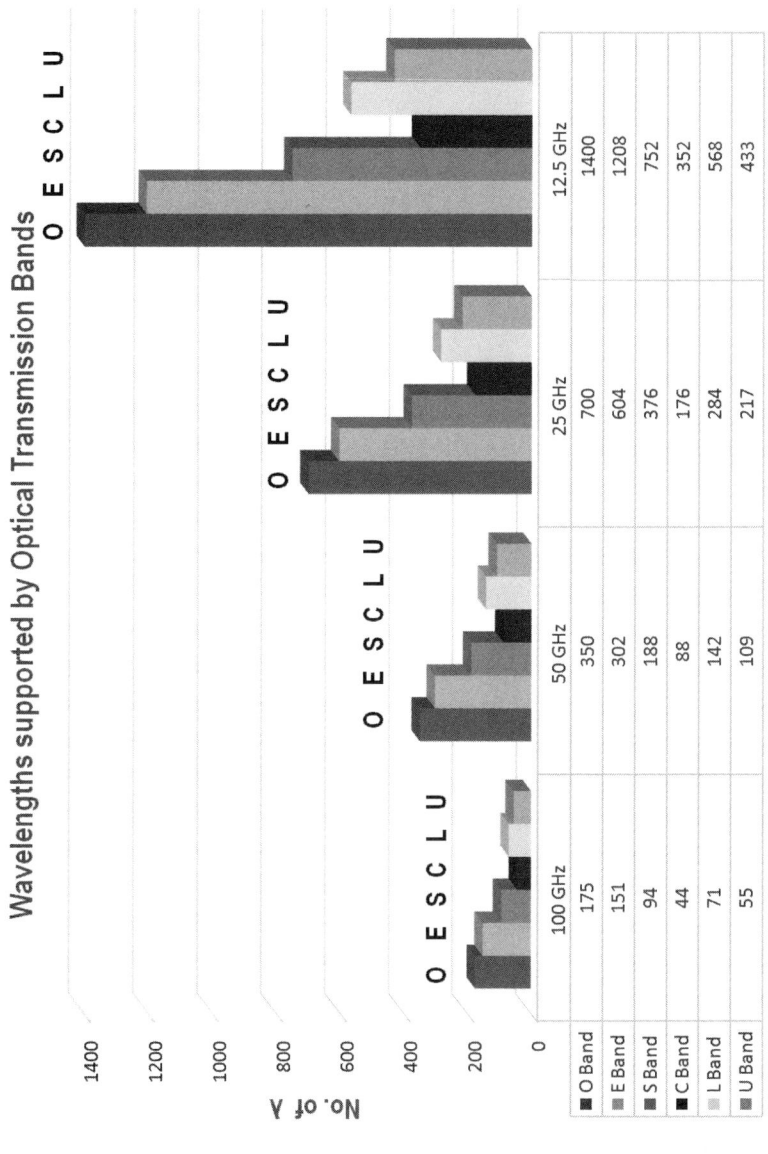

Figure 8.6 Optical transmission bands—lambda capacity.

Example 8.3

Some manufacturers provide for a higher number of wavelengths by increasing the spectral width of the C band (also referred to as the extended C band). Certain manufacturers like the erstwhile Nortel, now a part of Ciena technologies, offer wavelengths spread across four grids—grid 1 to grid 4. Grids 1 and 3 are the ITU-T–defined DWDM grids for the C and L bands respectively.

Table 8.5 lists the separation between wavelengths in nanometers for the common WDM deployments. As indicated earlier it is important to note that the CWDM frequencies are equidistant in the wavelength domain and hence will differ by exactly 20 nm while that of DWDM are equidistant in the frequency domain and hence will not be exactly equidistant in the frequency domain. As shown in Table 8.5 the separation between wavelengths for DWDM systems with 50-GHz spacing will be approximately 0.4 nm. Grids 2 and 4 are proprietary grids corresponding to the C and L bands respectively but with an offset of 25 GHz from the ITU-T–defined wave grids for the C and L bands respectively. This implies that the wavelength corresponding to grid 2 and grid 4 are spaced 50 GHz from each other. However as these wavelengths are not standardized they cannot be used for intercarrier communication.

Figure 8.7 lists the first three frequencies corresponding to the ITU-T G.694.1 C band DWDM wavelengths. The difference between the first and the second wavelengths is approximately 0.4 nm corresponding to a frequency separation of 50 GHz. The starting wavelength of grid 2 is offset by a frequency of 25 GHz or 0.2 nm from the standard first wavelength with the subsequent wavelengths also conforming to the same separation. Thus the difference between any two wavelengths of grid 2 as well as grid 4 is also 50 GHz. There are manufacturers that provide DWDM systems that operate in the 25-GHz spacing within the C band and the extended C band also. Additional details about specific vendor products can be found at their respective website using the URLs provided in this chapter.

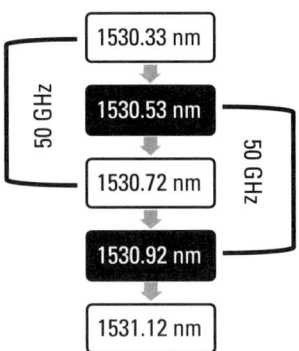

Figure 8.7 Proprietary WDM grids.

The development of chip-based integrated optical circuits or photonic integrated circuits (PICs), as they are commonly referred to, has led to advanced DSP techniques, resulting in the manufacture of advanced optical systems capable of working with wavelengths with reduced guard bands at lower costs. This has been supplemented by the deployment of FEC techniques as well as DCMs. An example of this is the proprietary triple FEC (TriFEC) as well as super-concatenated FEC (SFEC)[1]. Nortel has also come out with a revolutionary technique referred to as NGM, which is a per wavelength electronic dispersion compensation (eDCO) technique that reduces the number of regenerators used in long-haul networks. It is important to note that the actual number of wavelengths possible across the optical transmission bands would be less than the theoretical values depicted in Figure 8.6 due to the technology limitations, cost of equipment required, and the effect of attenuation owing to the optical fiber nonlinearities.

Flexible WDM Grids

Current WDM networks support optical line rates of 2.5G, 10G, 40G, and 100G within the 50-GHz channels. The ITU-T G.709/Y.1331 "Interfaces for Optical Transport Network" proposes a flexible (new) OTUCn (n * 100G format) that can be scaled to 25.6 Tbps. Similarly IEEE P802.3cd 400G (400 GbE) and 200G (200 GbE) standards are planned for release in December 2017. The transition to these higher-order bit rates has necessitated the evolution of a flexible WDM grid pattern capable of supporting wavelengths up to 1 Tbps. To support these high data rates (50-GHz channel spacing would not be sufficient), multiple subcarrier channels (referred to as superchannels) would be employed. These superchannels would be provisioned and switched across the network as a single channel using a flexible grid pattern [5]. The concept is illustrated in Figure 8.8.

The flexible grid pattern (ironically referred to as gridless channel spacing) is defined by the ITU-T G.694.1 (clause 7) standard. The objective of introducing the flexible architecture was to facilitate mixed bit rates or modulation formats to be transmitted optimally over frequency slots of different widths. This flexible grid pattern has a central frequency granularity of 6.25 Hz and a slot width granularity of 12.5 GHz [6].

$$\text{Nominal Centeral Frequency of Slots} = 193.1 + n * 0.00625$$

where

---

[1]. Developed by Nortel Networks.

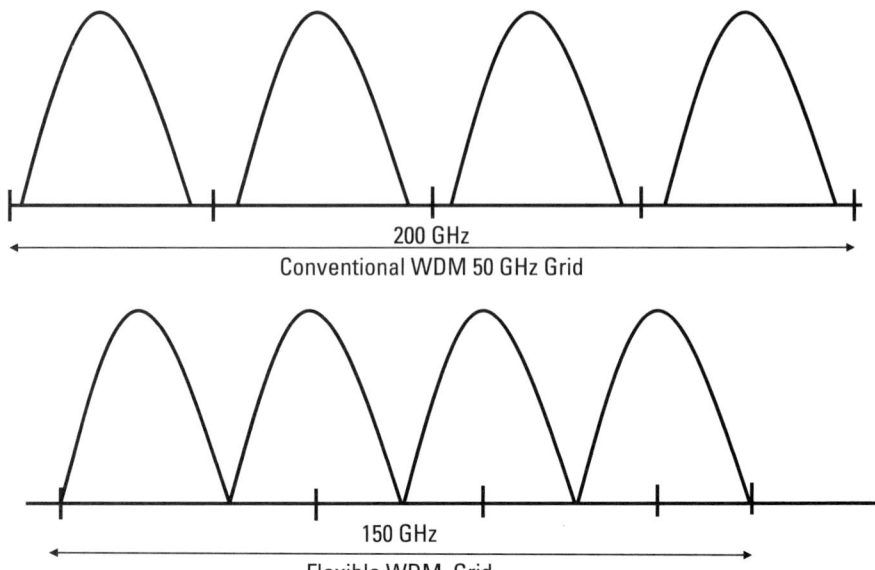

Figure 8.8 Flexible grid WDM transmission—superchannel.

n = positive or negative integer (including 0);
0.00625 is the nominal central frequency in terahertz.

The slot width = 12.5 * m.
Where m = positive integer, 12.5 is the slot width in gigahertz.
This arrangement facilitates multiple channels of different slot widths to be accommodated over a channel.

Note 8.1

It is important to note and understand that existing WDM networks can support 400G transmission, using the flexible grid, by tuning (transponders) to the existing 50-GHz grid pattern (subcarriers can be tuned to any wavelength using the formula mentioned earlier in the section), facilitating complete backward compatibility with existing WDM networks.

Note 8.2

The deployment of the flexible grid providers for closer subcarrier spacing results in capacity improvement of the network (over 30%).

Example 8.4

The transport of 4*100G channels over a WDM network using the existing 50-GHz grid would require a bandwidth of 4*50 GHz = 200 GHz. The use of flexible WDM grid results in enhanced spectral efficiency wherein only 150 GHz is required for transmitting the 4*100G signals (without guard bands) thereby enhancing bandwidth utilization by:

$$50 \text{ GHz}/200 \text{ GHz} = 25\%$$

The concept is illustrated in Figure 8.8.

Note 8.3

It is pertinent to note that the spectral efficiency (efficiency of transmission over a media) increases with higher bit rates. Spectral efficiency refers to the number of bits transmitted per hertz of the spectrum (optical). Table 8.6 presents the spectral efficiency at different optical transmission rates.

Note 8.4

PSK is a technique wherein the phase of a reference signal (carrier) is modulated in accordance with the data being transmitted. QSK is a type of PSK that modulates two bits simultaneously by using four phase shifts of the carrier at 0°, 90°, 180°, and 270°. Dual polarization QPSK (DP-QPSK) uses horizontal and vertical polarization along with QPSK to modulate information. This technique allows four bits to be represented by one symbol. DP-QPSK is the preferred method for high-speed transmission over a long-haul fiber-optic link. Using DP-QPSK a 100G transmission would require only 25G symbols to transmit.

8.7 Optical Transmission Challenges

The media for optical transmission is the optical fiber. There are different types of optical fibers available to suit different application needs. These fibers differ primarily in their dispersion characteristics. ITU-T recommendations have standardized and classified the different types of optical fibers. The most com-

Table 8.6
Spectral Efficiency of Optical Transmissions

S.N	Line Rate (Gbps)	Modulation	Spectral Efficiency (bits/Hz)
1	10	NRZ	0.2
2	40	DP-QPSK	0.8
3	100	DP-QPSK	2

monly used fiber for intracity networks is the standard SMF (SSMF), which is optimized for operation at 1,310 nm. DSFs, as per ITU-T recommendation G.653, optimized for 1,550-nm transmission corresponding to the low attenuation band were used for long-haul communication. These fibers were suitable for SDH transmission but could not be employed for WDM networks owing to their small core diameter. In the case of WDM networks more than one signal is transmitted through a single fiber and hence larger core diameters are beneficial. However it may be noted that NZ-DSF fibers developed for DWDM applications normally have core diameters and apertures that are smaller than standard SMFs. A commonly used fiber for long-haul DWDM transmission is ITU-T G.655 fiber. The transmission through an optical fiber is affected by its linear as well as its nonlinear characteristics. These characteristics are as described as follows [7]:

1. Linear characteristics: The major issues arising due to the linear characteristics of the fiber are attenuation and dispersion:
    a. Attenuation: Attenuation refers to the losses encountered as rays of light travel through the optical fiber. It is important to note that the amount of attenuation is also wavelength-dependent. There can be three common causes for attenuation (especially in SSMF). They are the following:
        i. Waveguide losses: The waveguide losses are primarily due to microbending and macrobending of the optical fiber. Microbending losses are a result of surface imperfections (perturbations) during the manufacture of an optical fiber. Macrobending losses are however due to the imperfect grooming, termination, and routing of fibers either at terminal or intermediate equipment sites or anywhere along the physical path. Macrobending losses can be reduced or even eliminated by the use of fiber-routing trays, fiber risers, slack storage trays, or other similar mechanisms. Macrobending losses are primarily due to the sharp bends (curves) that a fiber may be subjected to anywhere along its path.
        ii. Scattering losses: Scattering refers to loss of signal power due to diffusion of light as a result of the material structure of the optical fiber, which results in blisters, roughness, cracks, and other surface imperfections. This phenomenon is also referred to as Raleigh scattering.
        iii. Absorption losses: Absorption losses may be intrinsic or extrinsic in nature. Intrinsic losses are a result of the limited spectral

width of the optical fiber while the impurities in the glass or silica contribute to the extrinsic losses.

b. Dispersion: Dispersion basically refers to the elongation of the optical pulse as it travels over the optical fiber. Dispersion is one of the major reasons for the limited bandwidth capability of the currently available optical networks. The effects of dispersion increase exponentially with the increase in channel bandwidth. The effects of dispersion on a 40-Gbps transmission link is 16 times more pronounced than that of a corresponding 10-Gbps transmission link and 256 times that of a 2.5-Gbps link.

The light pulse components having various wavelengths propagate in the optical fiber at different velocities due to the wavelength dependency of the refractive index of the silicon oxide. This phenomenon is called CD. The total CD can be computed by the following formula:

$$CD = D_C * \sigma_\lambda * L$$

where D_c is the dispersion coefficient for the fiber (ps/nm.km), σ_λ is transmitter source spectral width (nm), and L is the total fiber span (km).

CD develops as the common result of several effects including the RI of a fiber. It is however interesting to note that waveguide dispersion can be influenced by shaping the profile of the optical fiber's RI, providing manufacturers an option of producing optical fibers with different dispersion characteristics. Normally zero dispersion coincides with 1,310 nm and the low-loss band corresponds to the 1,550-nm window. By adding suitable dopants we can have a fiber where the zero dispersion and the low-loss band coincide at 1,550 nm. Such fibers are referred to as DSFs.

2. Nonlinearities: The nonlinear characteristics of an optical fiber have a reflective impact on the functioning of optical networks, especially WDM networks. These include increased attenuation and consequently smaller span lengths leading to additional deployments of amplifiers and regenerators, distortion of the transmitted optical signal, increased spacing between adjacent wavelengths, and interchannel interference cumulatively leading to reduced channel bit rates. The main nonlinearities include the following:

a. SPM: caused by variations in optical signal power leading to corresponding variations in the signal phase. In PSK systems, these variations may lead to a degradation in system performance since the receiver relies on the phase information. Additionally, SPM causes spectral broadening of pulses because the variations in a signal's phase result instantaneous variations of frequency around the signal's central frequency. This may lead to spreading of pulses, thereby affecting the maximum bit rate.

    b. XPM: a nonlinear optical effect where one wavelength of light affects the phase of another wavelength thereby causing a shift in the phase of a signal. It is caused by a change in the intensity of light power of a signal propagating at different wavelengths. XPM leads to asymmetric spectral broadening, and a combination of XPM and SPM affects pulse shape.

    c. FWM: occurs when signals at two wavelengths operating at different frequencies combine to produce extra wavelengths or sidebands causing interference or distortions in the original signal.

    d. SRS: caused by the interaction of light with molecular vibrations. The scattering results in shifting of a portion of light from the high-frequency part of the spectrum to the low-frequency part. The lower frequency signal is referred to as the Stoke's wave. As the power of the input signal increases, the fraction of power transferred to the Stoke's wave increases rapidly. Under very high input power, SRS transfers almost all of the input signal power to the Stoke's wave.

    e. SBS: caused by the interaction of light with sound waves (as opposed to molecular vibrations in SRS). In SBS the resultant Stoke's wave propagates in the direction opposite to the input signal. However the intensity of the scattered light is much greater in SBS than in SRS.

Note 8.5

When a photon is incident on an atom/molecule (silica) it may get absorbed resulting in the atom/molecule transiting onto an excited phase. To return back to its normal state the atom/molecule has to release energy either in the form of photonic emissions or heat (energy loss). When the emitted photon has less energy in comparison to the absorbed (incident) photon, the energy difference is referred to as the Stoke's shift. If the emitted photon has a higher energy as compared to the incident (absorbed) photon, the energy difference is termed as the anti-Stoke's shift.

The inelastic[2] scattering of a photon to a lower energy photon induces nonlinear scattering effects in fiber-optic cables. The phonons[3] in the medium absorbs the energy difference. The energy difference is absorbed by the molecular vibrations or phonons in the medium. The optical wave displaces the silica atoms (in the fiber core) causing them to move in opposing directions at a higher amplitude in comparison to the acoustic wave, which causes the silica atoms to move in the same direction, amplitude, and phase.

SRS is triggered by optical phonons and the resultant stokes wave propagates in the same direction as the signal traveling through the fiber. In SBS, scattering is caused by acoustic waves, leading to contradirectional and frequency-shifted waves (Stoke's wave). This difference implies that SBS occurs only in one direction (backward) in contrast to SRS, which can occur in both directions (forward and backward).

8.8 WDM Network Components

This section presents the commonly used equipment[4] within a WDM network. The functionalities of some of the key system components are listed as follows:

1. ADMs/optical line terminals (OLTs):
   a. OADMs;
   b. ROADMs[5].
2. DCMs;
3. Multiplexers/demultiplexers (mux/demux) couplers;
4. Optical amplifiers;
5. Optical combiners;
6. Optical switches;
7. Optical translators;
8. Optical transmitters and receivers.

---

2. If the photons that are scattered from an atom/molecule have the same frequency/wavelength of the incident photon, the process is referred to as elastic scattering. When the scattered photons have a different frequency (usually lower) than the incident photon, the process is referred to as inelastic scattering.
3. Phonons can be visualized as particles that carry vibrational energy in a manner similar to that of photons (discrete and quantized). Further light traveling through the fiber (propagation wave) is split into two branches—optical branch (upper) and acoustical branch (lower).
4. Many of the functionalities offered by the components listed in this section may be combined within a single piece of equipment.
5. Includes wavelength selection (WSS) function.

9. Regenerators.
10. Wavelength selective switches (WSSs).

Optical Transmitters

Optical transmitters use laser to transmit data through an optical fiber. The information to be transmitted is encoded or modulated onto the laser signal using a variety of modulation techniques, both analog like amplitude modulation (AM), frequency modulation (FM), and phase modulation (PM), as well as digital including ASK, FSK, and PSK. Depending on the types of laser (tunable or fixed), optical transmitters can be either tunable or fixed.

Optical Receivers

Receivers can either be fixed or tunable. Fixed receivers use fixed filters or grating devices to filter out one or more wavelengths from a single fiber. Some of the most popular filters are diffraction grating, fiber Bragg grating, and thin-film interference filters. The diffraction grating is basically a flat layer of transparent material such as glass or plastic with a row of parallel grooves cut into it. The grating separates light into its component wavelengths by reflecting it with the grooves at all angles. At certain angles, only one wavelength adds constructively while all other wavelengths interfere destructively. Tunable lasers employ a variety of filters to choose specific wavelengths. An example is E/O filters, which use crystals whose index of refraction can be changed by electrical current.

Couplers

Couplers refer to devices that combine light into or split light out of a fiber. There are two basic types of couplers: active and passive. An active coupler requires power to function and it has some form of intelligence. A passive coupler simply combines light into or splits light out of a fiber, and it does not require power to operate. Couplers are also classified into single-mode couplers and multimode couplers corresponding to the two types of optical fiber applications.

Optical Amplifiers

Amplifiers impart a fixed gain or boost the strength of the weak input signal. An amplifier does not correct amplitude or phase distortion but merely increases the transmit power of the signal. Thus amplifiers do not require signal conversion and happen in the optical domain. The two basic types of optical amplifiers are semiconductor laser amplifiers and rare Earth doped fiber amplifiers

(DFAs). In addition, there are parametric amplifiers and Raman amplifiers. These types of optical amplifiers are described as follows.

- A semiconductor laser amplifier commonly referred to as a semiconductor optical amplifier (SOA) is comprised of a modified semiconductor laser that amplifies a weak signal via stimulated emission. SOAs can be of two types—Fabry-Perot amplifiers (FPAs) and nonresonant traveling wave amplifiers (TWAs). FPAs amplify light by back-and-forth reflections between mirrors. The process is similar to FPA with the exception that the end facets are either cleaved or coated with antireflection material to reduce internal reflection. A major limitation of optical amplifiers is the unequal gain spectrum. Optical amplifiers provide gain across a range of wavelengths, but they do not amplify all wavelengths equally.
- DFAs are coiled springs of fibers with suitable dopants added (e.g., Erbium). An external excitation is applied to the doped fibers causing the erbium atoms to go into a high energy or metastable state with the release of photons. The incoming signal photons collide with these excited erbium atoms and the resultant energy transfer produces photons with the same phase and direction as the photons of the incoming signal. Figure 8.9 illustrates the commonly used amplifier types.
- The distributed Raman amplifier (DRA) is based on the phenomenon known as SRS. The main advantage of using a DRA is the increased effectiveness of the amplification stage (due to the low noise figure). DRA used be very expensive in comparison with EDFA. However the cost has significantly decreased and is currently almost on par with EDFA. The use of EDFA in an optical network necessitates the deployment of a regenerator after five amplifier spans or sections. However if a DRA is being deployed in an optical network the ratio of regenerators and amplifiers is 1:15.

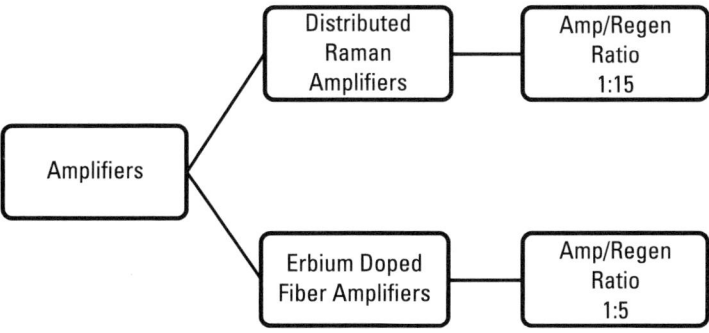

Figure 8.9 Types of optical amplifiers.

Regenerators

Optical regeneration is used to increase the length of the data path and ensure the stability of critical transmission parameters for long-distance spans (e.g., the length of the fiber and the power of the laser). Modern-day high-speed transmission systems mandate high-speed, reamplifying, reshaping, and retiming (3R) regeneration, which itself is responsible for bringing the optical signal back to a more readable form (after it has been impaired due to a variety of causes). The regeneration can be done in the optical or the electrical domain. Optical-level regeneration results in a higher span length and a significant reduction in power and equipment footprint. However the advantages are offset by a significant increase in cost for optical domain regeneration. Several regeneration types (back-to-back, unidirectional, and bidirectional) are used to carry out complex transformations: optical-to-electrical signal conversion (with the same bit rate), electrical regeneration, and conversion to the optical domain (with the same modulation and bit rate). This is done on a per-channel basis (for both CWDM and DWDM). The most commonly employed types of regenerators are depicted in Figure 8.10.

Passive Routers

Passive routers separately route each of several wavelengths existing on input fiber to the same wavelength on separate output fibers. Passive routers allow wavelength reuse (i.e., the same wavelength may be spatially reused to carry multiple connections through the router).

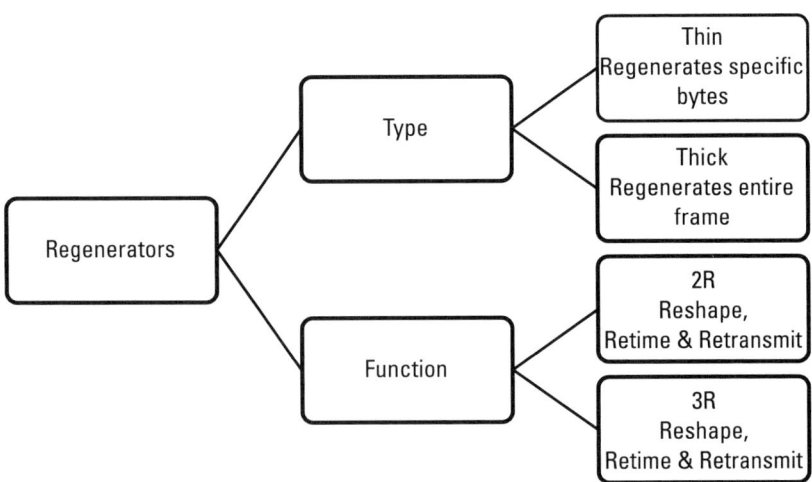

Figure 8.10 Types of regenerators.

Wavelength Selective Switch

WSSs use optical switches inside the routing element. The multiple wavelengths traveling on a fiber are separated using a grating demultiplexer whose outputs are directed to the optical switches. The multiplexed output is subsequently directed to an output fiber. An active switch is also referred to as a wavelength-routing Switch (WRS), wavelength selective cross-connect (WSXC), or cross-connect (NXN).

Wavelength Converters/Transponders/Muxponders

Wavelength converters convert data on an input wavelength onto a different output wavelength. Wavelength converters improve the efficiency in the network by resolving wavelength conflicts in the optical path. The techniques employed for wavelength conversion include optoelectronic wavelength conversion and all-optical wavelength conversion. Transponders are commonly used to covert non-ITU-T client signal wavelengths (for example 1,310 or 1,550 nm) to wavelengths compatible with the ITU-T wave grid. There may also be a case for multiplexing lower-order client signals to a higher order signal, which would be transmitted over a DWDM network. For example 4 * STM-16 client payloads may be multiplexed onto an STM-64 payload that is mapped to a specific ITU-T wavelength. The devices that facilitate this operation are referred to as muxponders.

WADMs

WADMs, also referred to as OADMs, consist of a demultiplexer followed by a set of 2X2 switches followed by a multiplexer. Although the design may sound fairly simple, the OADM is a complex[6] piece of equipment. OADM is a passive device that consists of (for the sake of simplicity) a set of mirrors aligned to the ports and a prism. For multiplexing the rays of lights coming in from the different ports (1: N) is focused onto the prism, which combines them onto a single or multiple ports (composite wavelength). The demultiplexing operation is the reverse of the multiplexing operation. In reality the complex arrangement and the precision needed to assemble the lenses and mirrors require state-of-the-art optical assembly and can be quite costly to fabricate. The various components include optical filters, fiber Bragg gratings, arrayed waveguide gratings, optical switches and circulators, microelectromechanical systems (MEMS), and liquid crystal and thermo optic switches[7].

---

6. A simple OADM design would consist of a fiber Bragg grating (FBG) and two circulators.
7. For complex OADM architectures.

Modern-day OADMs facilitate reconfiguration, referred to as ROADMs, wherein the mux configuration change can be effected electronically. ROADMs facilitate switching at the wavelength layer and provide the network with the flexibility of switching traffic from WDM systems. All OADMs currently in use are electronically reconfigurable and hence referred to as ROADMs. The configuration of the equipment, including the selection of the wavelength, can be done remotely. Previously such an operation would have required addition/deletion of wavelength-specific network cards being equipped at all the multiplexers and regenerators on the link followed by another complex procedure referred to as link equalization. This procedure ensures matching of the power levels of the optical signal at both ends of a link. [8].

In order to provide architectural flexibility along with operational simplicity and dynamic provisioning capabilities a new class of ROADMs has been introduced. They are referred to as colorless directionless/contentionless (CD/C) ROADMs. Their key features include the following:

1. Prepositioned bandwidth: Deployment of transponders and regenerators with CD/C ROADMs facilitating quick provisioning of wavelength services;

2. Bandwidth-on-demand: Ability to provide a new class of services—on-demand bandwidth provisioning (for specific period of time);

3. Optical layer protection: Ability to provide different QoS through the extension of a transport layer protection class to the optical layer. This includes the ability to support 1:1 (50 ms or less switching time for critical traffic), 1: n (shared protection for regular traffic) along with dynamic mesh restoration (capabilities);

4. Optical defragmentation: Ability to reoptimize the optical network, through a control plane, thereby compensating for optical path impairments (due to differing fiber types, modulation, and span length) dynamically. This feature is also referred to as optical defragmentation.

8.9 DWDM Links

A cross-section of a DWDM link is illustrated in Figure 8.11, which depicts only a small part of the link. This is done with the goal of maintaining simplicity. DWDM networks are usually designed with a mesh architecture. The configuration normally employed is a shared mesh wherein each node in the network has connectivity to a minimum of three other nodes in the network.

The input to the OADM could be SDH, OTN, IP, or any other compatible NE. Each of these input signals may be of varying bandwidths (2.5G,

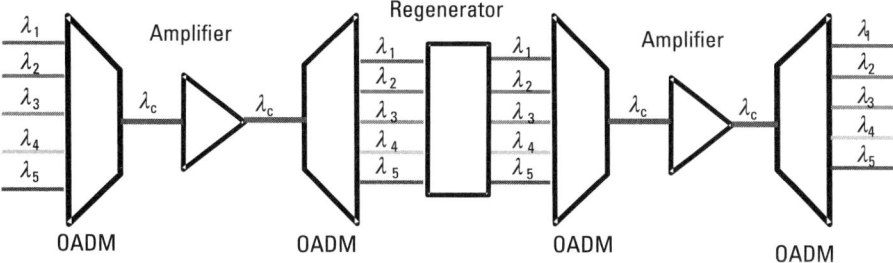

Figure 8.11 Sample DWDM link.

10G, 40G, and 100G). The OADM multiplexes these individual wavelengths ($\lambda_1, \lambda_2, \lambda_3 \ldots \lambda_n$) to form a composite signal comprising multiple wavelengths at the output (λ_c).

Note 8.6

The input signals (OADM) cannot be of the same frequency (wavelength). In case they are of the same frequency, wavelength converters have to be used to ensure unique wavelengths.

The input signals may have different levels of optical power. On multiplexing, the power of the composite signal will be a summation of the input power signals and may need to be amplified prior to transmission. The terminal amplifier (O-O) imparts a fixed gain prior to launching the signal over the long-distance fiber. Depending upon the link period amplification stages may be required using in-line amplifiers (ILAs). It may be noted that unlimited amplification stages cannot be permitted (refer to Figure 8.8). The number of stages (sections) that can be used depends upon the type of amplifier used in the network. A DRA would facilitate the use of 15 amplifier spans or sections prior to regeneration while an EDFA would limit the number of spans or sections to five or under.

Thus regenerators would need to be introduced at suitable sections (distances). A regenerator performs an O-E-E-O (OEO) conversion on a per-wavelength basis and hence the composite signal would have to be demultiplexed prior to regeneration. After regeneration the signals are multiplexed again and amplified (refer to the explaination of regenerators in Section 8.8) prior to re-transmission. The signals are finally demultiplexed and terminated at the end of the link or the drop locations (elements)

Figure 8.12 illustrates a branching DWDM link that uses a WSS and ROADM to branch wavelengths as per network requirements. The use of WSS and ROADM provides network planners the flexibility and scalability to build, interconnect, and expand networks as required.

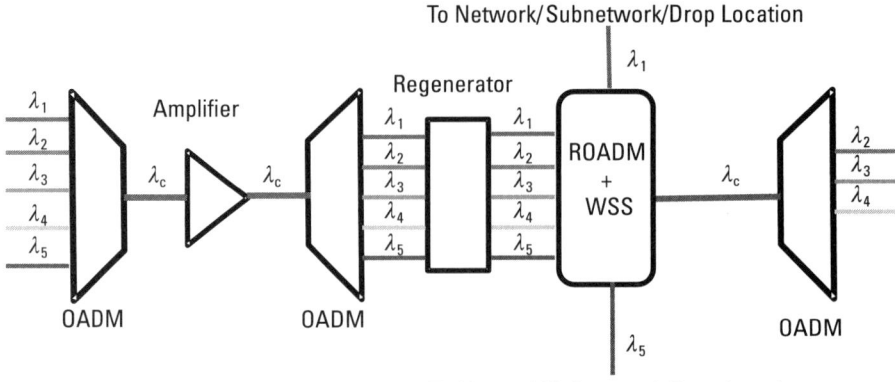

Figure 8.12 Branched DWDM network.

Note 8.7

In the simplified link section the used of DCMs has not been highlighted. The dispersion compensation is done at the intermediate stage (IS). Signal amplification would have to be performed after dispersion compensation. Newer OTN cross-connects would employ electronic dispersion compensation (e-DCM) techniques, eliminating or reducing the need for physical DCM. Physical DCM is normally carried out on the composite signal while e-DCM is done on a per-link basis and is more effective. It is possible to have sections in excess of 1,000 km without regenerators in the case of e-DCM systems.

8.10 Case Study: DWDM Deployment

8.10.1 Background

ESW, Inc., is a leading telecom service provider (tenth-largest by revenues) in South Asia with triple-play services provided to 29 states, 4,000 cities, and over 640,000 villages using a terrestrial fiber-optic network spanning over 200,000 kilometers. The network caters to over 70 million customers with over 30 million rural customers using only voice services. The backbone networks were-based on TDM SDH technology in a three-tier architecture, as illustrated in Figure 8.13:

1. Tier-1:
    a. Intracity links on a 10G STM-64 ring.
2. Tier-2:
    a. Intercity on 2.5 STM-16-Gbps rings.

DWDM 263

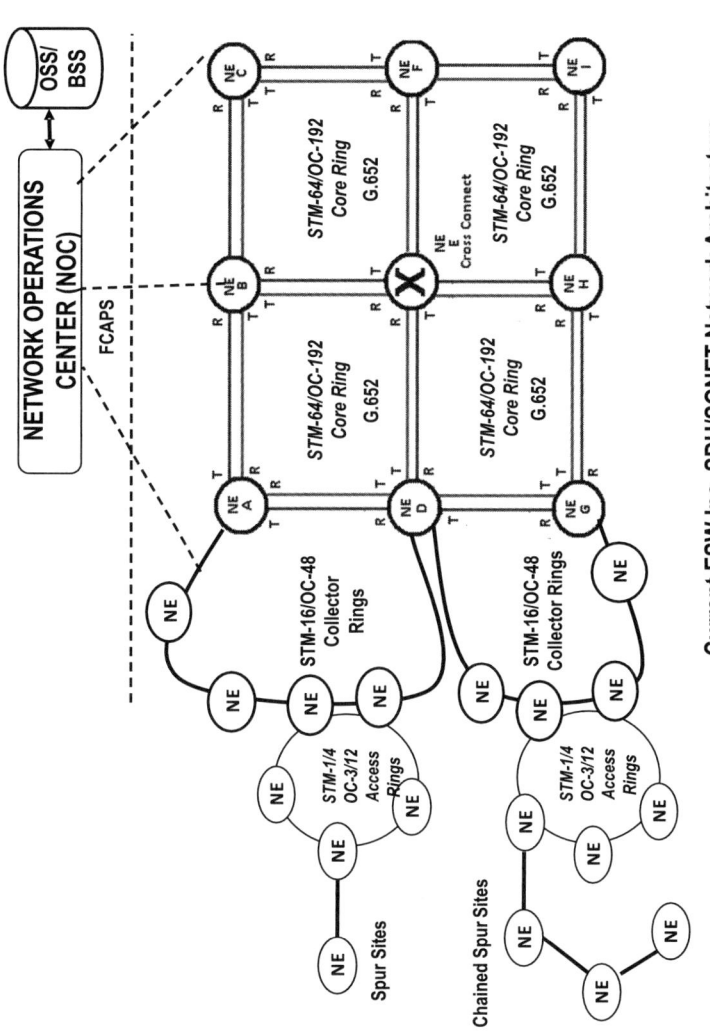

Figure 8.13 Current network architecture of ESW Inc.

3. Tier-3:
   a. Access rings of 622 Mbps STM-4/155 Mbps STM-1 rings.
   b. Spur sites on 155-Mbps STM-1 rings (unprotected).

The fiber consisted of 96-/48-/16-core G.652 cables for the core, collector, and access networks respectively. Figure 8.13 illustrates the network architecture of ESW.

8.10.2 Challenges

1. The maximum capacity of the core network, stagnating at 10 Gbps, is presenting a major traffic bottleneck.
2. Most of the metro locations in the network require additional bandwidth to cater to customer demands.
3. There has been a significant increase (nearly 60%) in customers requiring 10 GigE/1 GigE pipes between major metros.
4. In addition, the growing urbanization is fueling major infrastructure projects that are resulting in around 70% fiber cuts in urban areas (up to 100 cuts/day). This is severely impacting network up time and consequently noncompliance of SLA with major customers.
5. There is an anticipated five-fold increase in traffic in the next two years due to the deployment of the 5G ecosystem.

8.10.3 Requirements

ESW wants a solution that is cost-effective and that leverages existing infrastructure, while providing additional bandwidth for supporting the upcoming 5G ecosystem and improved network uptime.

8.10.4 Proposed Solution

The major highlights of the proposed solution for ESW are presented in Table 8.7 and illustrated in Figure 8.14.

As is evident from the background, ESW's network is based on purely SDH/SONET nodes. In order to keep the CAPEX at a minimum while providing for enhanced bandwidth at all the layers and improved network resiliency, the following changes were recommended:

1. *No change in fiber plants*—The proposed architecture does not envisage any change in the fiber plants at this point in time. In the future the core network fiber may be augmented to G.655.

Table 8.7
Key Features of the Proposed Network for ESW

S.N	Feature	Existing	New
1	Topology—core	Ring	Mesh
2	Topology—collector	Ring	Ring
3	Topology—access	Ring Unprotected spur	Ring Protected spur
4	Transport standard/technology	SDH	OTN (core) + DWDM OTN + SDH (collector/access)
5	Core bandwidth	10 Gbps	100 Gbps * 80λ
6	Collector bandwidth	2.5 Gbps	10 Gbps
7	Access bandwidth	155/622 Mbps	155/622/2.5 Gbps
8	10 GigE/1 GigE Transport	No native support GFP+VCAT+LCAS	Native support 10/1 GigE interfaces
9	Protection from fiber cut	Single (only two paths per NE)	Multiple
10	Protection mechanism	SNC-P MSP	Path protection Section protection
11	Control plane enabled	No	Yes

2. *Deployment of DWDM network*—The top nine metros (of the 29 states served) are to being configured over four DWDM rings with ROADM and OTN. The major junctions will employ colorless, directionless, and contentionless (CDC) ROADM (CD&C) with five directions or more. The DWDM network will support 80λ in the C band using a 50-GHz ITU-T grid.

The ROADMs facilitate an on-demand configuration with efficient and maximum resource utilization and a simplified network architecture. In addition it supports the following key feature sets:

a. It extends the flexibility of the mesh network architecture by facilitating the multiplexing of non-ITU-T (colorless) wavelengths onto the DWDM network by the use of transponders at DWDM nodes.

b. All tunable transponders have nonblocking access to all DWDM ports facilitating on-demand, dynamic network reconfiguration.

3. *Use of OTN in the core network*—The advantages of a photonic network based on OTN standard have already been detailed in this chapter. The use of OTN improves bandwidth availability at the core significantly by upgrading the 10-G rings to 100-G capacity. This would take care of the bottlenecks in the intercity links. The ONE

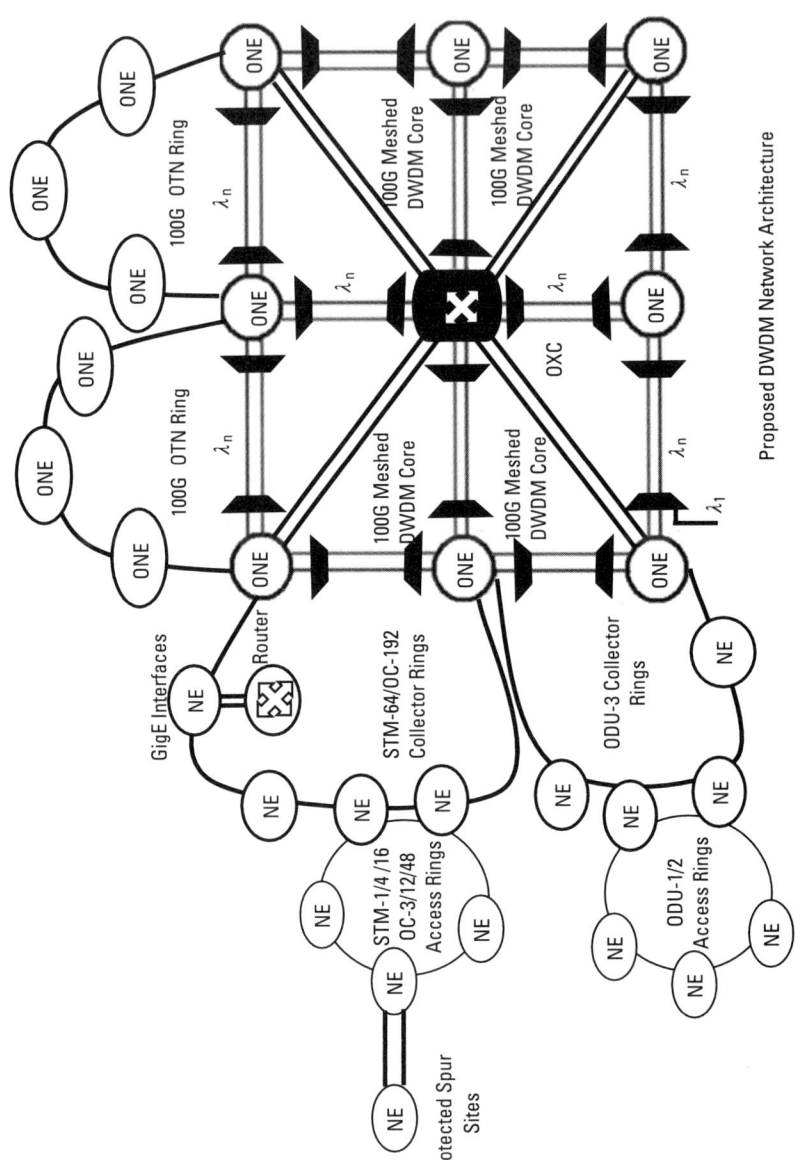

Figure 8.14 Proposed DWDM network architecture of ESW Inc.

will aggregate the lower-order traffic coming in from the collector rings into 100-G optimized traffic that can be switched on individual wavelengths. The use of OTN will facilitate the grooming of diverse lower-order client signals effectively.

4. *Mesh topology at the core*—In order to tackle the issue of multiple fiber cuts, implementing a shared mesh topology between the ONE in the core has been proposed. A 1:3 shared protection is proposed wherein every working path will have a minimum of three protection paths. Each of the ONEs has a minimum of three paths connecting them with neighboring NEs. This provides a cost-effective solution mitigating the issues related to multiple fiber cuts.

5. *Protected spur*—The spur sites can be extended protection (on a case-by-case basis) by extending another pair of fibers from the primary takeoff site. It should be noted that the pair of fibers needs to be route-diverse to ensure full protection.

6. *Control plane*—The full benefits of using OTN and a physical topology can be realized by provisioning a control plane with a hybrid protection mechanism. This ensure that dedicated protection paths can be defined for high-priority traffic/control channels resulting in lower switching times while providing automatic tertiary level protection for other traffic paths.

7. *Use of 10 GigE/1 GigE physical interfaces*—The deployment of OTN facilitates native support to Ethernet/IP traffic and direct interconnection of routing elements using physical interfaces. This eliminates the need for GFP protocol, LCAS, and VCAT techniques. The ability to support native IP traffic would cater to the requirement for provisioning additional 10 GigE/1 GigE pipes.

8.11 Summary

The optical fiber network is the backbone or the mainstay of telecommunication networks. Transmission speeds have evolved from 2.5 Gbps, a decade back, to 100 Gbps. In fact, speeds of 400 Gbps and 1,000 Gbps have already been tested. The backbone networks are based on the ubiquitous TDM technology, which has managed to hold its own. However, the development of high-bandwidth applications and the explosive growth of the internet have created capacity demands that exceed traditional TDM limits. The transport network based on SDH technology facilitates the transmission of only a single wavelength through the fiber at 10 Gbps, thereby creating a capacity bottleneck. This implies that the current technologies are able to harness only a minuscule amount of the fiber capacity. The cost of laying a fiber is significant and accounts for a major chunk of the network CAPEX. In order to meet the growing bandwidth demands a technology referred to as WDM was developed. WDM allows for the transmission of multiple rays of light or wavelengths on a single fiber, thereby increasing the capacity utilization manifold. This cutting-edge technology combined with the deployment of OTN networks and network management systems and ADM enables carriers to adopt optically based transmission networks that will meet the next generation of bandwidth demand at a significantly lower cost. WDM network components are currently expensive, but their cost is expected to lower significantly due to the technological advances in the field of optics and semiconductors. Going forward IP over DWDM is going to be the technology of choice for telecom operators.

8.12 Review

8.12.1 Review Questions

1. The ITU-T recommendation G.694.1 defines the CWDM wavelength grid.
    a. True
    b. False
2. Modern-day telecommunication networks employ DWDM 25GHz technology.
    a. True
    b. False
3. The CWDM wavelengths straddle the _____ optical bands.
    a. C
    b. S
    c. O-L
    d. L
4. eDCO is a technology that facilitates _____.
    a. Dispersion compensation
    b. Amplification
    c. Regeneration
    d. Multiplexing

5. The optimal ratio of regenerators to amplifiers in an optical network with EDFA amplifiers is _____.
   a. 1:15
   b. 2:30
   c. 1:8
   d. 1:5
6. A C band DWDM multiplexer operating on a 50-GHz grid allows simultaneous transmission of _____ wavelengths.
   a. 80
   b. 88
   c. 40
   d. 44
7. The _____ element facilitates branching on an optical network.
   a. WSS
   b. Amplifiers
   c. Regenerators
   d. DSCM modules
8. A regenerator and multiplexer can be used interchangeably on the WDM network.
   a. True
   b. False
9. An amplifier requires OEO signal conversion.
   a. True
   b. False
10. CD results in pulse elongation.
    a. True
    b. False

8.12.2 Exercises

1. What are the advantages of deploying ROADM in an optical network?
2. Explain the process of multiplexing in a WDM ADM. How is the composite wavelength formed?

3. Trace the evolution of WDM networks.

8.12.3 Research Activities

1. Trace the future developments in WDM networks.
2. Describe the architecture and the functioning of a ROADM.
3. What is meant by the term branching in an optical network?
4. Describe in detail the architecture of IP over DWDM (IoD) networks. Illustrate with a network diagram.

8.13 Referred Standards

G.652: Characteristics of a single-mode optical fiber cable

G.653: Characteristics of a dispersion shifted single-mode optical fiber cable

G.654: Characteristics of a cut-off shifted single-mode optical fiber cable

G.655: Characteristics of a non-zero dispersion shifted single-mode optical fiber cable

G.703: Physical/electrical characteristics of PDH interfaces

G.957: Optical interfaces for equipment and systems relating to the synchronous digital hierarchy

G.671: Transmission characteristics of optical components and subsystems

G.691: Optical interfaces for single channel STM-64 and other SDH systems with optical amplifiers

G.692: Optical interfaces for multichannel systems with optical amplifiers

G.959.1: Optical transport network physical layer interfaces

G.694.1: Spectral grids for WDM applications: DWDM frequency grid

G.694.2: Spectral grids for WDM applications: CWDM wavelength grid

8.14 Recommended Reading

8.14.1 Books

Mukherjee, B., *Optical Communication Networks,* McGraw Hill, July 1997.

Green, P. E., *Fiber-Optic Networks,* Prentice-Hall, 1993.

Lachs, G., *Fiber-Optic Communications,* McGraw-Hill Telecommunications, 1998.

Keiser, G., *Optical Fiber Communications,* McGraw-Hill, 1983.

Palais, J. C., *Fiber Optic Communications*, Fifth Edition, Prentice-Hall, 2004.

Bates, R. J., *Optical Switching and Networking Handbook,* McGraw-Hill, 2001.

Goff, D. R., *Fiber Optic Reference Guide,* Boston: Focal Press, 1999.

Ramaswami, R., and K. N. Sivarajan, *Optical Networks: A Practical Perspective,* Morgan-Kaufmann, 1998.

8.14.2 URLs

http://theory.lcs.mit.edu/~jacm/Abstracts/aggarwalbcrss96.ltx.

http://www.broadband-guide.com/lw/reports/report09985.html.

http://www.ciena.com/products/.

http://www.iec.org/online/tutorials/dwdm/index.html.

http://www.protocols.com/papers/wdm.htm.

http://www.sycamorenet.com.

http://www.webproforum.com/wpf_all.html.

http://www-bsac.eecs.berkeley.edu.

8.14.3 Journals

Doshi, B. T., et al., "Comparison of Next-Generation IP-Centric Transport Architectures," *Bell Labs Technical Journal,* Oct.–Dec. 1998, pp. 63–85.

Awduche, D., and Y. Rekhter, "Multi-Protocol Lambda Switching: Combining MPLS Traffic Engineering Control with Optical Crossconnects," Internet Draft, Work in Progress, 17 pages, 1999.

Chen, Y., et al., "Metro Optical Networking," *Bell Labs Technical Journal,* Vol. 4, No. 1, Jan.–Mar. 1999, pp.163–186.

Alferness, R., et al., "A Practical Vision for Optical Transport Networking," *Bell Labs Technical Journal,* Vol. 4, No. 1, Jan.–Mar. 1999, pp. 3–17.

Giles, R., and M. Spector, "The Wavelength Add/Drop Multiplexer for Lightwave Communication Networks," *Bell Labs Technical Journal,* Vol. 4, No. 1, Jan.–Mar. 1999, pp. 207–229.

References

[1] Schneider, K. S., *Primer on Fiber Optic Data Communications for the Premises Environment* [n.d.]. Retrieved June 10, 2008 from http://www.telebyteusa.com/foprimer/fofull.htm>.

[2] ITU-T G.694.2, *Spectral Grids for WDM Applications: CWDM Wavelength Grid,* from http://www.itu.int.

[3] Shannon, C. E., "A Mathematical Theory of Communications," reprinted with corrections from *The Bell System Technical Journal,* Vol. 27, pp. 379–423, 623–656, July, October, 1948 from http://web.mit.edu/persci/classes/papers /Shannon48.pdf.

[4] Lange, B., et al., *Deliverable DJ1.2.1: State-of-the-Art Photonic Switching Technologies*, 25-10-2010, p. 9.

[5] Eisenach, R., "Evolution to Flexible Grid WDM," *Lightwave* (2013), web document. Source URL<http://www.lightwaveonline.com/articles/print/volume-30/issue-6/features/evolution-to-flexible-grid-wdm.html).

[6] ITU-T G.694.1, "Spectral Grids for WDM Applications: DWDM Frequency Grid" (02/12) from http://www.itu.int.

[7] Warier, S., *The ABCs of Fiber Optic Communication,* Norwood, MA: Artech House, 2017.

[8] Ramaswami, R., and K. N. Sivarajan, *Optical Networks*, San Francisco: Morgan Kaufmann, 1998.

Part III

Next-Generation Photonic Networks: Applications and Architecture—The Future

9

Photonic Circuit-Switched Network Architecture

9.1 Chapter Objectives

The exponential growth in the number of users accessing triple-play services coupled with the explosive progression in the number and type of services have placed high demands on the transport layer, in terms of reliable high-speed throughput and six nines network availability. These stringent requirements coupled with service and application requirements have resulted in a complex architecture that provides for reliable transport of fixed granularity and dynamic bandwidths. The erstwhile TDM-based SDH/SONET transport networks are being replaced by high-speed packet-based network architectures, using the ubiquitous IP suite, with scalable and flexible capacity expansion achieved by integrating DWDM transport using a physical mesh architecture. The emergence of OTN has provided a fresh lease of life to TDM-based core networks with its ability to support native data rates, and it remains in contention to form the basis of the next generation networks.

The use of optical fibers was initially confined to the core or backbone networks. However with the growth of services and the emergence of wireless technologies like LTE, the bandwidth requirements at the access layer have increased exponentially. This has hastened the process of fiber deployment in the access network, along with wireless last-mile connectivity. This chapter provides a detailed listing of the current and emerging standards for cable- and fiber-based access networks. The chapter emphasizes the architecture of a FTTx network using PON technologies. It also includes a concise section on the emerging 5G ecosystem and its impact on next-generation architecture. The chapter also provides a concise outlook on the architectural choices for the photonic backbone networks and concludes with a framework for the converged backbone networks of the future.

This chapter provides a logical conclusion to the second part of the book, dealing with optical network architecture. Chapter 6 introduces the basis for a converged core or transport network using TDM technologies—SDH/SONET; Chapter 7 provides details the network architecture and functioning of an OTN along with the optical control plane; while Chapter 8 provides an the conceptual and deployment framework of a wavelength switched network.

Key Topics

- The architecture of broadband cable networks;
- The benefits of optical access networks;
- The architecture of FTTx networks;
- PON and its application in FTTx;
- The impact of 5G technologies and the proposed network architecture;
- The different choices for the photonic backbone networks;
- The architecture of the next-generation packet optical transport network;
- Generalized multiprotocol label switching;
- The impact of SDM techniques;
- The functionalities of agile optical networks;
- The need for and importance of CD&C ROADM architecture.

9.2 Introduction

Current trends are indicative of the fact that networks in the future are going to be all-optical E2E with DWDM technologies being deployed in the access layer. The full potential of photonic technology will be realized with the convergence of optical and wireless communication systems with support for a new generation of cost-effective services, including interactive video transmissions. However the launch of 5G technologies in 2020 could witness the decline of optical fiber-based access networks. With bandwidth capabilities ranging from 1 to 100 Gbps, 5G-based mobile broadband services would be prevalent in the last mile, replacing both the traditional copper-based connectivity and the newer optical PONs.

Network operators are faced with the dilemma of choosing between enhancing the current cable network standards and migration of high-bandwidth optical networks. While the broadband cable networks are still evolving there is no doubt that the future networks would be fully optical. This chapter is based on the above outlook and presents the architecture of cable networks based on DOCSIS standards and optical access networks based on PON technologies

The growth of converged network infrastructure has been exponential over the last decade. The developments in mobile broadband technologies, increasing teledensity (especially in developing nations), and the development of new services have led to a massive increase in traffic in the access network and necessitated the shift to optical broadband access solutions. Consequentially FTTx deployments have witnessed a manifold increase in the last decade. The bandwidth requirements at the access network have spurred the use of DWDM technologies in the metro and backbone networks and development of spatial multiplexing techniques. This trend is expected to continue in the foreseeable future.

9.3 Optical Access Networks

An end user connects directly to the access network to use required services. The development of new services (like video streaming) has led to increased bandwidth requirement at the access level. This increased demand for bandwidth, in the access network, has necessitated the transition from copper-based POTS to cable television networks to optical access networks. The investment required to set up fiber infrastructure is a major deterrent to this transition.

Copper-based technologies include the DSL family (ADSL and VDSL) while cable networks are based on DOCSIS and operate over coaxial networks or hybrid fiber coaxial (HFC) networks. LTE, mobile broadband service, broadband over power (BoP) line and satellite communication networks can also be used for providing access services. There are several factors that dictate the choice of technology to be deployed in the access network; primary among these factors are cost and technology future-proofing. The transition to fiber networks has been slow primarily due to the cost factor, though the concept of end-to-end optical networks was envisaged in early 1980s. Nevertheless, there has been a gradual transition from copper-based access networks to architectures that bring the fiber closer to the end user. This includes the many variations of hybrid fiber networks referred to as FTTx where x could be curb or cabinet, distribution point, street, home, office, or building. The network architecture has evolved from point-to-point topology to point-to-multipoint topologies.

9.3.1 Broadband Access Networks (Cable)

There are different technologies available for delivering broadband services to customers—residential and enterprise. These include the following:

1. Digital subscriber line (DSL) using copper cables of POTS;
2. Cable modem;
3. Fiber;
4. Wireless (LTE/4G services);
5. Satellite;
6. Broadband over power lines (BPL).

DSL-based services have been replaced by cable modem and/or LTE-based services. This focus of this chapter is fiber-based access networks. The chapter starts, however, by discussing the popular methods of providing cable services, since they continue to be used globally.

Access networks, also referred to as last-mile networks (or first-mile networks in telecom), refers to the portion of the network accessed by end users for using network services. In the case of the legacy PSTN networks this would refer to end-user connectivity with the CO in the United States or LECs in Europe using coaxial cables (employed for cable television systems) or wireless connectivity using wi-max. DSL and cable modems are used (especially in countries like India) as broadband solutions. These access systems with last-mile copper connectivity were designed to support data rates in multiples of 64 Kbps to 2 Mbps. Copper-based (last-mile) cable modems can support rates upto 50 Mbps. However the physical reach and channel capacity is limited wherever copper cabling is employed. The key technologies deployed globally for cable TV transport are described as follows:

1. DOCSIS [1]: Refers to an international standard for data transfer over cable TV systems. It specifies protocols and modulation schemes for bidirectional data transfer over cable. The standards were formulated by CableLabs, a nonprofit innovation and R&D lab setup by a consortium of cable TV companies in the United States in the late 1990s. The first version, DOCSIS, was released in 1997, and the current version is 3. 1. Table 9.1 lists the DOCSIS standards. Please note that there is likely to be some difference in the throughput between publication sources, depending on the reference to gross and net bit rates. Cablelabs also certifies that hardware like cable modems is DOCSIS-complaint. DOCSIS standards have been approved as international standards by ITU-T.

Table 9.1
DOCSIS Standards

Standard	ITU-T Standard	Downlink Speed (Mbps)	Uplink Speed (Mbps)
DOCSIS 1.x	ITU-T J.112	38	9
DOCSIS 2.0	ITU-T J.122	38	27
DOCSIS 3.0	ITU-T J.222	152	108
DOCSIS 3.1	ITU-T J.222	10,000 (10 Gbps)	1,000 (1 Gbps)

The frequency allocation plans for cable TV in Europe, conformant to PAL/DVB-C standards (8-MHz bandwidth), are different from those in the United States. The adaption of DOCSIS standards for the EU is referred to as EuroDOCSIS.

DOCSIS 3.1 supports full-duplex communication that provides the advantages of time-division duplexing (TDD) and frequency-division duplexing (FDD) with interference protection and intelligent scheduling capabilities. A DOCSIS 3.1–based network can provide for 10-Gbps (symmetrical) throughput using 1 GHz over HFCs with the upstream and downstream traffic sharing the common spectrum simultaneously. DOCSIS 3.1 uses orthogonal FDM (OFDM), efficient error correction [low-density parity check (LDPC)]–based FEC and smaller guard bands to improve spectral efficiency. It defines a new physical layer (PHY) technology that accommodates channels ranging from 24 to 192 MHz as opposed to the traditional 6-MHz channels.

2. Converged cable access platform (CCAP) [2]: Proposed by CableLabs by integrating the converged cable access platform (CMAP) and converged edge services access router (CESAR) into a single framework. CCAP integrates QAM and cable modem termination system (CMTS) technology into a unified device while supporting technologies like Ethernet PON (EPON), EPON over coaxial (EPoC), and DOCSIS [3]. The DOCSIS and CCAP architecture (integration of multiple functionalities onto a single node) is illustrated in Figure 9.1. CCAP is based on distributed architecture with three commonly used variations:

   a. Remote PHY: The CMTS/CCAP functionality is included in the remote node (from cable head end). This option is commonly used in the American, European, and Chinese markets.

   b. Remote MAC-PHY: This approach splits the CMTS among the MAC and PHY layers with the PHY layer functionality incorpo-

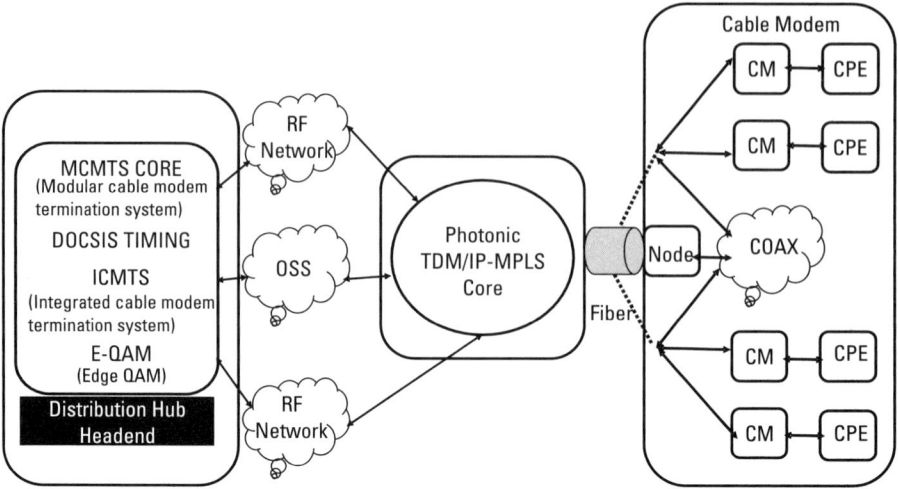

Figure 9.1 DOCSIS/CCAP architecture.

rated onto the remote node. This option is also used in the American, European, and Chinese markets.

 c. Split-MAC: This approach is a variation of the remote PHY and remote MAC-PHY with the MAC layer functionality split between the cable head end and the remote node, which includes the PHY layer functionality.

 The interface between the head end and the remote node (fiber node) is a digital optical/HFC plant that can be based on any of the following layer 2 technologies that support optical transmission:

 i. Ethernet;

 ii. EPON;

 iii. Gigabit PON (GPON).

The broadband access systems using cable access technologies meet the current requirements of customers but are expected to transition to fully optical networks, with a distributed architecture and communication over the IP (IP/DOCSIS). This move would facilitate the convergence of cable networks with telecom networks.

9.3.2 Optical Fiber Access Networks

The emergence of services like HD video streaming mandates the use of broadband access to residential users. Conventional HD has a resolution of (vertical pixel height) of 720p (1,920 * 720) or 1,080p (1,920 * 1,080). ITU-R Rec.2020 has defined two additional resolutions 3,840 * 2,160 commonly referred to as 4K or ultra HD and 7680 * 4,320 referred to as 8K. The 4K and 8K refers to the factor by which the resolution is increased as compared to HD standards. An internet speed (download) of 5 Mbps is recommended for HD quality transmission while a speed of 25 Mbps is recommended for ultra HD quality streaming [4].

Copper in the last mile restricts the bandwidth available to end users, necessitating the shift to optical fiber networks. The Full Service Access Network (FSAN) working group comprised of over 70 telecom service providers, equipment manufacturers, and test labs works toward defining, standardizing, and driving product and service offerings for deployment of broadband fiber access networks. There are three approaches to deploying fiber-based access networks:

1. Active optical network: AONs use powered devices like multiplexers or routers to manage signal distribution to end users.
2. PON: In contrast to AONs, PONs [5] use passive components like optical splitters to manage signal distribution to end users.

FTTx is a term used to describe a broadband access network-based wholly or partly on optical fibers, using any of the preceding approaches. Table 9.2 lists the variants of an FTTx network. Please note that the terms may be slightly different across various literature; the function, however, remains same.

Table 9.2
FTTX Variants

S.N	Type	Connectivity	Description
1	FTTB	Fiber-to-the-building	CO-to-building connectivity
2	FTTC	Fiber-to-the-curb	Fiber terminated in a street cabinet that is close to the subscriber
3	FTTD	Fiber-to-the-desktop	Fiber termination to a user terminal (within offices)
4	FTTDp	Fiber-to-the-distribution point	Fiber termination to a junction box located near the customer premises (e.g., a junction box provided on each floor of a building)
5	FTTF	Fiber-to-the-factory	Data communication network within factory premises
6	FTTH	Fiber-to-the-home	Deliver triple-play services to residential customers through in-house termination points
7	FTTN	Fiber-to-the-node/neighborhood	Fiber terminated in street cabinet, located at some distance from customer premises, with last mile on copper
8	FTTO	Fiber-to-the-office	Fiber distribution system for enterprise customers, within office
9	FTTP	Fiber-to-the-premises	Similar in concept to FTTH; may include small enterprise customers (can be thought of as a combination of FTTB and FTTH)

Figure 9.2 illustrates the FTTx architecture.

9.3.3 5G Mobile Wireless Networks

The development in the area of wireless technologies, more specifically the impending transition from 4G to 5G networks, would have a dampening effect on the deployment of optical access networks. 5G networks are expected to be deployed starting from 2020. The key features of 5G networks include the following [6]:

- Access bandwidth of 1–100 Gbps—Support for applications like virtual reality (VR);
- Ultra-low latency—Roundtrip delay of less than 1 ms;
- Significantly reduced energy footprint;
- Enhanced system availability[1];
- Support for massive machine communications—IoT .

The ITU-R initiated program "IMT[2] for 2020 and beyond," initiated in 2012, set the stage for the development of the 5G ecosystem. IEEE–SA[3], an

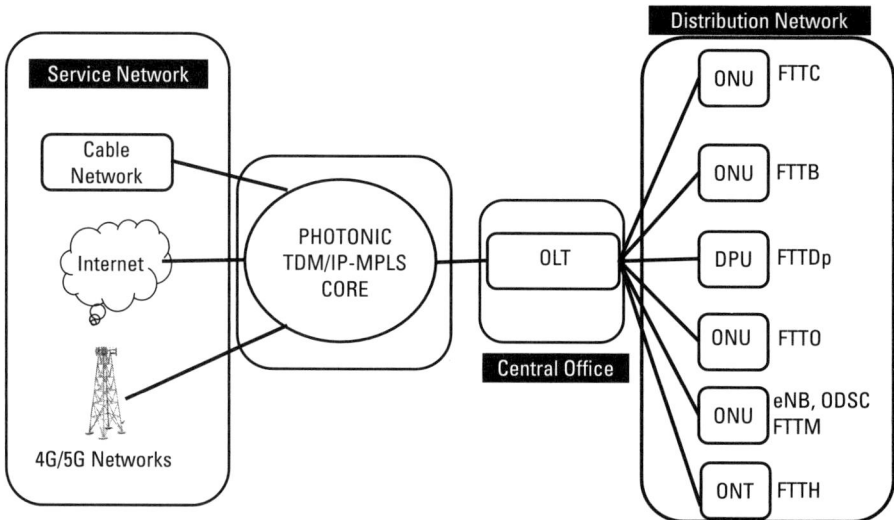

Figure 9.2 FTTx architecture.

1. Perceived 99.999 uptime—also referred to as five nines uptime. Translates to a downtime of five minutes 16 seconds annually.
2. International Mobile Telecommunications.
3. IEEE Standards Association.

open, global collaborative platform for wireless communities, is also engaged in building a platform based on 5G to develop IoT and smart cities [7].

Spectrum

5G services would be offered using three distinct frequency bands as outlined in Table 9.3.

The Asia Pacific Telecommunity (APT), an intergovernmental organization established by the United Nations Economic and Social Commission for Asia and the Pacific (UNESCAP) and the International Telecommunication Union (ITU), in a joint effort with telecom service providers, equipment manufacturers, and R&D institutions, is primarily engaged in the development and application of information and communication technologies (ICT). The APT wireless group (AWG) in its nineteenth meeting in 2015 agreed to explore the following frequency ranges for 5G services [8]:

1. 25.25–25.5 GHz;
2. 31.8–33.4 GHz;
3. 39–47 GHz;
4. 47.2–50.2 GHz;
5. 50.4–52.6 GHz;
6. 66–76 GHz;
7. 81–86 GHz.

The United States is exploring the following frequency ranges for 5G services [9]:

1. 27.5–29.5 GHz;
2. 37–40.5 GHz;
3. 47.2–50.2 GHz;
4. 50.4–52.6 GHz;
5. 59.3–71 GHz.

Table 9.3
Proposed Spectrum Allocation for 5G

S.N	Frequency Range	Description
1	< 1 GHz	Regular usage—urban, suburban and rural coverage, support for IoT
2	1 to 6 GHz	Enhanced capacity and coverage (3.3 to 3.8 GHz)
3	> 6 GHz	Ultrahigh broadband communication
4	30 to 300 GHz	Millimeter waves

The need for new spectrum has been necessitated by the scarcity of the existing bands. The range of frequencies between 30 and 300 GHz, also referred to as millimeter waves, since their wavelength is less than 10 mm, is being considered for supporting the 5G ecosystem. The coverage offered by systems working on millimeter waves is however limited besides facing attenuation by man-made and natural objects and phenomena like rain. A novel approach to get around this drawback is to reduce the cell size and use small cells to augment traditional macro transmitters.

As opposed to the existing mobile communication framework the 5G ecosystem additionally facilitates extensive machine communication with the internet of things (IoT) as the key enabling technology, besides the development and integration of smart cities (including smart offices and residences); new age transport like self-driven cars, personalized unmanned air transport taxis, and automated tubes; and support for exceedingly complex applications. The 5G applications can be broadly classified into the following three types [10]:

1. Extreme mobile broadband (xMBB);
2. Massive machine–type communications (mMTC);
3. Ultra-reliable machine-type communications (uMTC).

This includes new communication paradigms like network-controlled device-to-device (D2D) communication that supports point-to-point, multicast, and broadcast links. ETSI's multiaccess edge computing (MEC) provides cloud-computing capabilities and a service environment at the network edge that facilitates the deployment of cutting-edge services including the following:

1. Augmented reality (AR);
2. Data caching;
3. IoT;
4. Location services;
5. Optimized local content distribution;
6. Video analytics.

The use of cloud computing reduces the CAPEX of network service providers while the deployment of MEC will facilitate the enhanced utilization of the RAN by third-party application service providers thereby enhancing revenues. The emergence of SDN and network function virtualization (NFV) will be a crucial factor in extending cutting-edge services to the last mile.

Hardware Architecture

The architectural framework for a 5G ecosystem would be radically different from the current 4G evolved universal terrestrial radio access network (EU-TRAN). The major changes expected are outlined as follows [11]:

- The current passive RF antennas will be replaced by active antennas[4] with most of the functions (RF up/down conversion) of the remote radio head (RRH)—also referred to as the remote radio unit (RRU)—built into the radome[5]. This process eliminates an additional member in the RAN and facilitates the connection with the DU through an enhanced front-haul connection based on the common packet radio interface (CPRI). The use of AAS would offer the following additional features:
    1. Vertical sectorization;
    2. Virtual sectorization;
    3. 3-D beamforming;
    4. Adaptive beamforming.
- The DU would be a part of the cloud infrastructure (C-RAN) with support for full duplex communication with the user equipment (UE). The cloud-based radio network (C-RAN) architecture would support the creation of interconnected central DU pools based on open platforms with virtualization functions, serving multiple AAS and small cells.
- The antennas would incorporate multiple transceiver ports to support massive MIMO. The number of antennas could range from 128 to 512. The development of patch antennas with low power requirements would facilitate the creation of small cells using public infrastructure. The small cells would ensure the efficient use of the available spectrum. The use of massive MIMO mandates the need for adaptive beamforming wherein an advanced algorithm plots the best transmit path to individual users while reducing interference between parallel transmission streams.
- The backhaul could be based on an IP/MPLS core over fiber with or without DWDM infrastructure. The 5G systems could also witness the emergence of self-backhaul which essentially involves the DU connecting with the backbone over an enhanced air interface (sharing the channel with wireless access). Figure 9.3 illustrates the 5G network architecture.

---

4. Operating at frequencies less than 1 GHz.
5. Protective structure, transparent to radio waves, housing the antenna elements.

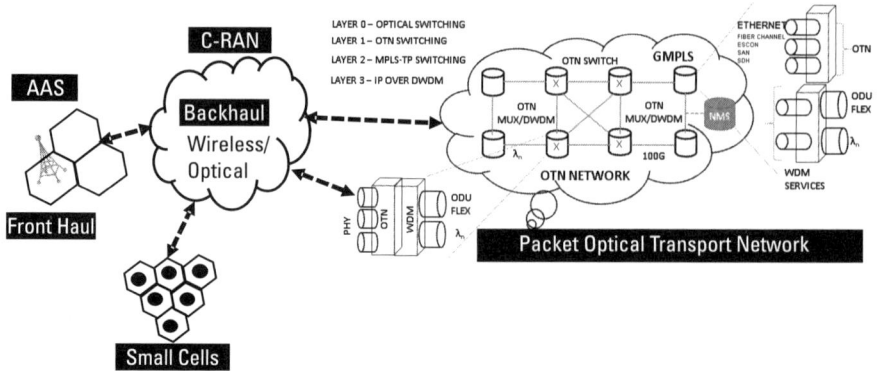

Figure 9.3 Architecture of a 5G network.

Software Architecture

The software architecture of a 5G network is completely different and complex compared to the previous generation of mobile networks. The architectural framework is illustrated in Figure 9.4, summarized in Table 9.4, and described as follows [11].

1. Converged data plane: The 5G network is based on a converged data plane that integrates TDM-based photonic networks and IP/MPLS core over a DWDM infrastructure. The converged data plane enables data transfer to and from a variety of users with the capability of handling multiple conversations through multiple protocols and peer elements over a distributed cloud infrastructure.

2. Software plane: The software plane is responsible for the provisioning and functioning of all the software and the associated services within the network. This includes the physical, logical, and cloud resources

3. Infrastructure control plane: The infrastructure control plane is responsible for the operation of all network devices including the data transmission and reception through the converged data plane. This includes the virtual network functions including mobility control, cloud control, and the mobile-edge computing framework.

4. Application and business service plane: The plane is responsible for the implementation of the service-specific business processes while providing interfaces to the entities of the other planes that need to use the services. A modular architecture facilitates the dynamic creation of additional services using a combination of one or more existing ser-

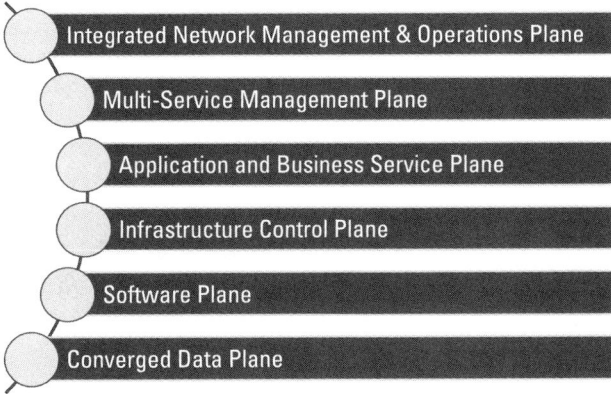

Figure 9.4 Software architecture of 5G systems.

Table 9.4
5G Systems: Software Architecture

S.N	Entity	Description
1	Converged data plane	Next-generation packet OTN
2	Software plane	Software functioning and provisioning
3	Infrastructure control plane	Operation of network nodes and devices
4	Application and business service plane	Implementation of the service-specific business processes and interface (service access) management
5	Multiservice management plane	Setup and management of network instances and nodes
6	Integrated network management and operations plane	Core management functions including fault management, configuration management, accounting, performance management, and security management (FCAPS), network monitoring, information management, segment and domain management

vices. The application layer can use the underlying services to implement various functions.

5. Multiservice management plane: The service management plane is responsible for the setup and management of network instances, for each node in line with the available logical and physical resources. It also implements a set of functions responsible for network management including fault, performance, and configuration management.

6. Integrated network management and operations plane: The core management functions including FCAPS, network monitoring, information management, and segment and domain management.

9.4 FTTx Networks

As indicated in Section 9.2, FTTx is a generic term for an optical access network. For a long time, connectivity to end users was supported by a copper network that was followed by the emergence of wireless networks and interfaces. However the new service offerings and the consequent bandwidth requirements have necessitated the deployment of a high-capacity medium like optical fiber in the access network. After considerable deliberation, cable service operators finally started to transition to a fiber-based access network, and this trend is now consistent globally. As indicated in earlier chapters deploying a fiber network is a capital-intensive activity.

Depending upon requirements, fiber access networks can support point-to-point (P2P) or point-to-multipoint (P2MP) architectures with active as well as passive elements in the distribution networks. Table 9.5 lists some of the popular deployment scenarios.

PONs have emerged as the preferred solution for high-speed access networks, in varied geographies including the United States, Europe, and Asia. The core network has been the focus of service providers as access requirements have been pretty static. The situation has changed in the last decade, due to the increased penetration of converged network infrastructure and wide acceptance of triple-play services. It is now understood that access networks have a significant cost impact due to ever-increasing end-user concentration and their unique service requirements. These developments have contributed to the increased acceptance of optical fiber–based access networks in general and PON in particular.

The initial work done on PON, by the FSAN group, was standardized by ITU-T as G.983. The initial version of G.983 was based on ATM and was

Table 9.5
FTTH Design Approaches

S.N	Link Type	Description	Number of Fibers	Number of Transceivers	Data Rate
1	P2P	P2P link between CO and N end users	N	2 * N	Speed of terminal equipment (OLT)
2	P2P	P2P link between CO and distribution switch and P2P links between switch and end users	1 (between CO and distribution switch), N between distribution switch and end users	2 * (N + 1)	Speed of terminal equipment (OLT)
3	P2MP	P2MP link between CO and N end users	N	2 * N	Speed of terminal equipment (OLT)
4	P2MP	PON in distribution network	1 between CO and distribution network and N between distribution network and end users	N+ 1	Speed of terminal equipment shared between end users

referred as APON. This was replaced with the current version (G.983.5), with significant improvements (service capabilities and enhanced survivability), which is referred to as BPON. BPON typically provides downstream bandwidth of STM-4/OC-12, 622 Mbps, and STM-1/OC-3 or 155 Mbps upstream bandwidth. To scale up the bandwidth to gigabit-per-second levels and meet industry demands, while enhancing transmission efficiency, the ITU-T came out with the G.984 Gigabit-capable Passive Optical Networks (GPON) standard, which supports STM-16/OC-48 rates of 2.488 Gbps in the downstream direction and 1.244 Gbps and 2.488 Gbps in the upstream direction [12]. Table 9.6 lists the popular types of PON.

9.4.1 PON: Architecture and Functioning

PON refers to a set of technologies conceived by the full service access network (FSAN) working group and standardized by ITU and IEEE. PON is a converged infrastructure that can support multiple technologies including POTS and VoIP and data, video, and telemetry information. A PON consists of the following components:

1. OLT: OLT provides an interface with the backbone network and is located at a collector/aggregation location on the service provider's network. The OLT may be a multiplexer in the case of TDM networks, an OADM in the case of WDM networks, or a router in the case of IP networks, located at a central location (referred to as the CO in legacy networks) and connects the distribution network to a SDH/OTN/WDM/IP Core [13].

2. ONU: An ONU is typically located at the edge of the distribution network. It can also be referred to as an ONT and can also be located at the customer premises. It may be noted that the term ONT is defined as per ITU-T standards, the IEEE equivalent is ONU.

3. ODN: The ODN refers to the network that interconnects the OLT and the ONU.

PON uses optical splitters, in a point-to-multipoint architecture, to serve multiple customers using a single optical fiber. PON eliminates the need to lay multiple fibers between a CO and individual customers and does not require any power source (hence the name). Figure 9.5 illustrates the concept of PON and its various flavors.

The passive elements used in PON include the following:

1. Fibers;

Table 9.6
Types of PON

S.N	Abbr.	Type	Standard	Downstream Speed	Upstream Speed	Description
1	APON	ATM PON	ITU-T G.983	155 Mbps	155 Mbps	PON with electrical layer based on ATM
2	BPON	Broadband PON	ITU-T G.983.5	622 Mbps 1.2 Gbps	155 Mbps 622 Mbps	Dynamic bandwidth distribution & protection
3	CPON	Composite PON	–	1 to 10 Gbps	1 Gbps	Supports combination of TDM and WDM technologies (including CWDM), WDM in downstream and TDM in upstream (using a pair of fibers)
4	EPON/ GPON	Ethernet PON	IEEE 802.3ah IEEE 802.3av	1 Gbps 10.3 Gbps	1 Gbps 10.3 Gbps	Triple-play services through single fiber access, based on Ethernet frames
5	DPON	DOCSIS PON	DOCSIS	10 Gbps	1 Gbps	DOCSIS 3.1 over PON
6	GPON	Gigabit PON	ITU-T G.984.1	2.4 Gbps	1.2 Gbps 2.4 Gbps	Carrier grade standards, asymmetric data rates support, supports ATM, GEM and Ethernet, TDM and complete OAM functions supported
7	TWDM-PON	TDM and WDM PON	—	2.5 Gbps 10 Gbps	2.5 Gbps 10 Gbps	Four downstream/upstream wavelengths with tunable laser support, standardized as NG-PON2
8	WPON	WDM PON	Nonstandard	1 to 10 Gbps	1 to 10 Gbps	WDM support, fixed virtual point-to-point bidirectional connections over long distances
9	XG-PON	10-gigabit-PON	ITU-T G.987	10 Gbps	1 Gbps	Coexist with GPON devices on the same network
10	XGS-PON	10-gigabit symmetrical-PON	ITU-T G.9807.1	10 Gbps	2.5 Gbps 10 Gbps	Fixed wavelength allocation and non-tunable optics
11	NG-PON2	40-gigabit-PON	ITU-T G.989	40 Gbps	10 Gbps	Composite time and wavelength-division multiplexed system
12	XG(S)—PON+	40-gigabit-PON	—	40 Gbps	10 Gbps	Use of DWDM, OFDM
13	NG-PON2+	n-Gigabit PON	—	n*40 Gbps	n*10 Gbps	Multiwavelength deployment with wavelength switching
14	(FOAS)*	Convergence of SDN, NFV, 5G, IoT, use of wavelength splitters				

*Future optical access System

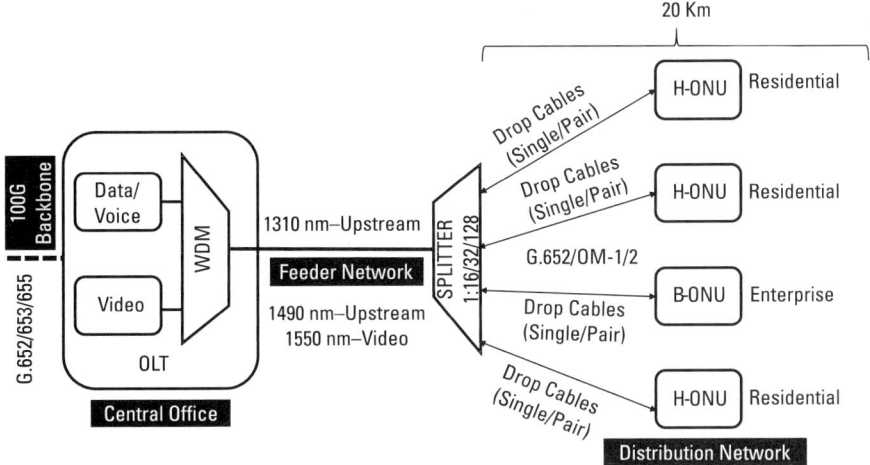

Figure 9.5 PON architecture.

2. Splitters (connectors);
3. Connectors.

The ODN refers to the interconnection of these passive components. In addition the PON consists of the following:

- A common splitting technology, power splitting, involves a power splitter connected to the OLT at the CO, with a feeder fiber and connecting to a number of ONUs via distribution fibers. The number of ports supported is dependent on the type of PON and can range from 16 to 256 ports with a distance ranging from a few kilometers to 40 km and beyond. The splitting technologies include the following:
  - Fused or fused biconical taper (FBT) couplers;
  - Planar splitter or planar lightwave circuits (PLC);
  - Fused coupler arrays;
  - Monolithic fused couplers.

PON provides a high-bandwidth, cost-effective solution for connecting end users to an optical core network. Its primary advantages include the following:

1. Support for longer distances on the distribution network as compared to DSL;
2. Reduced need for amplifiers;

3. Minimized fiber requirement;
4. Reduced CAPEX and OPEX due to the use of passive components;
5. High bandwidth to end customers (dependent on the splitting ratio);
6. Broadcast support;
7. Easy addition of end users.

PONs may be classified as follows:
1. TDM PON;
    a. ATM PON (APON);
    b. Broadband PON (BPON);
    c. Ethernet PON (EPON);
    d. Gigabit PON (GPON);
    e. 10G EPON;
    f. Next-generation PON (NG-PON);
2. WDM PON;
3. OFDM PON.

The APON/BPON, GPON, and NG-PON architectures were standardized by the FSAN, while EPON and 10G-EPON were standardized by the IEEE 802 study group.

9.4.2 WDM Infrastructure Integration

A further option to split/combine optical signals involves another passive device in the form of a WDM. WDM splitters use two wavelengths of either 980 and 1,550 nm; 1,310 and 1,550 nm; or 1,480 and 1,550 nm (common). The typical usage is voice and data operating at 1,490 nm and 1,310 nm, in both directions (downstream and upstream directions) with video signals operating at the 1,550-nm wavelength band. The combined signals are transmitted over a single fiber using WDM and distributed to end users via passive optical splitters.

The basis of NG-PON2 is the combination of TDM and WDM technologies and is referred to as TWDM-PON. TWDM-PON supports four symmetrical wavelengths of 40 Gbps with a provision to add additional wavelengths in the future. TWDM would enable an OLT to serve additional customers with GigE bandwidths. [14]

9.4.3 NG-PONs

The FSAN NG-PON task group is charged with evolving standards to meet the requirements of service providers, developing specifications that build on the current standards, and providing a cost-effective and efficient, scalable access mechanism that can cater to the developments of the future. The specifications from the task group include the following:

1. XG-PON1 (ITU-T G.987 series);
2. NG-PON2 systems (ITU-T G.989 series).

The NG-PONs would be based on OFDM modulation techniques and support 40-Gbps and 100-Gbps interfaces.

9.5 Photonic Core Networks

The backbone networks can be based on the TDM or packet-switched technologies. Each of the technologies has distinct advantages and disadvantages. This section presents a hybrid network architecture likely to be adapted by telecom service providers:

9.5.1 Packet Optical Evolution or the Packet Optical Transport Service

There are several advantages and disadvantages of the three approaches (OTN, IPoDWDM, and IP/MPLS over DWDM) discussed in this chapter for building next-generation core networks. As it stands the optimal architecture would involve a converged infrastructure capable of supporting diverse traffic types while providing the scalability and capacity to accommodate the dynamic traffic requirements required to support new-age applications. Packet optical evolution or the packet optical transport service (P-OTS) is the outcome of this approach.

In the current scenario, the network functionalities are implemented through separate nodes that are a part of the OTN, IP/MPLS, and DWDM networks, respectively. The new nodes, referred to as packet optical networking platforms (PONPs), integrate the functionalities and feature sets of SONET/SDH, WDM, and carrier Ethernet into a single cost-effective network node. The integrated node will lead to process and cost optimization (CAPEX and OPEX) and provide operators with a scalable solution with full OAM&P functionalities to meet their traffic demand in the near future.

MPLS-TP

MPLS-TP is a variant of the MPLS protocol that was jointly proposed by ITU-T (Study Group 15—SG-15) and IETF to support the capabilities and functionalities of a packet transport network. MPLS-TP is a connection-oriented packet-switched (CO-PS) network layer technology, with protocol extensions, designed for use in transport networks.

Provider Backbone Bridge Traffic Engineering (PBB-TE)

PBB-TE is defined by ITU-T G.8110.1/Y.1370.1 and IEEE 802.1Qay-2009. PBB-TE is an adaptation of Ethernet technologies to transport networks. PBB-TE extends the functionality of traffic engineering to transport networks enabling E2E path configuration (source to destination).

9.5.2 IP/MPLS Optical Core Networks

MPLS was originally conceived as a method of enhancing the speed or efficiency of routers in forwarding packets. It has subsequently emerged as a technology standard that provides traffic engineering and virtualization capabilities to large backbone IP networks [15].

Traditionally a router buffers the packets received on its ingress ports, extracts the headers to track the source and destination IP addresses, consults its routing table to find out the egress port, and moves the packet out on the specified interface. This action is repeated at all routers until the packet is delivered to its final destination.

MPLS works by generating a short fixed-length label that contains the forwarding information for the packet on the MPLS network. The IP packets are encapsulated with these labels by the first MPLS encountered on the edge of the network. The subsequent nodes within the network make the forwarding decision by checking the MPLS label as opposed to the IP header. The MPLS label is removed by the egress node (edge router). The nodes or routers that handle the packets are referred to as label-switched routers (LSRs). LSRs can be classified as MPLS edge routers and core LSRs capable of high-bandwidth packet-processing. The packet-routing tables are generated by the IP routing protocols RIP or OSPF. This routing control plane is independent of the MPLS label forwarding plane. This approach facilitates changes in the routing plane without any reconfiguration or impact on the label forwarding plane and vice versa.

As discussed earlier MPLS has two major components, the control plane and the data plane. The control plane, consisting of routing protocols, the routing table (RIB) and the label information base (LIB) is responsible for exchanging layer-3 routing information and labels using a variety of protocols including OSPF, EIGRP, IS-IS, and BGP, in addition to mechanisms for label exchange labels like TDP, LDP, and RSVP. On the other hand the data plane is responsible for the forwarding of the data packets. The primary advantages of an IP/MPLS-based optical core can be summarized as follows:

1. In an MPLS network, only edge routers fully process each packet while packet forwarding is based on label switches. This reduces the latency of the network in comparison with networks based on standard IP routing.
2. Separation of the control and data planes.
3. Enhanced network performance and support for QoS.
4. Traffic engineering capabilities.
5. Separation of customer-specific routing table.

MPLS is an integration of layer-2 and layer-3 technologies. The availability of layer-2 features to layer 3 facilitates traffic engineering. The use of MPLS traffic engineering facilitates establishment of label-switched paths (LSPs) across the core.

The integration of IP and DWDM networks was affected through SDH/SONET NEs with an overlaid DWDM core. This resulted in an additional layer, with its attendant issues of management, OEO conversion at the interface between the IP NEs (routers) and the optical layer. IP over DWDM architecture removes the additional transport layer, resulting in combining router interfaces directly with an optically switched DWDM layer. This streamlines the core architecture enhancing flexibility, supports efficient power control, and enhances reliability in additional to significant savings in CAPEX and OPEX.

IP over DWDM Architecture

The IP over DWDM approach creates a single agile network that integrates a dynamically reconfigurable optical transport layer with tunable optical interfaces. The approach minimizes the OEO conversion and creates a multiservice provisionable platform (MSPP) that provides bandwidth management capabilities at the wavelength level through the use of photonic switching devices. Figure 9.6 illustrates the IP over DWDM architecture:

Enhanced Router Interfaces

There are two interfaces that support the traditional method of transport of IP data over SDH/SONET networks [16], described as follows:

1. LAN-PHY 10 GigE interface: Maps Ethernet frames directly onto a 10 GigE optical circuit. The interface does not support OAM&P functions required for network management.
2. Packet over SDH/SONET (PoS) STM-64/OC-192 interface: The interface maps the IP-layer data with the STM-64/SONET TDM frames. The interface uses the overhead bytes of SDH/SONET to extend OAM&P features.

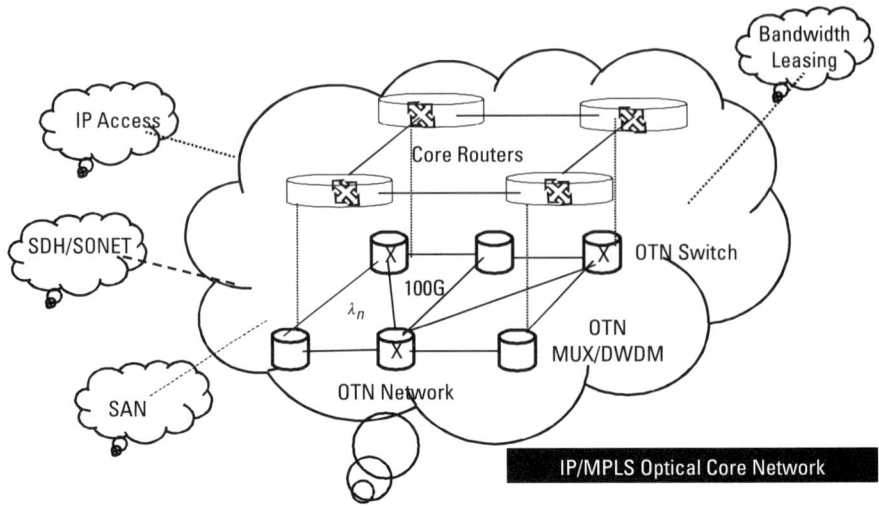

Figure 9.6 IP over DWDM architecture.

In contrast the IP over DWDM architecture uses a new generation of 10 GigE router interface that combines the features of the LAN-PHY interface with the ITU-T G.709 standard facilitating the digital wrapping of Ethernet frames with the added benefit of support for OAM&P function required by network operators. This interface is referred to as WDM-PHY. In addition, the IP over DWDM network model facilitates the integration of the user and the control plane using a common management framework resulting in simplified network management architecture.

9.5.3 NG-POTN: The Network of the Future

The NG-POTN [17] would be based on packet optical evolution or the P-OTS. This architecture proposes OTN as the transport mechanism for packet services with support for MPLS-transport profile (MPLS-TP), provider backbone bridge traffic engineering (PBB-TE), synchronous and connection-oriented Ethernet, or IP/MPLS. The end-to-end circuit management, monitoring, and protection functions, handled by SDH/SONET networks earlier, would be performed by WDM networks. This would result in the elimination of the transport layer resulting in better network efficiency.

P-OTS provides a flexible, scalable, efficient, service-oriented, and converged transparent architecture that can cater to the requirements of TDM and packet/IP-based services and extend the point-to-point wavelength services, while providing protection and restoration functions. The OPEX-optimized network would be capable of remote wavelength configuration, automatic power balancing, and end-to-end performance monitoring over the optical layer. The network architecture is illustrated in Figure 9.7.

The salient features of the proposed network architecture include the following:

1. Generalized MPLS (GMPLS), as specified by RFC 3945, supports multiple types of switching, including the following:
    i. TDM switching;
    ii. Lambda switching;
    iii. Fiber (port) switching.

    GMPLS extends the original architecture to include all LSRs whose forwarding plane cannot recognize cell or packet edge (restricting the ability to forward data based on the information carried in packet or cell headers). This includes NEs in the OTN where switching decisions are based on time slots, wavelengths, or physical ports. These LSRs can be classified as follows:

    a. PSC interfaces are interfaces capable of forwarding data based on contents of packet headers (i.e., interfaces capable of identifying packet boundaries). Examples include routers with interfaces forwarding data based on the IP header and MPLS headers.

    b. Layer-2 switch-capable (L2SC) interfaces are interfaces capable of forwarding data based on contents frame/cell headers (i.e., interfaces capable of identifying frame/cell boundaries). Examples include Ethernet bridges that switch data based on the content of the MAC header and the ATM of nodes.

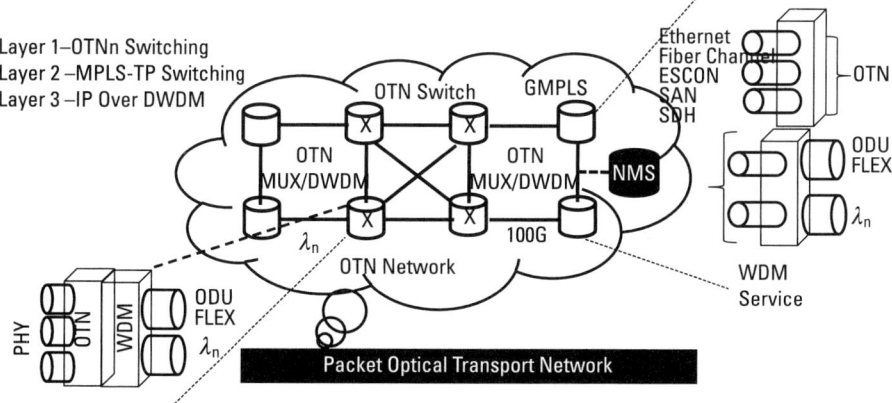

Figure 9.7 Architecture of the next-generation OTN.

c. TDM-capable interfaces include NEs with interfaces capable of switching data on a time slot basis. Examples include SDH/SONET G.707 nodes like DXC, terminal multiplexers (TMs), ADMs and G.709 digital wrapper–enabled OTN nodes.

d. Lambda switch-capable (LSC) interfaces are interfaces are capable of switching data on the basis of individual wavelengths. Examples include OXCs and photonic cross-connects (PXCs) as well as NEs with G.709-compatible interfaces with wavelength capabilities.

e. Fiber switch-capable (FSC) interfaces are interfaces can switch data based on a position of the data in physical space. Examples include OXC or PXC that can support spatial multiplexing with few mode fibers

2. SDM is a radical new approach to meet the exponential bandwidth requirements over fiber in the coming years, in which multiple spatial modes and/or multiple channels create parallel channels for transmission of independent signals. The developments in coherent technology and DSP provide a vastly improved receiver sensitivity and spectral efficiency and a transformation of optical signal to the electrical domain (adding reference carrier) with complete amplitude, phase, and polarization information [18]. The information-carrying capacity of a fiber can be increased by any or a combination of the following approaches:

a. Enhanced modulation schemes—Increase spectral efficiency;

b. Use of 25-GHz or a 12.5-GHz WDM grid—Pack more closely spaced wavelength channels over the fiber;

c. Use of dual polarization;

d. Spatial multiplexing

3. New types of optical fiber are described as follows.

a. Few-mode fibers: MMFs have a higher core diameter that permits the propagation of several independent modes within the same fiber. The numbers of modes are dependent on the diameter of the core and the RI profile of the fiber. The multiple transmission modes in an MMF can be suppressed by modifying the fiber characteristics, allowing only a few modes (typically 10 modes or fewer) to propagate. These fibers, referred to as few-mode fibers, represent a fundamentally new approach to increasing the information density over fiber.

b. HCFs: The HCF is another innovative concept for increasing the information bandwidth over fiber. Hollow fibers consist of a small hollow region (core) surrounded by a cladding (that acts as a grating) that confines the light to the core. The propagation losses in a hollow fiber are less than 0.1 dB/km with a nonlinear coefficient several times lower than a conventional SMF

c. MMFs: It is well known and understood that the available fiber bandwidth is often underutilized due to the limitations of the transmission technologies. The development of 100G interfaces and the imminent release of 400G Ethernet interfaces would help in enhancing the capacity utilization of the fiber. However efforts are being made, in tandem, in other related areas with a view toward enhancing the bandwidth-carrying capacities of the fiber-based backbone networks. One such area is spatial multiplexing, described in the previous section. Multicore fibers (MCFs), as the name suggests, are fibers that contain multiple cores—typically four to eight. The diameter of the cores is normally 8 μm with a cladding diameter of 130 μm. Generally the fiber cables have an acrylic coating and a diameter of 250 μm. MCFs are available in single and multimode types and find significant application in SDM and next-generation PONs.

Figure 9.8 illustrates the important characteristics of the new types of fibers.

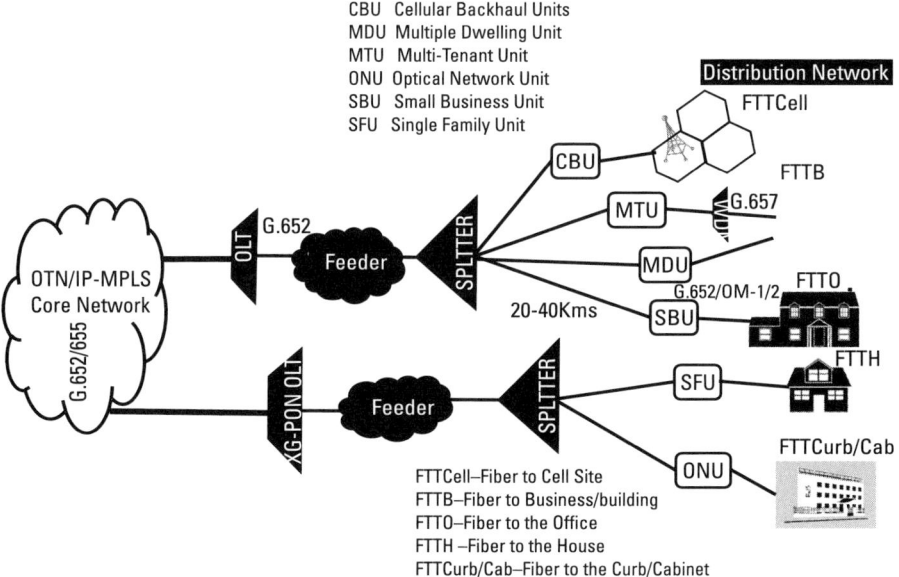

Figure 9.8 Next-generation access network architecture.

4. Agile optical networks: The developments in the DWDM front have led to the development of a new generation of ROADM architectures. This helps carriers provision an agile and cost-effective optical transport layer that supports a new feature set. These include the following:
   a. Transparent interconnectivity among nodes;
   b. Multiservice provisioning (2.5 Gbps, 10 Gbps, 40 Gbps, and 100 Gbps);
   c. Dynamic add/drop configurations;
   d. Support for physical mesh architecture;
   e. Enhanced control plane function;
   f. GMPLS;
   g. Remote establishment, breakdown, and reconfiguration of links.
5. Colorless, directionless, and contentionless ROADM architecture: The new generation ROADMs support the following architectures:
   - PXC-based architecture: The PXCs are also referred to as OXCs or transparent optical cross-connects, as they operate in the optical signal. The incoming optical DWDM signals are demultiplexed and switched by optical switching modules with data rate and protocol transparency.
   - WSS-based hierarchical OXC (HOXC) architecture: A HOXC can handle multiple hierarchical optical paths, wavelength paths, and wave bands (which consists of multiple wavelength paths) to provide a dynamic, scalable, and flexible cost-optimized design that can match new-age service and application demands. A HOXC consists of multiple hierarchy of waveband cross-connects (BXCs) along with wavelength path cross-connects using arrayed waveguide gratings (AWGs), thin-film filters, and concatenated AWGs along with 3-D MEMS[6], PLCs[7] in conjunction with mechanical fiber switches.
   - On-demand configuration: ROADMs facilitate an on-demand configuration with efficient and maximum resource utilization with a simplified network architecture. In addition, they support the following key feature sets:
   - Enhanced flexibility: Extends the flexibility of the mesh network architecture by facilitating the multiplexing of non-ITU-T (colour-

---

6. Manufactured using microfabrication techniques with physical dimensions ranging from less than 1 microns to millimetres.
7. Consists of a planar chip with passive optical fibers and a fiber pigtail.

less) wavelengths onto the DWDM network by the use of transponders at DWDM nodes.
- Nonblocking access: All tunable transponders have nonblocking access to all DWDM ports facilitating on-demand, dynamic network reconfiguration.
- CD&C ROADM: CD&C ROADM is a combination of PXC and HOXC architectures. It consists of nonblocking MEMS-based PXC along with multiport splitters and combiners. CD&C ROADMs support the entire feature set of PXC and HOXC systems and eliminate the need for deployment of WSS systems at local ADM sites, resulting in lower CAPEX.
6. Transparent optical transport with traffic protection: The integration of the IP and optical layers results in an efficient, optically switched DWDM layer that can service diverse traffic types irrespective of frame structures and bit rates. Carriers have the flexibility to use a mix of client signals of different rates with appropriate modulation schemes and the most suitable interfaces. The removal of the transport layers facilitate the provisioning of services like VoIP and wireless applications besides video (triple-play services) that are sensitive to network latencies.

The seamless integration of IP and DWDM layers also facilitates the adoption of IP-protection schemes like fast reroute (FRR) in addition to protection schemes specified by G.709 standards without additional hardware and its attendant costs. This outweighs the disadvantage of SDH/SONET networks, which cannot extend their protection schemes to the IP layer.

9.6 Summary

The term access networks, also referred to as last-mile networks (or first-mile networks in telecom), refers to the portion of the network accessed by the end users for availing themselves of network services. The erstwhile DSL-based services have been replaced by cable modem and/or LTE-based services. The architecture of the access networks will witness a sea change with the transition to all-optical networks and the consequent convergence of technologies—cable, data, and telecom.

DOCSIS is an international standard for data transfer over cable TV systems. It specifies protocols and modulation schemes for bidirectional data transfer over cable. DOCSIS 3.1 supports full duplex communication that provides the advantages of TDD and FDD with interference protection and intelligent scheduling capabilities. A DOCSIS 3.1-based network can provide for 10-Gbps (symmetrical) throughput using 1 GHz over HFC networks with the upstream and downstream traffic sharing the common spectrum simultaneously.

CCAP, a complementary architecture for providing cable access, integrates QAM and cable modem termination system (CMTS) technology into a unified device that supports EPON, EPoC, and DOCSIS. The DOCSIS and CCAP architecture integrates multiple functionalities onto a single node.

FTTx is a term used to describe a broadband access network based wholly or partly on optical fibers, using any of the preceding approaches. Depending upon the requirements fiber access networks can support P2P or P2MP architectures with active as well as passive elements in the distribution networks. PON uses optical splitters, in a P2MP architecture, to serve multiple customers using a single optical fiber. PON eliminates the need to lay multiple fibers between a CO and individual customers and does not require any power source. PON provides a high-bandwidth, cost-effective solution for connecting end users to an optical core network.

The next-generation access (NGA) network's future optical access system (FOAS) will support devices that provide multiple functionalities resulting from the convergence of cable and telecom networks with SDN, NFV 5G, IoT, and DWDM technologies.

However the launch of 5G technologies in 2020 could witness the decline of optical fiber-based access networks. With bandwidth capabilities ranging from 1 to 100 Gbps, 5G-based mobile broadband services would be prevalent in the last mile replacing both the traditional copper-based connectivity and the newer optical PONs.

The NG-OTN would be based on packet optical evolution or the P-OTS service. This architecture proposes OTN as the transport mechanism for packet services with support for MPLS-TP, PBB-TE, synchronous and connection-oriented Ethernet, or IP/MPLS. The end-to-end circuit management, monitoring, and protection functions, handled by SDH/SONET networks earlier, would be performed by WDM networks. This would result in the elimination of the transport layer resulting in better network efficiencies.

9.7 Review

9.7.1 Review Questions

1. The ITU-T recommendation J.222 maps to DOCSIS 3.1 standard.
    a. True
    b. False
2. DOCIS 3.1 supports the following data rates _____.
    a. 10 Gbps downstream/1 Gbps upstream

b. 1 Gbps downstream/10 Gbps upstream

c. 10 Gbps downstream/10 Gbps upstream

d. 1 Gbps downstream/1 Gbps upstream

3. EPON is a layer-2 technology.

    a. True

    b. False

4. Ultra-HD resolutions are defined by ITU-T Rec 2020.

    a. True

    b. False

5. A shared fiber network uses multiplexing technologies to ensure support of multiple users over a single fiber.

    a. True

    b. False

6. AONs use powered devices like multiplexers or routers to manage signal distribution to end users.

    a. True

    b. False

7. _____ is a generic term used to describe a broadband access network based wholly or partly on optical fibers, using multiple approaches.

    a. FTTB

    b. FTTH

    c. FTTN

    d. FTTx

8. Fiber access networks can support P2P or P2MP architectures.

    a. True

    b. False

9. The ODN refers to the interconnection of passive components.

    a. True

    b. False

10. WDM splitters can support any of the following wavelengths—980/1,550 nm, 1,310/1,550 nm, or 1,480/1,550 nm.

    a. True

    b. False

9.7.2 Exercises

1. List the primary differences between EPON and EEPON.
2. Describe TPON and CPON.
3. Discuss the convergence of cable, Ethernet, IP, and DWDM technologies.
4. Define the term P-OTS. How is it relevant in the future?

9.7.3 Research Activities

1. Describe in detail the FOAS.
2. Compare XG(S) PON+ and NGPON2+. Which of the two technologies is more likely to be adapted in the future?
3. List the architectural considerations for building next-generation optical access networks.
4. Discuss the impact of 4G/5G services on the deployment of optical access networks.

9.8 Case Study: Next-Generation Access Networks

ESW, Inc., is a leading telecom service provider in South Asia with triple-play services provided to 29 states, 4,000 cities, and over 640,000 villages using a terrestrial fiber-optic network spanning over 200,000 km. The network caters to over 70 million customers over a multitiered, integrated OTN and DWDM network. The company wishes to strengthen its presence in the optical access space by providing fiber access solutions to enterprises and the retail market.

9.8.1 Challenges

1. Diverse service requirements of its enterprise and retail customers;
2. High-speed access requirements;
3. Existing copper as well as fiber connectivity;
4. Making the network scalable to meet future service and customer requirements;
5. Leveraging existing installed fiber capacity on the access network;

9.8.2 Solution

9.8.2.1 NGAs

The proposed solution is-based on the NGAs. The key benefits of adopting NGA include the following:

1. Provisioning of high-capacity channels with up to 10-Gbps bandwidth;
2. Reduced CAPEX due to increased reach and the consequent reduction in hardware (COs);
3. Support for P2P and P2MP topologies;
4. Capability for integration of wireline and wireless nodes and the deployment of wireless infrastructure in customer premises. This feature can be an important factor in in the transition to 5G ecosystem;
5. Significantly reduced OPEX.

9.8.2.2 Depoloyment of 10-Gigabit-Capable PON (XG-PON) Systems

The XG-PON system is defined by the ITU-T G.987.x series of recommendations. The XG-PON may leverage WDM infrastructure as defined by the ITU-T G.984.x series of recommendations. There are two types of XG-PONs, based on the provisioned upstream bandwidth:

1. XG-PON1—upstream bandwidth—2.5 Gbps/downstream Bandwidth—10 Gbps;
2. XG-PON2—upstream bandwidth—10 Gbps / downstream bandwidth—10 Gbps. XG-PON2, being symmetrical, is also referred to as XGS-PON. Table 9.7 summarizes some of the key features of the XG-PON network.

9.8.3 Network Architecture

Optical Access Network

The local access optical network can be either active or passive with support for P2P and P2MP architectures. This can include the following:

1. FTTH;
2. Fiber to the cell site (FTTCell);
3. FTTB/C;
4. FTTCab.

Table 9.7
X-PON: Key Features

S.N	Parameter	Values	Remarks
1	Optical fiber support	G.652	Support for G.657 bending loss insensitive SMF
2	Wavelength provisioning	Upstream: 1,260 to 1,280 nm Downstream: 1,575 to 1,580 nm	
3	Power budget	16 to 31 dB	Scalable to 35 dB
4	Line rate	10 Gbps	40 Gbps downstream capacity—G.989.1
5	Split ratio	1:64/128/256	
6	Physical transmission length	20 km	
7	Logical transmission reach	60 km	Maximum value

9.8.4 ODN

The ODN (illustrated in Figure 9.8) is common to all the FTTx architectures mentioned above. The difference in the architectures lie in the services provided and the placement of the ONUs. There can be several options for the ODN architecture depending upon backward compatibility and the services to be offered.

9.8.5 Wavelength Provisioning

The XG-PON upstream signals are provisioned over a range of wavelengths between 1,260 and 1,280 nm while the downstream signals are provisioned over a wavelength range of 1,575–1,581 nm, also referred to as the basic band. In addition there are reserved bands also referred to as enhancement bands within the ranges 1,290–1,330 nm, 1,360–1,480nm and 1,480–1,560 nm and another band with the upper limit fixed at 1,625 nm. The wavelength range for video services is fixed by the ITU-T recommendation G.983. 3.

9.9 Referred Standards

BT.2020: Broadcasting service (television)

DOCSIS 1.1: DOCSIS standard for data transfer over cable TV systems

DOCSIS 2.0: DOCSIS standard for data transfer over cable TV systems

DOCSIS 3.0: DOCSIS standard for data transfer over cable TV systems

DOCSIS 3.1: DOCSIS standard for data transfer over cable TV systems

G.983.2 (2002): ONT management and control interface specification for B-PON

G.983.3 (2001): A broadband optical access system with increased service capability by wavelength allocation

G.983.4 (2001): A broadband optical access system with increased service capability using dynamic bandwidth assignment

G.983.5 (2001): A broadband optical access system with enhanced survivability

G.984: Gigabit-capable passive optical networks (GPON): General characteristics

G.984.6: Gigabit-capable passive optical networks (GPON): Reach extension

G.987: 10-Gigabit-capable passive optical network (XG-PON) systems

G.987.1: 10-Gigabit-capable passive optical networks (XG-PON): General requirements

G.987.2: 10-Gigabit-capable passive optical networks (XG-PON): Physical media dependent (PMD) layer specification

G.987.3: 10-Gigabit-capable passive optical networks (XG-PON): Transmission convergence layer specification

G.988: ONU management and control interface specification (OMCI)

G.989: 40-Gigabit-capable passive optical network (NG-PON2) systems

G.989.1: 40-Gigabit-capable passive optical networks (NG-PON2): General requirements

G.989.2: 40-Gigabit-capable passive optical networks 2 (NG-PON2)

G.989.3: 40-Gigabit-capable passive optical networks (NG-PON2)

G983.3: A broadband optical access system with increased service capability by wavelength allocation

G983.4: A broadband optical access system with increased service capability using dynamic bandwidth assignment

IEEE 802.1ag: Local and Metropolitan Area Standards

IEEE 802.3ae: 10-Gigabit Ethernet over fiber; 10GBASE-SR, 10GBASE-LR, 10GBASE-ER, 10GBASE-SW, 10GBASE-LW, 10GBASE-EW

IEEE 802.3ah: Ethernet in the First Mile

IEEE 802.3bs: 400 Gbit/s Ethernet over optical fiber using multiple 25G/50G lanes

IEEE 802.3bz: 2.5 Gigabit and 5 Gigabit Ethernet over Cat-5/Cat-6 twisted pair—2.5GBASE-T and 5GBASE-T

IG983.1: Broadband Optical Access System Based on Passive Optical Networks

ITU-T J.381: Requirements for advanced digital cable transmission technologies

J.112: Transmission systems for interactive cable television services

J.122: Second-generation transmission systems for interactive cable television services—IP cable modems

J.222: Third-generation transmission systems for interactive cable television services: IP cable modems.

9.10 Recommended Reading

9.10.1 Books

Lam, C. F. (ed.), *Passive Optical Networks, Principles, and Practice,* Academic Press, 2007, Ch. 1, p. 117.

Agrawal, G. P., *Fiber-Optic Communications System,* Third Edition, John Wiley & Sons, 2002.

Harstead, E., and P. H. van Heyningen, "Optical Access Networks," in *Optical Fiber Telecommunications,* IV B (ed. by I. Kaminow and T. Li), Academic Press, 2002 pp. 438–513.

9.10.2 URLs

www.nestgroup.net

https://en.wikipedia.org/wiki/IEEE_802.3

http://www.fpnmag.com

www.ftthcouncil.org

https://www.broadband-forum.org/

http://www.cablelabs.com/innovations/featured-technology/

9.10.3 Journals

Kaneko, S., et al., "Agile OLT-protection Method Based on Backup Wavelength and Discovery Process for Resilient WDM/TDM-PON," *Proc. of ECOC (European Conference on Optical Communication) 2014,* Cannes, France, September 2014.

Jackman, N., et al., "Optical Cross Connects for Optical Networking," *Bell Labs Technical Journal,* Vol. 4, No. 1, Jan.–Mar. 1999, pp. 262–282.

References

[1] CableLabs. DOCSIS 3.1 Statement of Compatibility. DOCSIS 3.1 Specifications Document. [Online] http://www.cablelabs.com/wp-content/uploads/specdocs/CM-SP-PHYv3.1-I05-150326.pdf.

[2] ETSI. (2011). Access, Terminals, Transmission and Multiplexing (ATTM); Integrated Broadband Cable and Television Networks; Converged Cable Access Platform Architec-

ture. From etsi.org: http://www.etsi.org/deliver/etsi_tr/101500_101599/101546/01.01.01_60/tr_101546v010101p.pdf.

[3] Sundaresan, K. (2015). From NCTA Technical Papers: www.nctatechnicalpapers.com / Paper/...evolution-of-cmts-ccap-architectures/download

[4] NETFLIX. (n.d.). Internet Connection Speed Recommendations. From NETFLIX: https://help.netflix.com/en/node/306.

[5] Stern, J. R., et al., "Passive Optical Local Networks for Telephony Applications and Beyond," *Electronics Letters,* Vol. 23, 1987, pp. 1255–1257.

[6] Mercer, C. (2017, March 13). From *Techworld:* http://www.techworld.com/personal-tech/what-is-5g-everything-you-need-know-about-5g-3634921/.

[7] http://5g.ieee.org/standards.

[8] Proposal on WRC-15 agenda item 10: A wrc-19 new agenda item for identification of IMT in the frequency band(s) above 6GHz, Document APG15-5/INP-35, 20 July 2015.

[9] http://reboot.fcc.gov/reform/systems/spectrum-dashboard.

[10] *5G Radio Access,* Ericsson White Paper, April 2016.

[11] 5G PPP Architecture Working Group. (2016, July 01). 5G Architecture. From 5G-PPP: Source URL< https://5g-ppp.eu/wp-content/uploads/2014/02/5G-PPP-5G-Architecture-WP-For-public-consultation.pdf>.

[12] Davey, R. P., et al., "DWDM Reach Extension of a GPON to 135 km," *Journal of Lightwave Technology,* Vol. 24, 2006, pp. 29–31.

[13] Senior, J. M., and S. D. Cusworth, "Devices for Wavelength Multiplexing and Demultiplexing," *IEE Proceedings,* Pt. J, 136, 1989, pp. 183–185.

[14] Davey, R. P., et al., "DWDM Reach Extension of a GPON to 135 km," *Journal of Lightwave Technology,* Vol. 24, 2006, pp. 29–31.

[15] Kompella, K., and P. Belotti, "Transport Networks at a Crossroads: The Roles of MPLS and OTN in Multilayer Networks," *Optical Fiber Communications/National Fiber Optic Engineers Conference,* Session OTuG, Anaheim, CA, March 6, 2011.

[16] Melle, S., et al., "Alien Wavelength Transport: An Operational and Economic Analysis," *Optical Fiber Communications/National Fiber Optic Engineers Conference,* Session NThF2, Anaheim, CA, March 22–27, 2009.

[17] Melle, S., D. Perkins and C. Villamizar, "Network Cost Savings from Router Bypass in IP over WDM Core Networks," *Optical Fiber Communications/National Fiber Optic Engineers Conference,* Session NTuD, Anaheim, CA, Feb. 24–28, 2008.

[18] Nokia Siemens Networks. [n.d.]. Space Division Multiplexing: A New Milestone in the Evolution of Fiber Optic Communication. From Modegap: http://modegap.eu/?publication=space-division-multiplexing&wppa_open=yes.

10
Packet-Switched Photonic Networks

10.1 Chapter Objectives

As discussed in most chapters of this book, it is a foregone conclusion that the bandwidth requirements at the access and consequentially the backbone continue to rise exponentially. The ever-increasing teledensity has brought low-cost internet access within the reach of billions of subscribers, globally. The enhanced competition in the converged services market and the resulting cost war has put service providers under tremendous financial strain. Cost effectiveness, efficient bandwidth utilization, and return on investment have become the commonly used mantras in the telecom industry. The scenario is exemplified by the two of the world's largest telecom markets, China and India. These developments have compelled service providers to look beyond the traditional TDM-based SDH/SONET technologies and embrace the ubiquitous packet-switched network model. Packet-switching technologies are most suited for networks that need to respond to traffic dynamically (statistical multiplexing), support multiprotocol traffic, and have the ability to scale up in tandem with an increase in traffic density. The support for statistical multiplexing techniques along with a scalable and flexible multiservice support architecture make it a value proposition for green field operators. Packet-switched networks offer high efficiency and low ownership costs for all services (non-TDM). The emergence of a new class of elements, referred to as packet transport switch, has brought about the integration of packet-switched and TDM architectures, amplifying the advantages of both and heralding a new converged architecture for backbone networks. This chapter presents the traditional and evolving architecture of packet-switched core networks besides the integration of MPLS components. Following Chapter 9's presentation of the various network models considered from the perspective of service providers currently operating TDM networks, this chapter presents the scenarios/perspectives conforming to the viewpoint of service providers operating packet-switched photonic networks. As discussed in Chapter 9, the current developments indicate the development and deployment of the NG-POTNs based on packet optical evolution or the P-OTS. This architecture uses OTN as the transport mechanism for packet services with support for MPLS-TP, PBB-TE, synchronous and connection-oriented Ethernet, or IP/MPLS.

Key Topics

- The business case for packet-switched networks;
- The architecture of packet-switched networks;
- The organization of ISPs;
- Approaches to building next-generation optical networks;
- MPLS;
- Virtual routers;
- Optical switching technologies;
- Generalized MPLS;
- The architecture of packet-switched networks.

10.2 Imperatives

The deployment of communication infrastructure, a highly capital-intensive activity, requires thorough understanding of the network requirements and supported services and a flexible and scalable architecture that can support the dynamic traffic requirements of the complex new-age applications. In addition the network entails a significant OPEX to maintain operational crews and tackle physical failures in the media and equipment that can dent network availability.

It is thus obvious that network operators are primarily concerned with a robust network architecture that helps them to do the following:

1. Enhance ROI;
2. Lower cost per bandwidth;

3. Ensure network availability;
4. Provide high turnaround times for dynamic deployment of services and/or applications in tune with market/customer demands.

The proliferation of the internet and allied technologies, down to the hinterland of developing countries, and the consequent rise of packet-switched data has put the spotlight on the deployment of packet-switching technologies at the core of the converged telecommunication network. The inability of the existing transport technologies, SDH/SONET, to ensure seamless and efficient transport of packet data over a circuit-switched core has also contributed to the movement toward building a packet-switched optical core. In fact it is estimated that nearly 70% of the tier-1 network traffic is packet-based data traffic. The architecture of the transport network, as prevalent, also led to overutilization of certain section of the network (primarily access and portion of the core) while the bulk of the network links were grossly underprovisioned. This lopsided provisioning is also based on the fact that while networks links are designed to support specified throughput, they need to be able to accommodate bursty traffic [1].

At the beginning of the last decade most of the long-haul networks were predominantly circuit-switched; currently, most of the green field networks are based on packet-switching technologies. In a bid to efficiently transport packet-switched data and meet customer demands, network planners added layers to the existing network, leading to a lack of synergies. Packet-switched technologies have advanced significantly over the last couple of decades and are now increasingly being deployed for backbone transport (carrier). The advent of 400-GigE interfaces have enhanced the value proposition, making packet-switched networks one of the preferred technologies of deploying converged photonic networks.

However, TDM-based core still has a place in the emerging next-generation network architecture. In fact, as detailed in Chapter 9, the next-generation architecture will be based on the seamless integration of circuit-switched and packet-switched networks complemented by the DWDM networks. The benefits of the deployment of OTN nodes, especially at the network edge[1], have been well established, and OTN would continue to be deployed in the core. The next-generation network architecture would be simpler (rid of multiple and often redundant layers) and flat, with minimal hierarchical layers with high levels of flexibility and scalability. Packet optical networking platforms (PONPs) would replace the conventional SDH/OTN nodes and provide the mechanism to integrate TDM and packet-switched technologies at the network core.

---

1. Refer to Chapter 9 for additional details.

10.3 Packet-Switched Network Architecture

Packet-switched networks provide a converged infrastructure that supports multiple user services, including voice, data, and video, using different types of access technologies, similar to TDM networks.

Example 10.1

The access technologies may include 4G (LTE), Wi-Fi, FTTH, cable access.

Packet-switched networks provide an E2E architecture that accords a high degree of scalability and flexibility and supports diverse service offerings in tune with business requirements. The design philosophy of a packet-switched network is similar to that of circuit-switched networks with multiple layers that facilitate aggregation and routing capabilities. Figure 10.1 illustrates the typical architecture of a packet-switched network.

10.3.1 Network Elements

In simple terms a packet-switched network is an interconnection of routers and/or switches, similar to the interconnection of multiplexers to form a TDM network. The routers can be classified based on their application context [2] as follows:

1. Access routers: Access routers, also referred to as customer premise equipment (CPE), connect a customer—individual/business to the

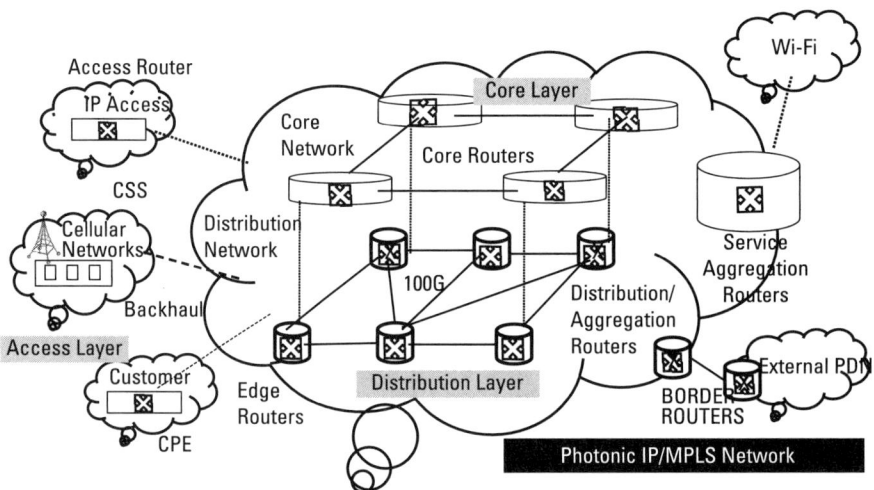

Figure 10.1 Architecture of a packet-switched network.

backbone and/or distribution network. Access networks provide for authentication, route and packet filtering, and/or classification.

Example 10.2—Access Router Specifications
Features/Services
- Flexible routing and switching integration;
- Inter-VLAN routing;
- Multiservice voice and data integration;
- Network analysis;
- Routing (bandwidth management capabilities);
- VPN access (firewall and encryption support);

Interfaces [2]
- Fast Ethernet;
- Gigabit Ethernet.

Protocols
- IP routing;
- IPv6;
- Ethernet;
- ATM.

2. Distribution routers: Distribution routers consolidate traffic from multiple access routers and provide region-specific routing capabilities and enhanced resiliency through physical ring topology and multi-homing and connectivity to backbone routers for interregion routing.

Example 10.3—Distribution Router Specifications
Features/Services
- Cloud-based services;
- Enterprise network function virtualization (NFV);
- HD video;
- Integrated switching;
- Intelligent path selection;
- Native application support;
- VPN;

---

2. Field-upgradable for future-proofing.

- WAN optimization.

Interfaces

- Gigabit Ethernet;
- Wireless WAN (3G/4G-LTE);
- Serial.

Protocols

- Multiprotocol support.

3. Backbone routers: Backbone routers facilitate regional connectivity to backbone networks and provide reachability information for internetwork, intranetwork, and outside world connectivity. Backbone networks provide for enhanced packet-forwarding capabilities.

Example 10.4—Backbone Router Specifications
Services

- Multi-terabit per second switching capacity;
- Scalable IP routing, internet peering, MPLS switching;
- Low-latency data center interconnection;
- P2P Ethernet VLL, multipoint Ethernet VPLS, IP VPN.

Interfaces

- Gigabit Ethernet;
- 10 GigE;
- 100 GigE/400 GigE.

Protocols

- Multiprotocol support;
- Routing protocols—BGP, IS-IS, OSPF, PIM;
- MPLS—LER/LSR;
- Signaling—LDP, RSVP;

4. Service aggregation routers (SARs)s: SARs provide highly scalable service adaptation, routing, and aggregation functionalities over an IP/MPLS infrastructure with support for multiple access protocols suited for a variety of enterprise applications.

Example 10.5—Service Aggregation Router Specifications
Services

- Flexible routing and switching integration;
- Inter-VLAN routing;
- Multiservice voice and data integration;
- Network analysis;
- Routing (bandwidth management capabilities);
- VPN Access (firewall and encryption support).

Interfaces

- Fast Ethernet;
- Gigabit Ethernet.

Protocols

- IP routing;
- IPv6;
- Ethernet;
- ATM;
- Multicast (protocol independent);
- Generic routing encapsulation (GRE).

Routers used for connectivity to an ISP (enterprise connectivity) can also be classified as follows:

1. Edge routers/switches: Edge routers (similar to access routers), also referred to as provider edge routers, are located at the edge of an ISP network to connect to a subscribers or an enterprise network using external border gateway protocol (EBGP). It is preferred to switch at the edge and route at the core.

2. Border routers: Border routers interconnect ISPs using border gateway protocol (BGP). In order to limit the link state updates, the concept of areas was introduced. Routers within an area have a similar link state database. The interconnection of multiple areas to the backbone is facilitated through area border routers. A router that has all its interfaces within an area is referred to as an internal router.

3. Core routers: Core routers are located within an autonomous system (AS) and facilitate traffic flows to/from edge routers. Core routers are similar to backbone routers.

RFC 4098 defines the following classes of routers using BGP:

1. Provider edge router: A router at the edge of the service provider network that connects with other service providers edge routers and/or subscriber edge routers.

2. Subscriber edge router: A router placed at the edge of the subscriber network and connected to the service provider edge routers using eBGP.

3. Interprovider border router: A class of routers operating on BGP that interconnect with routers from other service providers.

4. Core router: A core router is internal to a service provider network and is interfaced with edge routers, intraprovider core routers, or interprovider border routers using the iBGP protocol.

10.3.2 Physical Topology

The different layers of a packet-switched network (see Figure 10.1) include the following:

1. Access layer: The access layer consists of interconnected layer 2 devices—cell site routers (CSRs) and/or cell site switches (CSSs) at the network edge, in a ring topology with optical fiber (and/or microwave links) as the media.

Example 10.6—Access Layer Functionalities
- Address resolution protocol (ARP) inspection;
- Collision domain control;
- Increased network uptime;
- Layer 2 switching;
- QoS classification;
- Security;
- Spanning tree;
- Virtual access control lists (VACLs);
- VLAN support.

2. Aggregation/distribution layer: The aggregation/distribution layer represents the network boundary between the access and the core networks and consists of layer 3 routers and/or switches interconnected in a partial mesh topology.

Example 10.7—Distribution Layer Functionalities.
- Access control policy enforcement;
- Application gateways support;
- Control broadcast and multicast;
- Packet filtering (firewalling);
- QoS;
- Route summarization.

3. Core layer: The core layer, also referred to as the backbone layer, is a fully meshed (core elements)/ partial mesh network that is instrumental in providing high-speed, low latency, and reliable data transport across the network. It also supports connectivity to the internet, extranets, VPNs, and WANs.

Example 10.7—Distribution Layer Functionalities
- High-speed, low-latency links;
- Enhanced interior gateway routing protocol (EIGRP) and open shortest path first (OSPF) routing protocol support;
- Link redundancy;
- Load balancing;
- Traffic prioritization.

10.4 ISPs

ISPs, as the name suggests, provide users with access to the internet along with the related set of services. The different classes and types of ISPs are illustrated in Figure 10.2 and described in Tables 10.1 and 10.2, respectively. In the age of converged services telecommunication service providers also offer ISP services to retail customers and business enterprises. The common services provided by an ISP include mailbox services and internet access.

ISPs also provide the entire gamut of hosting, domain management, and e-commerce platforms and services to smaller ISPs as well as retail customers.

10.5 Design Considerations

As discussed in Section 10.3, the core network can be based on TDM or packet-switching technologies. Historically cost was one of the key factors dictating the choice of a network. TDM-based core networks provided a lower cost per

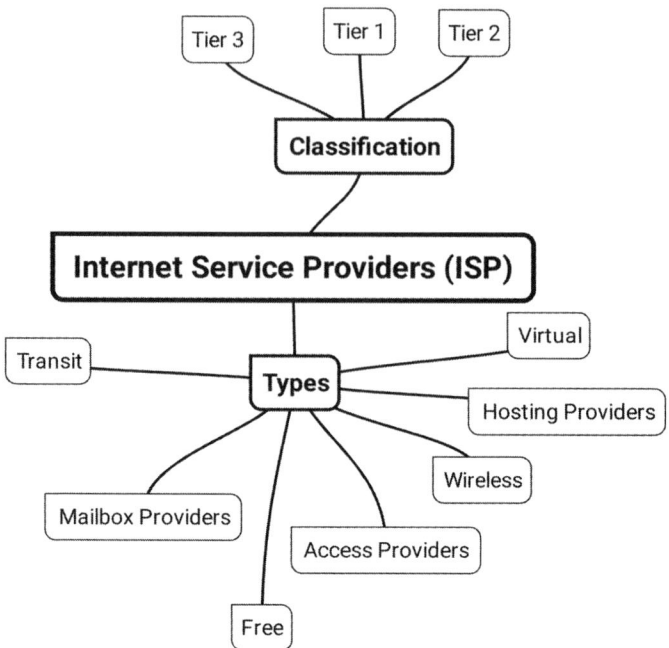

Figure 10.2 Types of ISPs.

Table 10.1
ISP Tiers

S.N	ISP Type	Description
1	Tier 1	Large international/national ISP with direct high bandwidth, reliable connectivity to the internet backbone
2	Tier 2	Regional ISPs (resellers) that buy services from tier-1 ISPs
3	Tier 3	Regional, retail ISPs that buy services from tier-1 ISPs

bandwidth in comparison with the packet-switched networks. The emergence of traffic engineering schemes like MPLS has neutralized the cost benefits of TDM networks. There are several factors that impact the choice of the network architecture. A few important ones include the following [3]:

1. In the current scenario the aggregation layer of the network (packet-switched and TDM) is static (based on the assumptions of the nature of user traffic, ingress points, and perceived traffic density and/or distribution scenarios). The future networks need to be built to

Table 10.2
Types of ISPs

S.N	ISP Type	Description
1	Access	End-user connection (retail/enterprise) (e.g., DSL/ADSL, fiber, cable, wireless, and satellite)
2	Free	Free services to user (with display of advertisements)
3	Hosting	Provides web hosting services—e-mail, FTP, virtual machines using physical servers and/or cloud infrastructure
4	Mailbox	E-mail hosting services
5	Transit	Provides bandwidth for interconnecting access and hosting providers
6	Virtual	Provides access services to end users through purchase or lease from other ISPs
7	Wireless	Provides end user connection through access points (hot spots) using wireless technologies like Wi-Fi

handle diverse services, traffic types, and bandwidth requirements dynamically.

2. The access networks need to be scaled to provide 10-Gbps throughput to cater to the bandwidth requirements demanded by new applications including reality technologies[3]. Service providers are interested in network architectures that provide the lowest cost per bandwidth coupled with management systems that reduce the amount of nonrevenue generating traffic.

3. The network needs to accommodate the backhaul requirements of the mobile broadband networks. As service providers move to LTE release 13 and deploy MIMO technologies, the bandwidth requirements are expected to increase by at least a factor of 10. The imminent launch of the 5G services and the resultant emergence of IoT and reality technologies is expected to further fuel the demand. Network planners need to understand the nature and the unique requirements (especially bandwidth and latency) of these emergent services and develop an architecture that facilitates the integration of the mobile broadband networks.

There are three strategies that are currently[4] being employed by network planners in their quest to build a network that can meet/exceed the burgeoning bandwidth and stringent latency requirements of emergent applications:

---

3. Discussed in detail in Chapter 11.
4. This chapter reflects the viewpoint of service providers operating packet-switched optical networks. As discussed in Chapter 9 the next-generation optical networks would be based on P-OTS.

1. Augmenting existing packet optical core: The simplest option is to enhance the packet-switched core by integrating additional high-capacity routers (IP) within the existing topology. This action would augment the bandwidth capability of the core, with the least time and effort, to meet the bandwidth requirements of the edge networks (increases due to additional users, new services/application, and/or expansion of access networks. However the approach entails high CAPEX, in comparison to the CAPEX required to bring about a similar increase in TDM networks. However the OPEX could be expected to be within predictable limits. Another drawback is that, while this option takes care of the short-term requirements, it limits the flexibility/scalability of the network to meet the dynamic requirements of the application in the future. The high number of layers, catering to different types/classes of services, makes convergence challenging, enhances latency, and increases power and space requirements.

2. Using a TDM core: This option envisages TDM-based OTN in the core with routing and its associated complexities at the network edge. However the P2P static connections at the core reduce network flexibility and scalability. In addition this architecture increases the complexity of network planning and OAM&P, due to the different types of nodes involved at the core and the edge. The OTN network does not support dynamic traffic provisioning and hence has to be provisioned for specific traffic capacities. Peak provisioning may result in bandwidth utilization, while lower capacity provisioning will not support bursty traffic.

3. Multilayered networks: This architecture is-based on the age-old adage "switch at the edge, route at the core." This architecture envisages the need for multiple layers that help in bypassing the core routers, whenever possible, for all transit traffic. However the flip side is that the network complexity is higher, that multiple management systems need to be used, and that the traffic provisioning is static for the OTN section. This model is suitable for networks with a high amount of transit traffic.

10.6 MPLS

A router typically buffers the incoming packet stream, extracts the header (source and destination IP), performs a routing table lookup, and maps the packet to the port servicing the destination address. This process is repeated at all the intermediate routers in the network path leading to inefficient and time consuming packet transport. MPLS, an IETF technology, is a switching technology that provides a method to regulate data traffic and provide a simplified packet forwarding mechanism with multiprotocol support for large and complex networks. MPLS is a scalable protocol-independent mechanism wherein data packets are analyzed at the ingress of the MPLS network and assigned a label, which is subsequently used by all the intermediate routers in the MPLS chain, to make forwarding decisions [4].

MPLS provides a scalable, protocol-independent transport that facilitates the creation of E2E circuits over any transport mechanism using a variety of layer 3 and layer 2 (L2) protocols ranging from ATM, FR, Ethernet, token ring, FDDI, PPP, IPX, Apple Talk, SDH, and SONET. MPLS operates in between the data link layer[5] (layer 2) and network layer (layer 3) and hence is also referred to as the layer 2.5 protocol. MPLS thus eliminates the need for multiple layer 2 networks for handling different traffic types and provides unified mechanism for supporting circuit- and packet-switched clients. Figure 10.3 illustrates the MPLS label and its field, while Figure 10.4 presents a simplified view of an MPLS network. The composition of the MPLS label is listed and described in Table 10.3 while the physical and logical components of an MPLS network are listed and described in Table 10.4.

10.6.1 Packet Forwarding Through the MPLS Domain

1. The packets are forwarded from one MPLS enabled router to another, through a predetermined path known as LSPs.

2. The router (MPLS-enabled) at the headend (ingress) of the LSP is known as the ingress LSR while the router at the tailend (egress) is referred to as the egress LSR.

3. There may be zero or multiple intermediate routers in an LSP. They are referred to as transit LSRs.

4. A MPLS-enabled router can function as an ingress LSR, transit LSR, and egress LSR in an LSP.

5. In addition a router can simultaneously function as an ingress LSR, transit LSR, and egress LSR for other LSP.

LABEL	CoS	S	TTL

Figure 10.3 MPLS label format.

---

5. OSI seven-layer model.

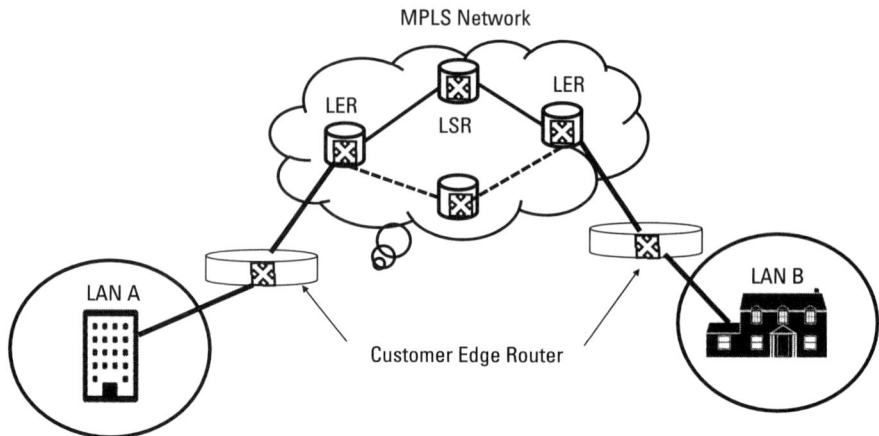

Figure 10.4 Architecture of an MPLS network.

Table 10.3
MPLS Label Description

S.N	Label Field	Bits	Description
1	Label	20	Used for FEC classification. The label value is an integer in the range 16–1,048,575 with values 0–15 reserved by IETF. For signaled LSP the label values are dynamically assigned by the device in the range 1024–499999.
2	CoS (experimental (EXP) usage field)	3	Class-of-service—The three bits define eight distinct classes of service that prioritize the packets traveling through the LSP.
3	S	1	Bottom-of-the-stack—A value of 0 indicates that the label is the bottom of the stack while a value of 1 indicates that the label stack implementation is in use. There can be more than one labels applied to packet.
4	TTL	8	Time-to-live (TTL)—The IP packet TTL is copied onto this field by the ingress LER and subsequently decremented by 1 at each transit LSR hop. The packet is discarded when the TTL value becomes 0.

The packet-forwarding process in an MPLS domain is illustrated in Figure 10.5.

10.6.2 Label Switching Process

The label switching process in an MPLS network is detailed as follows:

1. Ingress LER—label push:
   a. Packet arrives at the ingress of the MPLS-enabled router (LER);

Table 10.4
MPLS Elements (Physical and Logical)

S.N	Element	Description
1	Ingress Label edge router (LER)	Positioned at the edge of the MPLS cloud, receives source packets, applies label (PUSH), sets LSP, initiates packet forwarding through MPLS domain
2	Label switch router (LSR)	Intermediate node that receives labeled packets and performs label removal (POP), label addition (PUSH) or label swap, traffic segregation as per forward equivalence class, and forwarding of packets to appropriate interface
3	Egress LER	Positioned at the edge of the MPLS cloud, acting as the exit point for the packet and performing label removal (POP) prior to packet forwarding to destination
4	Label-switched path (LSP)	Path traversed by a packet from the source to its destination over an MPLS network, through intermediate routers
5	Forward equivalence class	Cluster of packets with similar service requirements (specific application) and common path (destination)

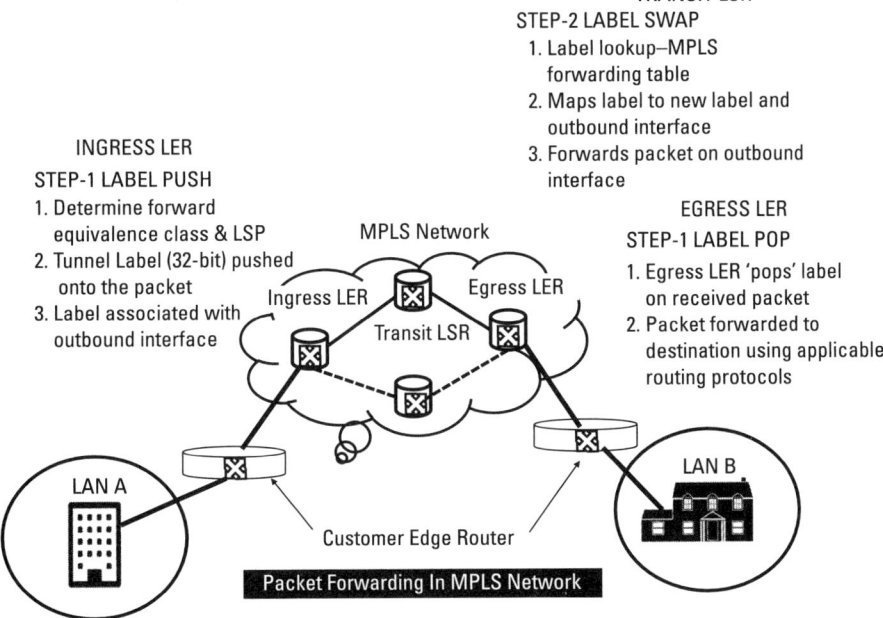

Figure 10.5 Packet-forwarding process in an MPLS network.

b. Router determines the forward equivalence class and LSP (if available).

  i. Forward equivalence class is a group of packets that are to be forwarded in a similar fashion.

  ii. Example—Packets from a given LAN/VLAN/VLL.

iii. Forward equivalence classes are mapped to LSPs (packets from a forward equivalence class with an LSP mapping are assigned to the LSP).
   c. A tunnel label (32-bit) is pushed onto the packet.
   d. Label is associated with an outbound interface on the ingress LER. The packet is forwarded to the transit LSR or egress LER (as the case may be) on the specified outbound interface.
   e. Packet forwarding by the transit LSR on the basis of this label (IP header need not be examined).
2. Transit LSR—label swap:
   a. On receipt of an MPLS packet, the transit LSR looks up the label on the MPLS forwarding table.
   b. The MPLS forwarding table maps the label and the inbound interface to a new label and outbound interface.
   c. The transit LSR swaps the received label with a new label and forwards the packet on the specified outbound interface.

Example 10.8—MPLS Forwarding Table

Table 10.5 presents a sample MPLS forwarding table.

3. Egress LSR—label pop:
   a. The egress LER pops the label on the received packet.
   b. The packet is forwarded to the destination using the applicable routing protocols.

10.6.3 LSP Types

1. Static LSP: Static LSPs are similar to static routes. They have to be manually configured (explicit configuration) at every LSR in the LSP. The national advantage in using a static LSP is the reduced load on the LSR. However network changes will require manual reconfiguration.

Table 10.5
Sample MPLS Forwarding Table

Inbound Interface	Label	Outbound Interface	Label
3/1	367	4/2	437

2. Signaled LSPs:
   a. Signaled LSPs are dynamically configured, using the RSVP protocol, at the ingress LER.
   b. RSVP signaling messages are used for resource reservation, at each LSR in the LSP, and to create dynamic association of labels and interfaces.
   c. A packet assigned to a signaled LSP follows a pre-established path from the ingress LSR to the egress LSR.
   d. The path chosen could be one of the following:
      i. Traversing a set of explicitly configured routers;
      ii. Shortest path (over the MPLS network) determined on the basis of local routing tables by IGP;
      iii. Traffic-engineered path (based on constraints—topology, administrative groups, and bandwidth reservations).

10.6.4 Penultimate Hop Popping

1. At the egress LER a dual lookup has to be performed on the MPLS forwarding table and the IP forwarding table to exit the incoming packets.
2. This process can reduce the forwarding efficiency of the egress LER.
3. In order to maintain the forwarding efficiency of the egress LER, the MPLS label popped at the LSR that is immediately prior (penultimate) in the LSP. This process is referred to as penultimate hop popping.
4. The packet is forwarded to the egress LER without any MPLS label attached.
5. The egress LER processes the packet assuming that it is not a part of the MPLS domain.

Example 10.9—Penultimate Hop Popping
Table 10.6 is a sample MPLS forwarding table.

10.6.5 Traffic Engineering

Traffic engineering (TE) helps in ensuring efficient use of the available network resources and reducing bottlenecks in the network. MPLS facilitates the creation of traffic-engineered LSPs. The process involves the following:

Table 10.6
Sample MPLS Forwarding Table (Penultimate Hop Popping)

Inbound Interface	Label	Outbound Interface	Label
3/1	367	4/2	

1. Gathering network topology information;
2. Charting optimal network paths;
3. Setting up and maintaining identified paths.

Figure 10.6 illustrates the TE process in an MPLS network.

10.6.5.1 TE Database

The traffic engineering database (TED) is used to store the MPLS domain topology information. This information is garnered from IS-IS domain and OSPF areas using TE [4]. The TED includes include constraints (TE), bandwidth reservations, administrative group memberships, and MPLS enabled interfaces. Some of the important parameters include the following:

1. Destination address of the egress LSR;
2. Path to be used (explicit);
3. LSP bandwidth requirements;
4. LSP priority;
5. LSP metrics;
6. Links belonging to specific administrative groups.

10.6.5.2 Traffic-Engineered Paths

As discussed earlier LSP definition can be static or dynamic (signaled LSP). A signaled LSP uses the constrained shortest path first (CSPF) at the ingress LER to dynamically calculate a traffic-engineered path between the ingress and egress LER. CSPF is an advanced version of the IGP routing protocols based on the shortest path first (SPF) algorithm [4]. The CSPF process is detailed as follows.

1. The CSPF process (on the ingress router) uses the user-specified path information (static LSP) or the information in the TED (signaled LSP) to compute a traffic-engineered path through the MPLS network.

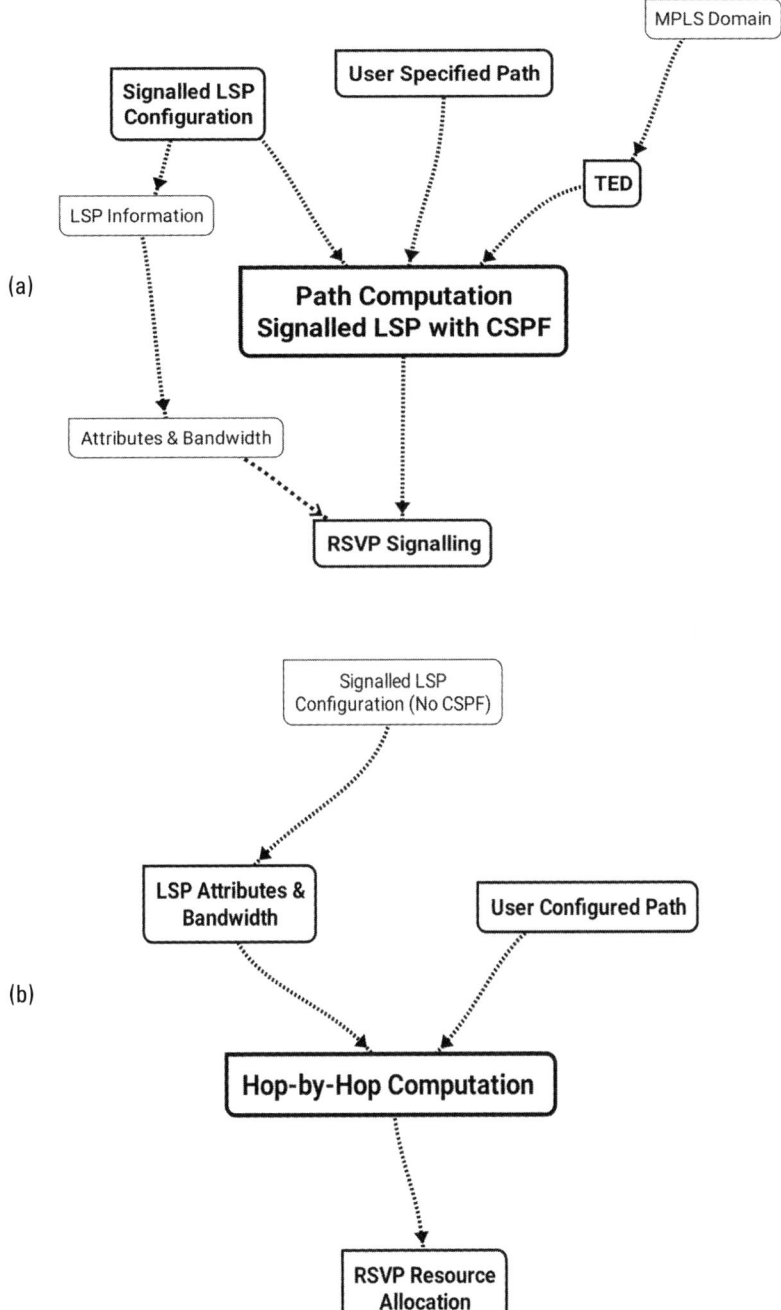

Figure 10.6 TE in an MPLS network.

2. It prepares a sequential list of interfaces that packets assigned to an LSP must take from the ingress to the egress router.

3. The process takes into account resource availability, network topology, and any user-specified constraints to come out with an optimal traffic-engineered path.

4. It may be noted that the CSPF-computed network path may be different from the SPF computed by the standard IGP routing protocols.

5. CSPF is enabled by default on signaled LSPs, but can be disabled (without CSPF, signaled LSPs can span multiple IS-IS levels and/or OSPF areas). If disabled, the shortest path is computed by the standard hop-by-hop routing mechanisms.

6. Following the path computation, resource reservation is done by the RSVP protocol and labels are assigned to the LSR.

10.6.5.3 CSPF Functioning

1. Path computation (using TED, LSP attributes, and requirements): LSP setup priority and bandwidth requirements are analyzed to select an LSP (in case of multiple LSP) for path computation. In case of similar priorities, path computation for the LSP with higher bandwidth requirement is initiated.

2. Removal of unsuitable links: The TED information is parsed by the MPLS device and eliminated from traffic-engineered paths based on the following criteria:
    a. Half-duplex links;
    b. Insufficient bandwidth;
    c. Administrative group mismatch.
        i. An include link statement in the LSP does not have a corresponding administrative group.
        ii. An exclude link statement (link belongs to an administrative group specified in the exclude group and/or does not belong to any administrative group.

3. Shortest path computation (for the MPLS domain):
    a. MPLS device computes the shortest path between the ingress and the egress routers, dynamically. The path can be comprised of a maximum of 255 nodes.
    b. The device computes the shortest path between specified nodes in case of explicit path assignments.

4. Path selection (in case of equal cost multiple paths):
   a. Path selected wherein the last node is the physical address of the destination interface;
   b. In case of multiple path options, selection based on hop count (least);
   c. In case there are multiple options still available, path selection is done on a random basis;
   d. Users have the option to choose a path with either of the following:
      i. Highest bandwidth availability;
      ii. Least bandwidth availability.

The CSPF process results in a sequential list of interfaces assigned to an LSP to enable packets to the egress LER. Subsequently RSVP signaling will establish the LSP on each LSR in the path [4].

10.6.5.4 RSVP Functioning

TE extensions are defined for RSVP. These extensions, as defined by RFC 3029, include the following [4]:

1. EXPLICIT_ROUTE;
2. LABEL;
3. LABEL_REQUEST;
4. RECORD_ROUTE.

The reservation process, described in Figure 10.7, is as outlined.

1. RSVP path message (ingress LER to egress LER):
   a. Admission control (checks LSP setup priority and mean rate settings) on the outbound interface (to ensure sufficient bandwidth availability on the interface);
   b. Path message consists of CSPF-engineered path (specified as an EXPLICIT_ROUTE Object (ERO);
   c. Message includes the following descriptions:
      i. Traffic type;
      ii. Bandwidth.

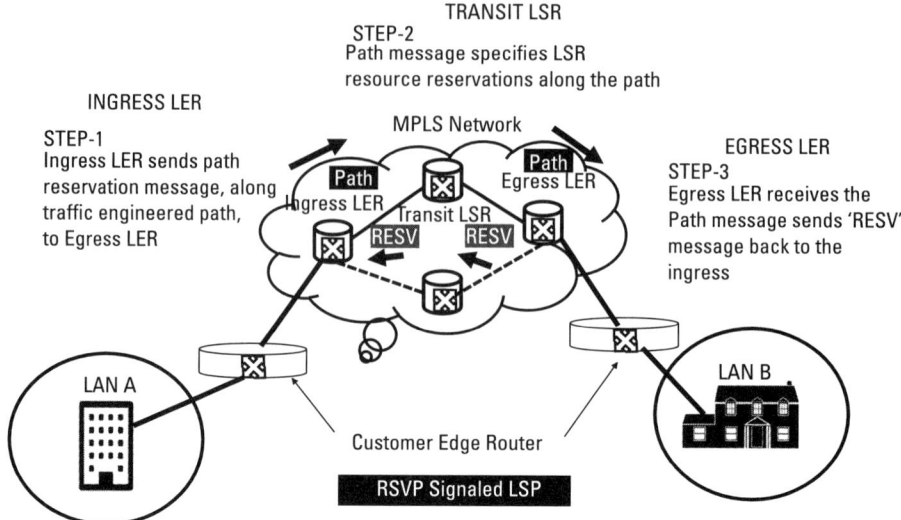

Figure 10.7 RSVP-signaled LSP.

    d. LABEL_REQUEST object for label allocation on the LSR and to request the egress LER to include a LABEL object in the reservation message send to the ingress LER.
2. RESV message:
    a. On clearing admission control ingress LER initiates path message to the address at the top of the ERO list (physical interface of the next LSR in the path).
    b. LSR performs admission control to ensure bandwidth availability on the outbound interface.
    c. The LSR removes its address from the top of the ERO list (on l=clearing admission control) and initiates path message to the new address (currently on top of the ERO list).
    d. The process continues until the last node in the ERO list (egress LER).
    e. The egress LER receives the path message and transmits a RESV message to the ingress LER (reverse path).
    f. The RESV message includes the LABEL object (in response to the original LABEL_REQUEST object in the path message).
    g. The LABEL object is used to associate labels with interfaces that are a component of the LSP, on the LER
3. Resource reservation:

a. Resource reservation is initiated as RESV messages travel upstream (from egress LER to ingress LER).

b. On receipt of RESV messages, the LSR performs admission control (again), on the interface which received the message (outbound interface for packets flowing through the LSP), culminating on bandwidth allocation (on clearing admission control).

c. Bandwidth is allocated on the basis of LSP mean rate setting.

d. LABEL object (in the RESV message) used to associate labels with interfaces (within LSRs MPLS forwarding table).

10.6.5.5 P2MP TE

P2MP is a key feature that facilitates forwarding of packets (information) to multiple destinations from a single source, along an MPLS (optimized) path. This allows the efficient transport of multicast data while enabling forwarding of information from a single source to multiple destinations along an optimized MPLS path. The P2MP feature is ideal for efficiently (with optimal bandwidth utilization) transporting multicast data traffic leveraging MPLS links. The following RFCs are supported [4]:

1. RFC 2205: Resource reservation protocol (RSVP)—Version 1 Functional Specification;
2. RFC 2961: RSVP Refresh Overhead Reduction Extensions;
3. RFC 3209: RSVP-TE: Extensions to RSVP for LSP Tunnels;
4. RFC 4461: Signaling Requirements for Point-to-Multipoint Traffic-Engineered MPLS Label Switched Paths (LSPs);
5. RFC 4875: Extensions to Resource Reservation Protocol—Traffic Engineering (RSVP-TE) for Point-to-Multipoint TE Label Switched Paths (LSPs).

The major advantages of configuring P2MP, illustrated in Figure 10.8, include the following:

1. Optimal network bandwidth utilization;
2. Network capacity improvement;
3. Efficient transport of unicast, multicast, and broadcast packets.

10.6.5.6 Label Distribution Protocol

When used to create tunnel LSPs, LDP allows a set of destination IP prefixes (known as a forwarding equivalence class) to be associated with an LSP. Each

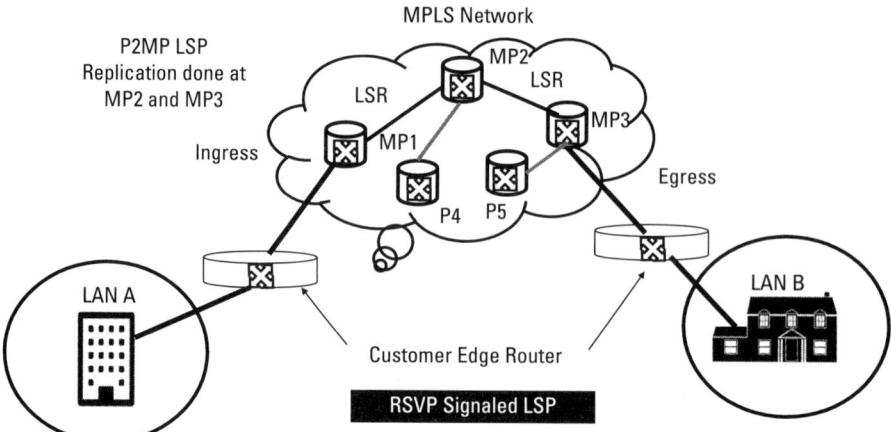

Figure 10.8 P2MP architecture.

LSR establishes a peer relationship with its neighboring LDP-enabled routers and exchanges label-mapping information [4]. This label-mapping information is stored in an LDP database on each LSR. When an LSR determines that one of its peers is the next hop for a forward equivalence class, it uses the label-mapping information from the peer to set up an LSP that is associated with the forward equivalence class. It then sends label-mapping information to its upstream peers, allowing the LSP to extend across the MPLS network. The devices advertise their loopback addresses to their LDP peers as a 32-bit prefix-type forward equivalence class. When an LSR installs a label for a forward equivalence class, it also creates an MPLS tunnel route, which is then made available to routing applications. This allows each router to potentially be an ingress LER for an LSP whose destination is the device's loopback address. The result of an LDP configuration is a full mesh of LSPs in an MPLS network, with each LDP-enabled router a potential ingress, transit, or egress LSR, depending on the destination. The system supports LDP for the configuration of non-traffic-engineered tunnel LSPs in an MPLS network. LDP is described in RFC 3036.

10.7 IP over MPLS

In the normal scenario the BGP builds a routing table using IP routes. In an MPLS network, BGP must consider MPLS tunnels as feasible routes, when used to propagate routes. A feature, referred to as BGP shortcuts, mandates the BGP to use an MPLS tunnel, when available, as a preferred route to the destination. Additionally BGP can be forced to use LSP metrics for best-route computations (user-configured). Destination routes through other edge routers are used by BGP for route configuration, when configured on the LER. The

BGP shortcut feature gives directions to BGP to include an MPLS tunnel as a preferred route in the routing table, on determining that a route through an edge router is reachable through the MPLS tunnel. The BGP shortcut feature can be enabled globally[6] along with an option to include LSP metrics. A device, with the feature enabled, attempts route resolution using the MPLS tunnel. In case of failure the device defaults to its IPv4 routing table [4]. Three key algorithms and the corresponding context specific system behaviors are described as follows:

1. Next-hop MPLS-disabled: Only IP routing tables are used to resolve routes.
2. Next-hop MPLS-enabled:
    a. Route resolution by LSP with a fixed metric of one.
    b. Routing table is used by the system for unresolved routes.
3. Next-hop MPLS With LSP metric consideration:
    a. The LSP metric is used as the IGP cost for the next hop, by BGP, for route resolution.
    b. Only paths through an LSP are chosen, if available, by BGP for next hop resolution.
    c. Paths with the lowest IGP (by comparison of IGP cost of each next hop) cost values are considered for equal-cost multipath routing (ECMP).

BGP uses the default decision process and native IP forwarding to build ECMP routes when next-hop MPLS is disabled.

10.7.1 Next-Hop MPLS

When next-hop MPLS is enabled on a device, BGP initially determines if an LSP can be used for route resolution, for each unique BGP next hop. The process is described as follows.

1. The native IP routing table is not checked in case BGP can resolve routes.
2. All LSPs with the lowest metric values (same) are selected for each BGP next hop, for all routes resolved by LSP.

---

6. Device-specific feature.

3. BGP sets the next-hop IGP cost to one (overriding the actual LSP metrics) in order to force it to be the preferred hop (as opposed to the hop resolved by native IP).
4. The IGP cost is compared, for each BGP next hop, and the least value IGP cost [for the next hop(s)] are entered into the routing table.

The MPLS service basket includes the following:

1. IP/MPLS;
2. Layer 3 VPN;
3. Layer 2 VPN;
4. Virtual private wire service (VPWS) (pseudowire)[7];
5. Virtual private LAN service VPLS);
6. Hierarchical VPLS;
7. MPLS-TP (transport profile).

10.7.2 Next-Hop MPLS: Comparing LSP Metrics

BGP resolves next hop using the LSP metric as the IGP cost for the next hop if the compare LSP metric feature is enabled. Subsequently only the IGP cost paths with the lowest values are considered for ECMP. The option to use a native IP path over an LSP path (in case of different BGP next hop) is user-configurable.

10.8 Virtual Routing

The stringent demands of the new-age services and applications have forced network planners and equipment vendors to alter their strategic outlook, honed with their experience in designing, deploying, and maintaining largely static networks and bringing about radical architectural improvements. Some of the key changes include the following [5]:

- Automated FCAPS support;
- Enhanced performance;
- High QoE;
- NFV;
- Reduced total cost of ownership (TCO);

---

7. Layer 2 P2P emulation over PSN.

- Service segmentation;
- SDN.

Virtual routing and forwarding (VRF) refers to a technology that facilitates multiple independent instances of the routing table to coexist on a router, at the same time.

The IP layer processing requires high CPU bandwidth. This requirement can be reduced if some of the processing can be done at the lower layers. Virtual routing is based on this concept. A virtual router combines the functionality of a MPLS edge router and an SDH/OTN ADM along with the versatility of a wavelength selector switch (WSS). It provides the capability to move packets from the ingress to the egress, thereby bypassing the IP layer (pass-through/cut-through), using a combination of the following:

1. Wavelength switching (lambda switching) on WDM/OTN/SDH/SONET nodes at layer 1,
2. Label switching at layer 2.

The key features include the following:

1. Carrier class services;
2. Cost/resource optimization;
3. Deployment/leveraging of cloud infrastructure;
4. Distributed control plane architecture with multiple dedicate processors (hot standby—hitless switchover);
5. Enhanced data plane—use of multiple virtual forwarders;
6. Enhanced service management;
    a. Rapid service deployment and automated FCAPS support;
    b. Routing configured as service with configurable CoS.
7. High network availability;
8. Increased agility;
9. Reduction of human-induced errors associated with manual configuration and provisioning;
10. Reduced latency (pass-through/cut-through);
11. Simplified network architecture;
12. Use of common platforms and standard management interfaces to manage multivendor network leading to cost optimization and lower CAPEX;

13. Virtual redundant backplane—control and data interconnection planes over cloud infrastructure.

The use of TE methodologies facilitates manual and/or automatic optimization, congestion management, and load balancing resulting in enhanced utilization and dynamic management of network paths, as opposed to the shortest-path methodology adopted by traditional routing protocols.

10.9 Optical Switching

It is evident that optical-switching technologies are mandatory to meet the bandwidth requirements of the future. Fortunately, optical-switching technologies have reached a fair level of maturity, and the vision of a fully optical network should be realized in the near future. In general there are three types of switching—circuit, packet, and message [6]. This section discusses the switching architecture and fabrics employed in circuit switching and packet-switched nodes with the focus on packet switching. Figure 10.9 illustrates the various types of optical-switching technologies. As discussed at several points in this chapter, optical switching eliminates the OEO conversions resulting in switching efficiencies, reduced latency, and OPEX.

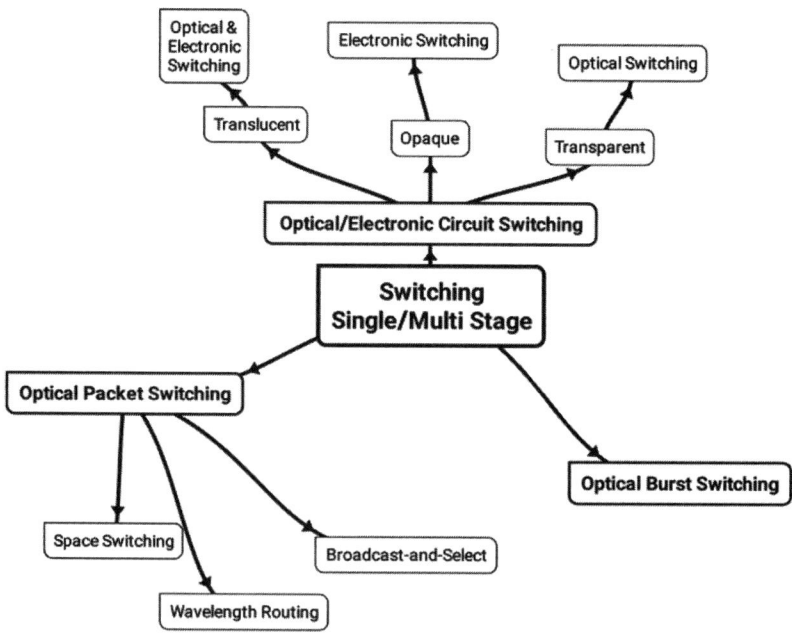

Figure 10.9 Switching techniques.

10.9.1 Optical/Electronic Circuit Switching

Conventional circuit switches employ MEMS and/or nanoelectromechanical systems (NEMS)–based cross-connects, which have slow configuration times (milliseconds). MEMS-based switching fabric works efficiently only with traffic that is statistically multiplexed. MEMS-based optical two-dimensional (2-D) (cross-bar) and three-dimensional (3-D) (mirror-tilt) switches are also available. These switches employ microactuators to position the mirror/lens assembly to realize optical switching. Switches are usually classified as follows [6]:

1. Opaque OXCs: Opaque OXCs also referred to as DXCs employ electronic switching, wherein the received optical signals are converted into electrical, switched by the switching fabric, and transmitted as optical signals directly through the output ports or through the multiplexing module. The process involves OEO conversion that is similar to the operation of a 3R regenerator. Thus the cross-connect also mimics the regen function during the cross-connection process. This is one of the advantages of using a DXC. DXCs are normally used at the junction of the core and aggregation layers or at the point of interconnection with the WDM network to groom the 10G traffic that comes in from the TDM network. The use of electronic switching limits the bandwidth that can be handled by the switching fabric, increases latency [1], and lacks protocol transparency. The monitoring of signal quality is simple due to the nature of the signals (electrical) being handled.

2. Transparent OXCs: Transparent OXCs, also referred to as transparent cross-connects or photonic cross-connect (PXCs), employ optical switches for signal cross-connection or in other words perform optical switching. The incoming composite wavelength (optical signals) is demultiplexed, using a standalone WSS or reconfigurable ROADM, switched by the optical switch and multiplexed (WDM multiplexer), and transported over the output port (all-optical operation). An OXC can support native data rates and provide protocol transparency and fast switching.

3. Translucent OXCs: Translucent OXCs represent a hybrid switch that employs both optical and electronic switching mechanisms. It offers the advantages of the opaque and translucent OXCs—regeneration facility of the opaque OXC and protocol transparency of the transparent OXC. It also facilitates inclusive optical signal monitoring.

10.9.2 Optical Packet Switching Techniques

Optical packet switching (OPS) provides for a flexible, reconfigurable optical bandwidth-efficient layer that can function independently, be integrated with WDM networks, or complement TDM networks. TDM-based optical transmission and switching technologies require OEO conversion and the resultant need for regenerators (increased latency). The integration of WDM technologies has provided a transparent layer that provides high-bandwidth and dynamic-switching capabilities. The use of ROADMs has enhanced the response time of the network to meet demanding and dynamic multiprotocol traffic. The introduction of OPS has led to the development of a new generation of elements referred to as OPS nodes. The OPS node consists of six functional blocks, listed as follows:

1. Input interfaces;
2. Mux/demux unit;
3. Space switch fabric;
4. Switch control unit (SCU);
5. Wavelength transponders;
6. Output interfaces.

A packet consists of the payload (including IP header) and an optical header used for routing/switching in the optical domain.

The following are three techniques-based on WDM that can be used in packet-switched networks [7]:

1. Broadcast-and select networks: The concept of broadcast-and-select networks is simple (multicasting) and was introduced as long ago as the early 1980s. A physical star topology is best suited for broadcast applications. However other physical/logical topologies including bus and ring and interconnection of star and/or ring topologies are possible. Some important aspects of broadcast-and-select networks are described as follows.

    a. The broadcast-and-select networks do not need the routing function, and full connectivity is maintained through broadcasts.

    b. The network consists of nodes equipped with one or more fixed or tunable SFPs.

    c. The nodes are connected to two fibers providing transmit and receive paths (one a single-WDM channel).

d. Packets are transmitted simultaneously by nodes, on different wavelengths, combined by the couplers, and broadcast onto the network.

e. The nodes choose the packets on their tuned wavelengths. Nodes have full visibility of the network traffic but receive packets only on the configured receiver wavelength.

f. Broadcast-and-select networks can be single- or multihop. In a single-hop network, optical messages are received directly by the receivers. Nodes need to be able to simultaneously transmit and receive over multiple wavelengths, necessitating rapid tuning of transmitters and receivers. Media access protocols (MACs) for collision (two or more nodes transmit simultaneously on the same wavelength) and contention resolution (destination receives packets simultaneously on multiple packets and has only one tunable interface) are also required.

g. Some of the issues associated with single-hop broadcast-and-select networks can be overcome using a multihop network that does not require tuning.

h. Each node will have a specified number of fixed/tuned transmitters and receivers configured for different wavelengths.

i. Each node transmits on one fixed wavelength and receives on another fixed wavelength.

j. This arrangement results in the creation of a virtual topology over the fixed topology with packets requiring multiple (more than one) hops to travel from the source to destination (increased delay).

Figure 10.10 illustrates single- and multiple-hop along with the physical and virtual topology and wavelength assignment. For all the simplicity, broadcast-and-select networks do not support wavelength reuse and require wavelengths corresponding to the number of nodes in the network. Hence these networks are not scalable.

Example 10.9—Design of Star Couplers for Broadcast-and-Select Networks

    i. An $n*n$ star coupler can be designed using multiple stages ($\log_2 n$ stages—$2*2$ couplers and $n/2$ couplers per stage with similar power losses for every pair of nodes).

    ii. $2n$ $2*2$ couplers are required to support a bus topology.

2. Wavelength routing: Wavelength-routed networks facilitate wavelength reuse and have better power margins (since there is no power

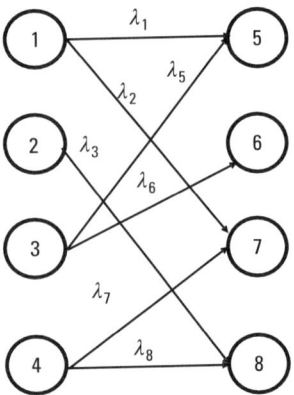

Figure 10.10 Broadcast-and-select networks.

splitting) and hence can be scaled to provide WAN connectivity [8]. The key features of broadcast-and-select networks include:

a. End-user nodes are optically linked (optical fiber link) to routing nodes (WXCs), collectively referred to as network nodes, which are interconnected by P2P links in a freewheeling topology.

b. Each node consists of a set of wavelength tunable transmit/receive interfaces.

c. The network employs wavelength routing to establish a lightpath (continuous optical route using a single wavelength (wavelength continuity constraint) at all nodes in the route) between source and destination nodes (without need for intermediate OEO conversion and buffering at any of the intermediate nodes).

d. It is to be noted that two wavelengths cannot be assigned the same wavelengths on any fiber (distinct wavelength continuity constraint).

e. Lightpath using separate links can use the same wavelengths (wavelength reuse).

f. A single-hop or multihop approach can be employed to support packet switching (conceptually similar to broadcast-and-select networks).

g. A multihop approach results in the creation of a virtual topology, overlaid on the physical network, and requires multiple intermediate nodes between the source and the destination nodes.

A wavelength-routed network is illustrated in Figure 10.10, and its functioning is highlighted in Example 10.10.

Example 10.10—Wavelength-Routed Networks

A network with seven nodes and two wavelengths provisioned over the pair of fibers interconnecting the nodes (with transmit and receive interfaces). The key features of wavelength-routed networks include:

   a. Nodes are interconnected using a single physical path (transmit/receive) over which two separate lightpaths are configured.
   b. The wavelength continuity constraint restricts the use of the same wavelength over shared links.
   c. This implies that some of the wavelengths can be reused while others cannot, based on the network topology.
   d. The use of wavelength converters (transponders) can be used to reduce enhance bandwidth utilization (reduce bandwidth loss due to wavelength continuity constraint).

3. Space switching: A fully functional optical switch requires a combination of space switching, WSSs, and Wavelength converters (transponders) [9]. A space switch is a nonfabric (cross-bar) consisting of semiconductor optical amplifiers (SOAs). All types of optical switches have associated losses (optical). The use of SOA, in a space switch, helps overcome these losses in addition to offering faster switching. Commonly used switch architectures, illustrated in Figure 10.10, include the following:
   a. Matrix-vector multiplier crossbar switch;
      i. Distributed gain matrix-vector multiplier crossbar switch (DGMVM);
      ii. Lumped gain matrix-vector multiplier crossbar switch (LGMVM).

      A centrally located switching plane provides the switching functionality. In DGMVM SOAs are placed after every splitter (2*1) and combiner module (1*2) while SOAs are placed only in the switching plane in the case of LGMVM.
   b. Benes switch: A nonblocking interconnection is provided through a switching plane of 2*2 switches.

10.9.3 Optical Burst Switching

Optical burst switching (OBS) refers to a mechanism that facilitates dynamic subwavelength switching of data. OBS refers to a mechanism that is the midpoint between static OCS and the dynamic OPS [10]. OBS differs from the other switching technologies in the fact that it reserves a control channel to transmit control information (using a reserved channel) in advance of the incoming payload. The control information is used to initiate a lightpath for the transport of the incoming payload and the technique is referred to as delayed reservation. OBS leads to better resource management in comparison with OCS mechanisms.

Mechanism

1. Packet aggregation (burst aggregation) happens at the ingress node (for a specific time period) facilitating packets with common QoS to be transmitted in a burst.
2. Control information (burst-specific, containing routing values including destination address, burst size, and timing information) precedes the burst transmission. The control information typically uses a reserved wavelength using the out-of-band optical supervisory channel (OSC), if available on the network [11].
3. The time difference between the receipt of the control information and the burst is referred to as the offset time.
4. The control packet, a part of the control information channel, is subjected to OEO conversion at each intermediate node to facilitate the configuration of the optical devices.
5. The bursts (multiple IP packets) arriving at the egress are stripped and routed to their respective destinations.
6. OBS can be configured depending on the underlying architectural framework. The different mechanisms include the following:
    a. Burst reservation;
    b. Burst cut-through;
    c. Burst store-and-forward.
7. In the case of burst reservation, the preceding control packet ensures that the resource reservation, through acknowledgments, and in the process avoids bus contention.
8. At the egress node, the burst is disassembled into constituent packets and routed to the destination.
9. The OBS mechanism is responsible for the following:
    a. Adaptation of IP traffic to the burst format;
    b. Burst switching;
    c. DWDM Transmission;

d. Burst wavelength conversion (BWC);
   e. TDM;
   f. Control packet processing;
10. The optical burst forwarding methods include the following [12]:
   a. Burst reservation: Resource reservation using control packets; avoidance of contention link-based time reservation of burst transmission on a per link basis.
   b. Burst cut-through: The control packet, on a hop basis, is followed by the burst on the specified wavelength.
   c. Burst store-and-forward: Provision for fiber delay line (FDL) buffering at each node; the burst is stored partially or in its entirety on each OBS node prior to forwarding to the next node to mitigate the effects of burst dropping and/or contention.

10.10 GMPLS

GMPLS extends the functionality of MPLS to TDM networks including SDH/SONET/OTN/DWDM and spatial switching technologies. GMPLS focuses on the routing (routing protocols) and signaling (signaling protocols) functions of the control plane. GMPLS extends the MPLS features to include devices supporting physical ports, TDM, and WDM within LSRs. The following classes of interfaces (through the extension of the control plane) are supported by the LSR:

1. Packet switch-capable (PSC) interfaces: Data forwarding on the basis on the packet header.
2. Layer-2-switch capable (L2SC) interfaces: Data switching on the basis of the frame and/or cell headers
3. TDM-capable interfaces: Data switching based on the time slot (repetitive).
4. Lambda switch-capable (LSC) interfaces: Data switching on the basis of the incoming wavelengths (λ).
5. Fiber switch-capable (FSC) interfaces: Data switching base on the physical fiber.

GMPLS does not support the MPLS features such as penultimate hop popping, equal cost multipath (ECMP), and label merging.

10.11 Packet Transport Networks

Packet transport networks (PTNs) refer to a technology that facilitates the integration of IP-oriented services and services supported by legacy SDH networks. It supports the transport of TDM/ATM over IP infrastructure. The PTN networks are based on the MPLS-TP standardized by IETF and ITU-T. PTN integrates the advantages of MPLS networks with the OAM&P benefits of SDH/SONET and OTN networks. GMPLS is the alternative to PTN. MPLS-TP is explained in detail in Section 10.12.

10.12 MPLS-TP

IP/MPLS implementation necessitates—and this was especially true when the technology was introduced initially—higher CAPEX and OPEX (in comparison with non-MPLS-enabled networks. It also suffers from the lack of a full-fledged FCAPS system (as compared to SDH/SONET systems. MPLS-TP is an IETF-defined network layer technology for use in conjunction with transport networks. MPLS-TP is a connection-oriented application with integrated mechanisms for supporting transport functionalities. It provides separation between the control (static or dynamic) and data planes leading to enhanced network resilience and flexibility and lower OAM costs.

10.13 Packet Versus Data Flows

Packet-switched networks transmit a stream of data originating from an application through a local/remote session over the transport layer (TCP) to the network layer (IP), which segments the stream into packets for transmission over the network. The packets (stateless, independent segments) can be easily transmitted over a resilient network infrastructure. However, most applications require a stream of data rather than independent segments. Technologies like virtualization OpenFlow and SDN are designed to enable applications and data flows. The network environment that we are so familiar with is undergoing a paradigm shift. In a virtualized environment, applications exchange information over virtual and dynamic interfaces. The user environment is also changing with a shift from desktop systems to mobile and smart devices (handsets) triggering a shift in the network service requirements. Future networks need to evolve to support the need for data flows as opposed to packet flows.

10.14 Summary

Packet-switched networks offer high efficiency and low ownership costs for all services (non-TDM). The emergence of a new class of elements, referred to as packet transport switches, has brought about the integration of packet-switched and TDM architectures, amplifying the advantages of both and heralding a new converged architecture for backbone networks. This chapter presents the traditional and evolving architecture of packet-switched core networks along with the integration of MPLS components. The next-generation network architecture referred to as future packet-based networks (FPBNs) would be simpler (removal of multiple and often redundant layers) and flat or with minimal hierarchical layers and offer high levels of flexibility and scalability. PONPs would replace the conventional SDH/OTN nodes and provide the mechanism to integrate TDM and packet-switched technologies at the network core. Packet-switched networks provide an E2E architecture that accords an high degree of scalability and flexibility and supports diverse service offerings in tune with business requirements. The design philosophy of a packet-switched network is similar to that of circuit-switched networks with multiple layers that facilitate aggregation and routing capabilities. It has been long understood that future networks must be fully optical. However the adoption of optical switching technologies has been relatively slower. The introduction of optical packet-switching technologies has uncovered a lot of issues related to routing, control, and packet transport mechanisms that were developed for mostly opaque electronic networks. As these issues get sorted the OPS algorithms will also evolve. Network planners need to consider strategies for migrating to optically switched networks. Field engineers working in the circuit-switched domain need to acquire competencies to maintain and troubleshoot packet-switched networks and vice versa. This chapter provides a solid grounding for novice as well as experienced engineers needing to develop or update their competencies on packet-switched networks. This chapter also provides food for thought—packet versus data flows.

10.15 Review

10.15.1 Review Questions

1. The _____ mechanism is responsible for the following:
   a. Adaptation of IP traffic to the burst format
   b. Burst switching
   c. DWDM transmission
   d. BWC

2. OBS can be configured depending on the underlying architectural framework. The different mechanisms include:

    a. Burst reservation

    b. _____

    c. Burst store-and-forward

3. A fully functional optical switch requires a combination of space switching, WSS, and wavelength converters (transponders).

    a. True

    b. False

4. The network employs wavelength routing to establish a _____ (continuous optical route using a single wavelength).

5. Wavelength-routed networks facilitate wavelength reuse and have better power margins and hence can be scaled to provide WAN connectivity.

    a. True

    b. False

6. The concept of broadcast-and-select networks is based on _____.

    a. Unicast

    b. Multicast

    c. Broadcast

    d. Telecast

7. OPS provides for a flexible, reconfigurable optical bandwidth-efficient layer that can function independently, be integrated with WDM networks, or complement TDM networks.

    a. True

    b. False

8. Opaque OXCs, also referred to as _____, employ electronic switching, wherein the received optical signals are converted into electrical, switched by the switching fabric.

    a. DXC

    b. PXC

    c. PXC

    d. CXC

9. Conventional circuit switches employ MEMS and/or NEMS-based cross-connects, which have a slow configuration times (milliseconds).

   a. True
   b. False

10. _____ refers to a technology that facilitates multiple independent instances of the routing table to coexist on a router.

    a. VRF
    b. Virtual forwarding
    c. Sequential routing
    d. Sequential forwarding

10.15.2 Exercises

1. Describe the architecture of a PTN. Discuss the advantages and disadvantages of PTN in comparison with other circuit- and packet-switched network architecture.
2. List the advantages of MPLS networks.
3. Describe in detail the MPLS extensions for use on circuit-switched networks.
4. List the advantages of wavelength-routed networks over broadcast-and-select networks.
5. Summarize the essence of this chapter in 200 words.

10.15.3 Research Activities

1. Describe the architecture of an all-optical network. Illustrate suitably.
2. Describe OBS in detail.
3. Describe in detail the architecture of optical switches.

10.16 Referred Standards

RFC 3031: Multiprotocol Label Switching Architecture

RFC 3032: MPLS Label Stack Encoding

RFC 3036: LDP Specification

RFC 2205: Resource Reservation Protocol (RSVP)—Version 1 Functional Specification

RFC 2209: Resource Reservation Protocol (RSVP)—Version 1 Message Processing Rule

RFC 3209: RSVP-TE

RFC 3270: MPLS Support of Differentiated Services

RFC 4090: Facility backup and Fast Reroute

RFC 3630: TE Extensions to OSPF v2

RFC 3784: Intermediate System to Intermediate System (IS-IS) Extensions for Traffic Engineering (TE)

RFC 4875: Extensions to Resource Reservation Protocol—Traffic Engineering (RSVP-TE) for Point-to-Multipoint TE Label Switched Paths (LSPs)

RFC 4461: Signalling Requirements for Point-to-Multipoint Traffic-Engineered MPLS Label Switched Paths (LSPs)

RFC 2961: RSVP Refresh Overhead Reduction Extensions

RFC 3209: RSVP-TE: Extensions to RSVP for LSP Tunnels

RFC 2205: Resource Reservation Protocol (RSVP)—Version 1 Functional Specification

10.17 Recommended Reading

10.17.1 Books

De Luc, G., *MPLS Fundamentals,* Cisco Press, 2006.

Antoniom, M. S., and K. G. Szarkowicz, *MPLS in the SDN Era,* O'Reilly Media, 2016.

Pepelnjak, I., and J. Guichard, MPLS and VPN Architectures, John Wiley and Sons, 2010.

Raza, K., and M. Turner, *Cisco Network Topology and Design,* Cisco Press, 2002.

Chikama, T., H.Onaka, and S. Kuroyanagi, "Photonic Networking Using Optical Add/Drop Multiplexers and Optical Cross Connects," *Fujitsu Sci. Tech. J.,* 35, 1, 1999, pp. 44–64.

Tsuda, T., T.Ohta, and H. Takeichi, "R&D for the Next-generation IP Network," *Fujitsu Sci. Tech. J.,* 37, 1, July 2001.

10.17.2 URLs

http://citeseerx.ist.psu.edu/viewdoc/download?doi=10.1.1.408.5296&rep=rep1&type=pdf.

http://community.fs.com/blog/the-principle-of-the-optical-switch-and-types.html.

http://www.hit.bme.hu/~jakab/edu/litr/wdm/OXC_4_dedo.pdf.

http://www.mplsinfo.org.

https://optiwave.com/resources/applications-resources/optical-system-optical-cross-connects/.

https://pdfs.semanticscholar.org/ee5f/b92c6fc7eeb209ee790636bcfaf3e24523c9.pdf.

https://www.google.co.in/?gfe_rd=cr&ei=9gFmWaLEBrCl8wfskqfQAQ&gws_rd=ssl#q=space+switch&spf=149985739997.

https://www.juniper.net/us/en/local/pdf/whitepapers/2000392-en.pdf.

https://www.netlab.tkk.fi/opetus/s38165/k04/Luentomat/L13_slides.pdf.

10.17.3 Journals

Coulibaly, Y., et al., "QoS-Aware Ant-Based Route, Wavelength and Timeslot Assignment Algorithm for Optical Burst Switched Networks," *Transactions on Emerging Telecommunications Technologies,* 2015.

Xin, Y., G. N. Rouskas, and H. G. Perros, "On the Physical and Logical Topology Design of Large-Scale Optical Networks," *IEEE/OSA Journal of Lightwave Technology,* Vol. 21, No. 4, April 2003, pp. 904–915.

Dorren, H. J. S., et al., "Optical Packet Switching and Buffering by Using All-Optical Signal Processing Methods," *Journal of Lightwave Technology,* 1(1):2–12, 2004.

O'Mahony, M. J., et al., "An Optical Packet-Switched Network (WASPNET)—Concept and Realization," *Optical Networks Magazine,* 2(6):46–53, 2001.

Yang, Q., et al., "WDM packet Routing for High-Capacity Data Networks," *Journal of Lightwave Technology,* 19(10):1420–1426, 2001.

Clos, C., "A Study of Non-Blocking Switching Network," *Bell Syst. Tech. J.*, Vol. 32, Mar. 1953, pp. 406–24.

References

[1] Rouskas, G. N., and L. Xu, "Optical Packet Switching," *Emerging Optical Network Technologies: Architectures, Protocols and Performance,* Springer, 2004, pp. 111–127.

[2] Kahlid, R., and M. Turner, *Cisco Network Topology and Design* [2002], Cisco Press. Retrieved from: http://www.ciscopress.com/articles/.

[3] ITU-T Recommendation Y.2611—High-Level Architecture of Future Packet-Based Networks.

[4] Brocade NetIron Multiprotocol Label Switch (MPLS) Configuration Guide 53-1003830-01, Pages 25-54, 395-398.

[5] Blumenthal, D., et al., "Special Issue on Photonic Packet Switching Technologies, Techniques, and Systems," *Journal of Lightwave Technology,* 17(12).

[6] Chu, P. B., S.-S. Lee, and S. Park, "MEMS: The Path to Large Optical Cross-Connects," *IEEE Communications,* 40(3):80-87, 2002.

[7] Blumenthal, D. J., "Photonic Packet Switching and Optical Label Swapping," *Optical Networks Magazine,* 2(6):54-65, 2001.

[8] Baldine, I., G. N. Rouskas, and D. Stevenson, "JumpStart: A Just-in-Time Signaling Architecture for WDM Burst-Switched Networks," *IEEE Communications,* 40(2):82-89, 2001.

[9] Callegati, F., G. Corazza, and C. Raaelli, "Exploitation of DWDM for Optical Packet Switching with Quality of Service Guarantees," *IEEE Journal on Selected Areas in Communications,* 20(1):190-200, 2002.

[10] Baldine, I., , et al., "Just-in-Time Optical Burst Switching Implementation in the ATDnet All-Optical Networking Testbed," in *Proceedings of Globecom,* 2003.

[11] Blumenthal, D. J., et al., "Optical Signal Processing for Optical Packet Switching Networks," *IEEE Communications Magazine,* 41(2):23–29, 2003.

[12] Callegati, F., "Optical Buffers for Variable Length Packets," *IEEE Communications Letters,* 4(9):292–294, 2000.

11

Virtualization and the SDN Ecosystem

11.1 Chapter Objectives

The telecommunication industry is traditionally capital-intensive with a focus on hardware platforms capable of delivering huge bandwidth to support converged applications. It is only over the last decade that the industry has indicated a marked shift by exploring existing developments in the software industry and customizing them to support telecom services. This shift has resulted in new paradigms of virtualization and SDN that promise to revolutionize the telecom market in the yeto come.

As per the latest report from HIS Markit, the NFV market, which includes hardware, software, and services, will be worth $15.5 billion by 2020 with a compound annual growth rate (CAGR) of 42% with the software market accounting for a percentage of the total industry spending. As the technology matures (by 2020) the software segment will account for only 11% of the industry revenues; 16% will be from NFVI (including servers, storage systems, and optical switches) while the bulk of the revenues, 73%, will stem from NFV applications—VNF[1]. In 2020, the services market for NFV will be larger than that for SDN, as this service provides focused service agility and operational efficiency.

During the period of 2017 to 2020 service providers will enhance their portfolio of software-based services to customers using virtualized infrastructure that virtual customer premises equipment (vCPE) and enterprise vCPE. Spend-

---

1. VNF can also be abbreviated as vNF.

ing on consumer and enterprise hardware and software is forecast to reach $1.5 billion worldwide by 2020.

NFV will lead to the virtualization and consolidation of service provider network elements onto industry standard server, storage, and switching systems that can be shared with multiple users. NFV and SDN technologies are being increasingly adopted in most of the telecommunication network domains including (but not limited) to the core network [evolved packet core (EPC)] of mobile broadband systems (4G), IP multimedia systems (IMSs), evolved multimedia broadcast multicast services (eMBMS), policy control and enforcement mechanisms, CPE, content delivery network (CDN), and backbone networks. This chapter presents the conceptual basis and the detailed framework for deploying SDN and NFV functions. Readers need no prior knowledge of these technologies. However, they should have a general understanding of the functioning of telecom networks.

Key Topics

- The need and imperatives for SDN and NFV;
- Use cases for NFV and SDN;
- The architectural framework for NFV;
- The architectural framework for SDN.

11.2 Introduction

Telecom networks are evolving to meet the high-bandwidth, dynamic traffic, and low-latency requirements of new-age services. The evolution however has been slow, as it involves changes in hardware that necessitate installation and commissioning (I&C) activities and the associated network reconfiguration. This is besides the need for additional space, CAPEX, power, and field personnel with the requisite competencies. Historically the telecom industry has been focused on hardware that cannot evolve to meet the dynamic requirements of new-age services. In addition, many of the hardware components are proprietary, albeit with standardized interfaces, and hence need additional effort to maintain and configure. The reliance on hardware mandates on-site interventions for I&C and OAM&P activities. A case in point is that of the optical interfaces, SFPs, whose form factor and interfaces are standardized. SFP modules (with the same data rates and supported wavelength) of one manufacturer generally cannot be used on another since the internal hardware is proprietary. [Some modules may include clock recovery (8-KHz) functions.] In summary, the hardware-based telecom systems are slow to evolve; difficult to install, configure, maintain, and reconfigure; and do not provide the dynamic reconfiguration facility that is the hallmark of the new-age service applications.

In a manner similar to the levels of support to applications that is extended by the dynamically configurable and fully automated cloud environments, VNFs enhance the agility of telecom networks and provide them with the capability to respond automatically to the needs of the traffic and services running over them. The key enabling technologies that support network virtualization are two complementary but mutually supporting technologies—SDN and NFV. Together they provide the supporting framework to fully realize the benefits of virtualization.

NFV [1] is an architecture that borrows (but is distinct) from the virtualization function technologies of IT, in order to create an entire class of virtualized network node functions that forms the building blocks of the telecommunication networks. These fundamental blocks may be stacked together, in unlimited permutation and combinations, to dynamically create new classes of services VNF-based on application, network, and customer demands. A VNF consists of one or more virtual machines that work on specific software and distinct processes that are hosted on high configuration server farms with multiple, redundant high-capacity storage systems connected to a network with high-speed switching devices and/or cloud infrastructure. This concept is radically different from the traditional network architecture, which relies on dedicated hardware resources for every function implemented on the network.

The NFV Group, initiated by global operators under ETSI and also referred to as the Industry Specification Group (ISG), is responsible for the evolution, adoption, and standardization (providing relevant information and feedback for standards but not directly participating in standards formulation) of the NFV ecosystem. Figure 11.1 illustrates the virtualization ecosystem.

Example 11.1

VNF is the software representation of the common network functions. These functions can be interconnected to create new services dynamically. Some common examples including the following:

1. Routers;
2. Tunnel gateways;
3. Signaling elements;
4. Optimization elements (load balancer);
5. Policy control;

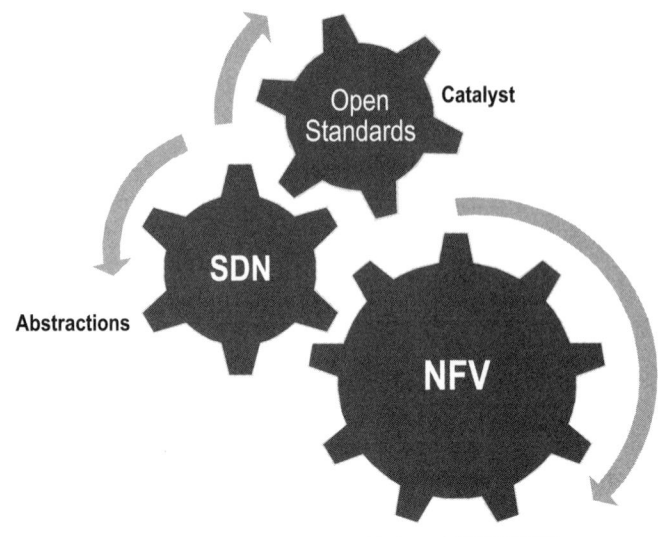

Figure 11.1 Virtualization ecosystem.

6. Authentication;
7. Security mechanisms:
   a. Firewalls;
   b. Antivirus and antispam protectors;
   c. Intrusion detection systems.
8. Mobile broadband networks (3G/4G):
   a. Home location register (HLR);
   b. Home subscriber server (HSS);
   c. Mobility management entity (MME);
   d. Radio network controller (RNC);
   e. Serving GPRS support node (SGSN);
   f. Gateway GPRS support node (GGSN);
   g. PDN Gateway (PDN-GW).

SDN technology is a radical approach that facilitates remote configuration, control, updating, and management of network functions and behaviors (low-level functionality abstraction) through open interfaces [1]. SDN infuses agility to the traditional, static network architectures making them dynamically respond to the computing and bandwidth needs of the applications, system, and users. The improved agility is due to the separation of the control (SDN controller) and data transport functions through the implementation and separation of control and data planes. SDN works with multiple industry standard communication protocols.

11.3 NFV Business Case

It is expected the number of connected devices will increase from its number of 12 billion to more than 100 billion by 2025 [2]. Existing architectures will not be able to meet these huge traffic and data flows through the network. Accordingly, new integrated service architectures will be required to leverage the cloud infrastructure to support the data requirements of this huge interconnected mass of devices.

The adoption of virtualization and SDN technologies will definitely bring about a long-term reduction in CAPEX and OPEX, but these are just byproducts of the evolved network architecture. The two primary benefits of this exciting transition are the following:

1. Lower total cost of ownership: The highly scalable and agile network environment will result in economies of scale leading to significantly lower total cost of ownership (TCO) as compared to the fixed behemoth networks of today.

2. Opportunity cost: Business enterprises that do not scale up their networks to adapt to the dynamic service requirements of the near future will lose their market share and fade away within the next decade.

The adoption of virtualization and SDN will result in incremental changes that will bring down the TCO and enhance revenue generation (the ability to dynamically provide and configure new services) for telcos as the effort of these incremental changes take shape. One of the earliest use cases for NFV has been the virtualization of CPE, leading to logistic efficiencies and the associated reduction in OPEX.

11.4 NFV Use Cases

The entire set of network nodes and functions, as we understand them today, are candidates for NFV. The high-level objectives of NFV include [3] the following:

1. Express service innovation and delivery (software implementation of network functions and E2E services);
2. Enhanced operational efficiency (automation and SOP consolidation);
3. Power savings (hardware resource optimization);
4. Use of open standardized interfaces (multivendor interoperability);
5. CAPEX and OPEX savings due to VNF implementation.

The following sections present ETSI-specified NFV use cases.

Use Case 1: CPE Virtualization

Need

A standard enterprise network consists of the HQ, zonal/regional offices, and branches integrated through a technology-centric networking architecture. The interconnection requires a dedicated, feature-specific device [access routers/provider edge (PE) routers/enterprise CPE] at the points of interconnection (POI). These static devices require on-site installation, configuration, and maintenance support that adds to the CAPEX, OPEX, and general logistic complexity. The enterprise routers are based on proprietary hardware and routing engines, increasing the complexity of network configuration. In addition an increased mobility of users, spurred by evolving organizational people's practices has resulted in an increased deployment of smart phones and BYOD that needs an additional layer of network security to ensure the safety and integrity of corporate data. These challenges present a compelling use case for virtualization of access hardware with the maximum benefits being realized by the virtualization of CPE, which is unique for each connection, as opposed to the PE, which is shared by multiple users.

Virtualization Scope and Function

Figure 11.2 illustrates an enterprise network configuration highlighting the interconnections between the CPE (can be owned by the enterprise customer or service provider) and the PE. The virtualization scope includes the following:

1. CPE function virtualization (vCPE);
2. PE function virtualization (vPE).

The two functions can be independently and separately virtualized and in tandem represent the complete NFV. The virtualization can be partial or complete. In case of a partial virtualization some of the functions (for example VPN, PE core functions) may not be virtualized. In case of a complete virtualization

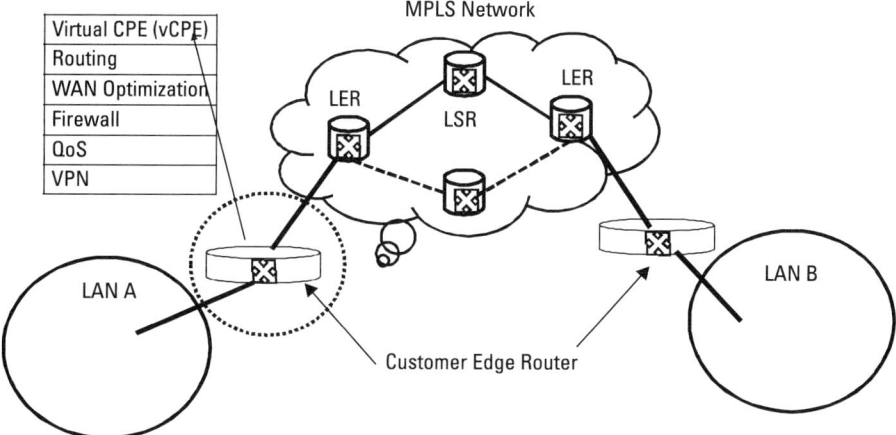

Figure 11.2 CPE virtualization use case.

a virtual machine would provide the equivalent functionality of the CPE and/or PE equipment.

Use Case 2: Mobile Broadband Networks

Need

Figure 11.3 illustrates the 4G ecosystem. The EUTRAN consists of a single node known as eNodeB (eNB). The eNodeB is connected to the remote radio unit (RRU) through an optical fiber cable using an industry standard (3GPP) CPRI interface that forms the front-haul network. The connectivity with the

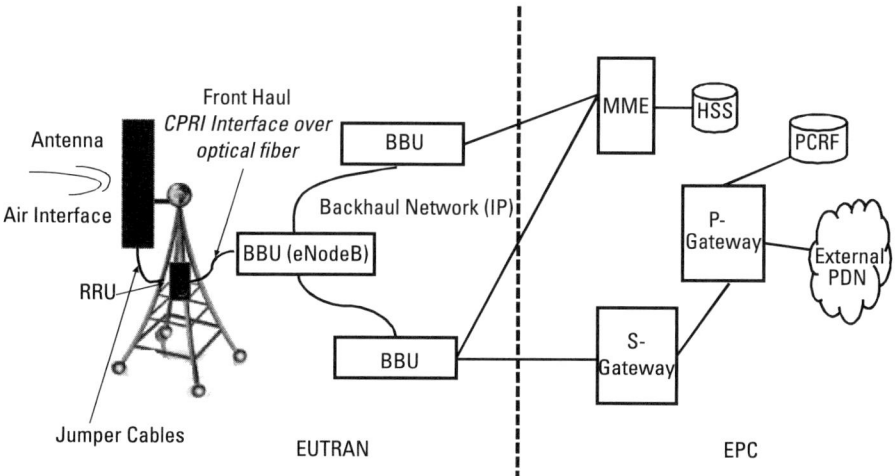

Figure 11.3 4G network architecture.

other eNBs and the core network is through a backhaul network that is based on IP/MPLS. Each site contains an individual BBU that is interconnected using the backhaul network. The architecture is illustrated in Figure 11.4.

In the conventional 4G network a stand-alone baseband unit (BBU) is located at each site with support for at least three bands (LTE-FDD and/or LTE-TDD) (frequencies—example—2,500/2,300 MHz; 1,800 MHz; and 900/850/700 MHz). The BBU provides the channel control and element modules that provides radio resource management (RRM), call processing, baseband T/R signal processing, O&M capabilities, time synchronization, alarm management (internal/external), and data flow management. Passive infrastructure is required for hosting the BBU unit. This includes cabinet/shelter, power supply (and grounding including earth pits), and backup (DG set/battery) and cable management units. The BBU does not operate at its peak capacity on a 24*7 basis. The current problem areas include the following:

1. Increased CAPEX (due to redundant BBU and the associated infrastructure);

2. Increased OPEX (O&M of a large number of BBUs with the associated infrastructure);

3. Low QoS (unequal load distribution and interference due to overlapping coverage areas);

Virtualization Scope and Function

As discussed earlier, most of the RAN and the core functions can benefit from virtualization. The virtualization function can be extended to 3G/2G networks

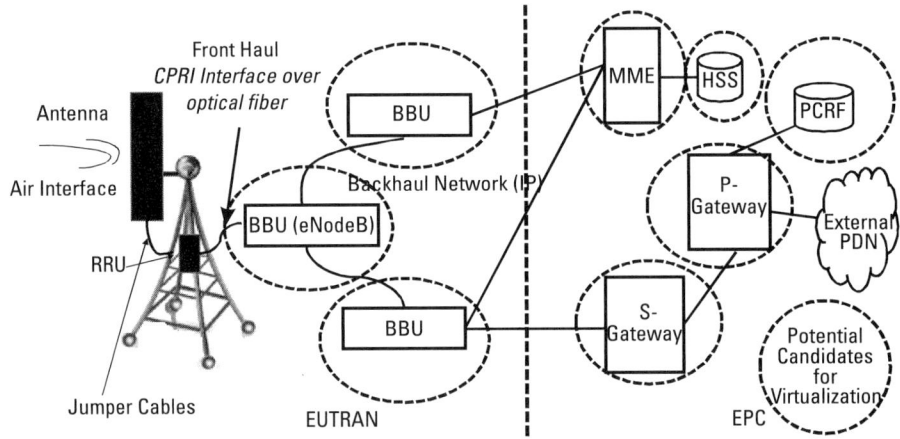

Figure 11.4 Potential candidates for virtualization in 4G network.

also. However the discussion, in this chapter, has been limited to 4G networks due to its current relevance. The potential areas of virtualization in 3G/4G networks, as illustrated in Figure 11.4 include the following:

1. Home location register (HLR);
2. Home subscriber server (HSS);
3. Mobility management entity (MME);
4. Radio network controller (RNC);
5. Serving GPRS support node (SGSN);
6. Gateway GPRS support node (GGSN);
7. PDN gateway (PDN-GW);
8. Load balancing: The BBU operates at a much lower capacity than its maximum for most of its operating cycle. C-RAN, also referred to as cloud RAN or centralized RAN, is a cloud-based architecture that allows multiple RRUs to connect to a centralized BBU pool over a fiber link (a front-haul network that can span upto 80 km). The important benefits include the following:
    a. Creation of a BBU pool that supports multiple RRH connections;
    b. High-bandwidth (10-Gbps) communication channel for inter-BBU communication (backhaul network)—collaborative radio[2].

The above scenario is a compelling case for real-time virtualization capability.

1. Traditional BBUs are based on vendor-specific proprietary hardware and software systems, which support standard physical and logical interfaces.
2. The C-RAN BBU pool can be based on open hardware-server farms with interface cards for interconnectivity.
3. The hardware hosts the software stacks representing the BBU functionality.
4. Real-time virtualization facilitates dynamic resource allocation and traffic management.
    a. Virtualized self-organizing network: The C-RAN is a perfect platform for centralized SON (C-SON) deployment. The C-SON would encompass the following:

---

2. This concept is different from BBU hoteling wherein BBUs are stacked at a common location but are not physically interconnected.

i. Self-configuration function;
   - Plug-n-play;
   ii. Self-optimization function;
   - Automatic neighbor relations (ANR);
   - Random access parameters optimization;
   - Automatic switch-off (energy-saving feature—A percentage of BBUs in dormant state during off-peak hours).
   iii. Self-healing functions;
   b. Coordinated multipoint transmission/reception: Coordinated multipoint transmission/reception (CoMP) enhances system performance through coordinated transmission/reception between user equipment (UE) and geographically separate BBUs.

11.5 NVF Framework and Services

Traditional network elements consist of an integrated set of, often proprietary, hardware and software (with standardized interfaces) that perform specific functions. New functionalities, or upgrades of existing functionalities, may require the upgrade of software, hardware, or both. The primary objective of NVF is to separate the software and hardware components of a network node to facilitate their organic development, independent of each other. The decoupling also implies that dedicated resources can now be shared or pooled amongst a larger set of entities resulting in elimination of redundancies and realization of economies of scale and process and product efficiencies. Figure 11.5 illustrates the broad-level NFV framework.

The framework consists of three main elements whose functions are as listed as follows [4]:

1. VNFs: the software representation of the key network functions that can be arrayed on a NFVI [5].
2. NFVI: encompasses the set of hardware and software components that provide the environment for deployment of VNFs. The NFVI can span multiple geographically diverse locations and includes the interconnectivity links.
3. Network functions virtualization management and orchestration (NFV-MANO): refers to the assembly of all the functional blocks including the associated data storehouses, interfaces for information interchange, and reference points that are required to support VNF lifecycle management.

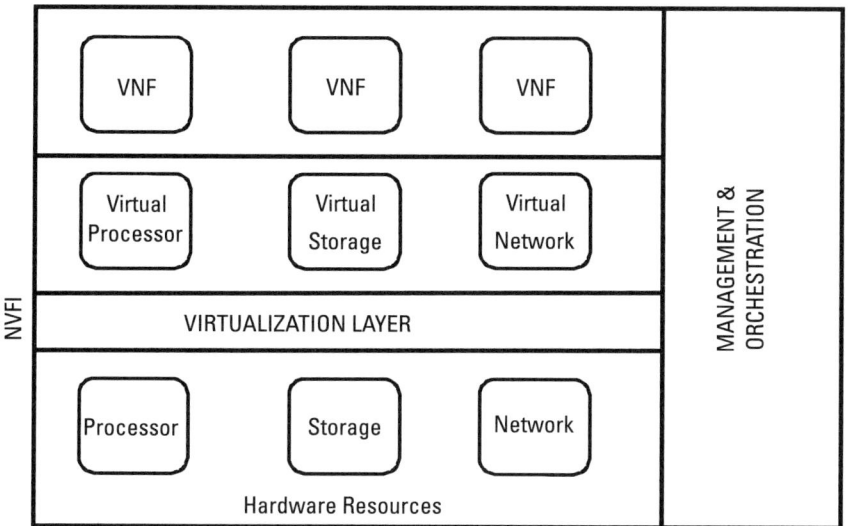

Figure 11.5 NFV framework.

The framework provides the virtual environment that enables dynamic creation and management of multiple VNF instances along with the associated data management tasks, aligning dependencies and other attributes that are crucial to the functioning of the VNF and the underlying physical and logical components. The life-cycle management of the VNFs include the following context-specific perspectives:

1. VNF running as virtual machine;
2. Deployment of a vendor-specific package that consists of a set on interconnected VNFs that handles specific network functions;
3. Operations and management of the deployed VNFs by the service provider.

11.5.1 Functional Blocks

The primary functional blocks for setting up, operating, and managing virtualized functions include the following:

1. VNF;
2. VNFI:
    a. Related hardware and software blocks;
    b. Virtualization layer;

c. Orchestrator;

d. Services.

3. Management systems;
4. Element management systems (EMSs);
5. Virtualized infrastructure manager;
6. VNF manager;
7. Operations/business support systems (OSS/BSSs).

11.5.2 VNF Services

NFV provides an E2E service whose behavior(s) matches that of the original nonvirtualized service. A VNF service[3]: implementation is through a forwarding graph of network functions and end/terminal points. The concept is illustrated in Figure 11.6. The implementation of complex functions is realized through network service chaining (VNFG)[4] or in simple words a combination of several VNFs.

1. VNF forwarding graphs (VNF-FGs): refers to a chain of VNFs (as described by GS NFV 001) with specified network connectivity requirements. The VNFs are connected by logical links that can be one of the following:

   a. Unidirectional;

   b. Bidirectional;

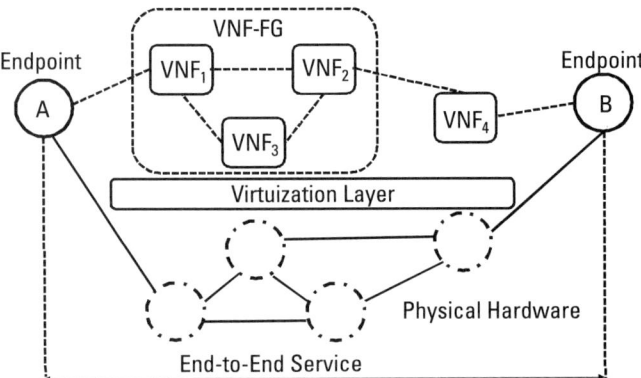

Figure 11.6 NFV service implementation.

---

3. Services provided by NSPs to end customers.
4. RFC 7665 defines the service function chaining (SFC) architecture.

c. Multicast;

   d. Broadcast.

2. VNF set: refers to a chain of VNFs where network connectivity requirements are not specified (as described by GS NFV 001).

3. NFVI point-of-presence (NFVI—PoP): refers to a (operator-specific) geographic location that hosts the physical devices (processors, storage, and network interface entities). A data center is an example of a PoP.

The following are some of the major characteristics of VNF services:

1. The VNFs are executed over the virtualization layer that forms part of the NFVI.

2. The exact physical deployment of the VNF over the NFVI is transparent from the end users.

3. The policy constraints are however guaranteed.

4. A network service is implemented as a forwarding graph of network functions (VNF-FG).

Figure 11.6 illustrates an example of an E2E virtualized network service, described as follows.

1. The E2E service consists of VNFs between end points over multiple NFVI-PoPs.

2. The VNF is executed over the virtualization layer (part of NFVI).

11.6 Design Considerations

Section 11.4 discusses the benefits of virtualization. Now we turn to some of the key issues that need to be included within the design framework, described as follows.

1. Backward compatibility/interoperability: The NVFI environment includes hypervisor, which integrates hardware, software, and firmware to create an environment to host, create, and maintain virtual machines. There can be two types of hypervisors:

   a. Type-1—Referred to as bare metal, embedded, or native hypervisors and run directly on the system hardware;

   b. Type-2—Run on the host system.

There are no formal standards that demarcate the hypervisor functions and types resulting in subtle differences between hypervisors of different vendors. This aspect should be given detailed consideration when planning for network service chaining. The compatibility issues can cause significant erosion in performance of a VNF. In addition the VNF implementation must be compatible with all the systems as the physical device that is being virtualized and issues related to backward compatibility need to be well thought out at the design stage.

2. Network performance: It has been observed that a single instance of a VNF may not be able to replicate the performance levels of dedicated hardware systems (context- and function-dependent). In order to circumvent the above issue, a network plannerin may instantiate multiple instances of a VNF. The multiple VNFs will add to the management overhead, leading to a performance dip of the NFV. The above mentioned factors and their impact on the SLAs (that were in existence) must be carefully worked out at the design stage. It is pertinent to note that the use of cloud infrastructure can further impact (reduce) the performance of the virtualized network.

3. Network reliability: A hardware-based network generally provides five nine availability or 99.999% availability or a downtime of five minutes and twenty six minutes in a year. A hybrid NVFI developed with COTS solutions may be unable to meet five nine availability requirements. The use of open standards and/or generic hardware may seriously impact network reliability.

4. Network security: The dynamic changes in the NVFI coupled with the liberal use of software components makes the NVF vulnerable to security breaches and the associated challenges. The dynamic deployment of VNF and resource optimization creates hurdles with the enforcement of network security policies.

5. Provisioning of services: In a conventional (nonvirtualized) set-up the service provisioning was directly effected at the NE. There are distinct changes in a virtualized network. These include the following:

   a. Service provisioning is extended to cloud-deployed VNFs.

   b. Complexities due to the different hardware and software systems (including the probable use of cloud infrastructure). This makes manual provisioning a difficult proposition and calls for the deployment of intelligence in the NVFI environment.

   c. Frequent service reprovisioning due to the movement of VNFs (migration across cloud services).

d. On-demand, specialized monitoring (e.g., vTAPs) and self-diagnostics of key systems supporting NVFI. This includes network performance management and monitoring.

11.7 Distributed NFV Architecture

During the initial stages of NFV deployment it was presumed that the virtualization environment (vPoP) would be data centers. This approach does work well for many use cases, but it would not be universally true. The objective of NFV was to provide unmatched flexibility in deployment including the location of the physical and/or virtual resources. VNFs should be cost-effective and deployed with a view of maximizing their efficient usage. The service provider should have the freedom to place the NFV at all locations including data centers, CPE equipment, network nodes, and the cloud infrastructure. This approach is referred to as distributed NFV architecture. In fact there can be three approaches to VNF deployment, as illustrated in Figure 11.7. In a centralized deployment model the VNFs are deployed at a single (centralized) location or a PoP-like data center (economies of scale). In a distributed deployment (D-NFV) model the VNFs are deployed at the customer premises (network edge), and in case of the hybrid model the VNFs are deployed at the network edge as well as a centralized location. The primary benefit of D-NFV is the enhanced TTM (time-to-market) for the service providers, which helps them to launch innovative and dynamic services based on market demand.

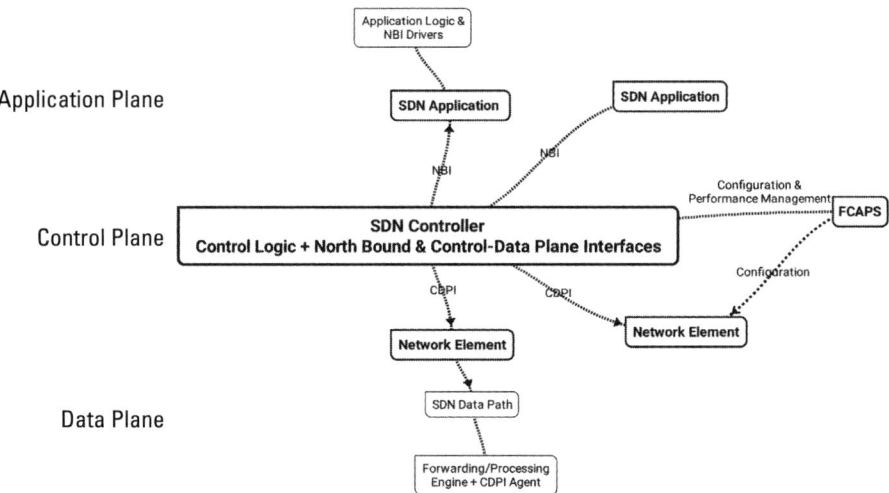

Figure 11.7 SDN architectural framework.

11.8 NFV and SDN: The Linkage

NFV and SDN are interrelated concepts that have value independently but complement each other to bring out a radically new value proposition when deployed in tandem. The origins and the objectives of both these technologies are different with SDN being developed (by researchers) to introduce programmability to network devices, while NFV is a service provider–initiated concept developed to facilitate the rapid development and deployment of services. The following are some of the key characteristics of NFV and SDN:

1. The crux of SDN is to build network devices where the abstract layers are clearly defined and separated (linked by standard interfaces).
2. The abstraction is realized by separating the control and the traffic functions resulting in the deployment of the control and the date planes.
3. The control plane is centrally located while the forwarding plane components are distributed.
4. There are northbound and southbound interfaces associated with the control plane with the functionality provided using application programming interfaces (APIs)[5].
5. The northbound interface provides network abstraction to the higher level applications/programs.
6. The southbound interface provides a channel (based on open standards or proprietary) for use by the SDN controller to communicate with the network devices facilitating centralized control and real-time dynamic response to network demands.
7. In contrast NVF creates a virtual environment that allows key network functions to be virtualized.
8. It is evident that SDN can be effectively deployed to create software representation of the virtualized functions.
9. This allows service providers to maximize the impact of virtualization and realize effective ROI.
10. The NFVI uses a centralized orchestration and management system that performs the processing, storage, and OAM&P functions required to run the VNFs. Monitoring is done by using vTAPs (virtual taps or monitoring points).

---

5. API refers to a set of functions/procedures that facilitates the development of programs/services that access a common set of features provided by an operating system, application, or service.

The above discussion implies that NFV is a potential use case for SDN. The implementation of VNF can be done using the SDN ecosystem. Table 11.1 summarizes the relationship between SDN and NFV.

11.9 Network Management and Orchestration

Network management and orchestration (MANO) provides the management framework (ETSI-defined) for the NVFI. NFV MANO is a working group of the ETSI ISG. MANO facilitates the rapid creation, maintenance, and deployment of the virtual functions within the NFVI. MANO consists of three functional blocks:

1. NFV orchestrator:
   a. Network services (NS), VNF on-boarding;
   b. Service lifecycle management;
   c. Resource management (global);
   d. Validation and authorization of NFVI resource requests.
2. VNF manager:
   a. VNF instantiation life-cycle management;
   b. Coordination/adaptation—NFVI and EMS configuration and event reporting.
3. Virtualized infrastructure manager (VIM):
   a. Management of NFVI resources (processor, storage and network resources).

Table 11.1
SDN and NFV

Parameter	NFV	SDN
Objective	Virtualization of network functions	Centralized control, separation of control and data planes, flexibility in network configuration
Target function	Service provider NEs	Cloud/campus infrastructure and data centers
Hardware platforms	Servers/switches	Servers/switches
Use cases	Content distribution network (CDN), routers (CPE/edge), firewalls, mobile RAN/core network elements/functions	Service Orchestration (including cloud, mobile network), NFV, converged storage systems
Standardization	ETSI NFV Working Group	Open Networking Forum (ONF)

The NFVI MANO framework is integrated with APIs for it to function. It provides users with the functionality to choose NVFI resources for deployment.

11.10 Open-Source NFV Platforms

The Open Platform for NFV Project (OPNFV) is an open-source reference platform aimed at providing a platform for rapid deployment of carrier-grade products and services. The project is initiated by the Linux Foundation and in coordination with ETSI, service providers, and equipment vendors.

11.11 SDN: Making a Compelling Business Case

Like most technological developments, SDN was necessitated by the ever-increasing bandwidth requirements of latency-sensitive real-time applications that support the dynamic deployment of services in tune with enterprise and industry requirements. SDN is another milestone in the movement toward the reduction of proprietary (albeit with standard interfaces) hardware and associated software systems in the ICT industry. The imminent arrival of the 5G systems and the related growth in technologies like IoT, tactile Internet development of artificial intelligence (AI), and related areas like man-machine-interface (MMI) all contribute to the growth of SDN and virtualization technologies.

The present networks are hierarchical and generally based on a tree structure. However SDN and NDV work better with flat structures. The flattening of the traditional networks is one of the strongest business cases for SDN with the resultant lowering of TCO, CAPEX, and OPEX. It is estimated that the flattening of the network architecture can result in savings of over 95%—space, power, cabling, and cooling systems[6] [5]. (See Table 11.2.)

The following are some of the drivers as well as key issues that need to be addressed during the design phase:

1. *Mobility*—Increased deployment of mobile devices;
2. *Increasing bandwidth requirements*—Reality technologies and big data have a huge requirement for bandwidth;
3. *Cloud infrastructure*—Increased system deployment over cloud infrastructure;

---

6. Actual savings in TCO, CAPEX, and OPEX are dependent on the specific case under consideration.

Table 11.2
Making a Compelling Business Case for SDN

Benefits	Impact
Service agility	Rapid deployment of new services based on business needs
Enhanced operational efficiencies	OPEX reduction by 80% Manpower cost reduction by 72% On-the-fly updating/modification of network configuration
Hardware reduction	Resource consolidation—CAPEX savings Reduction in server hardware by 65% Reduction in switching hardware by 80% Reduced OPEX—power savings by 80%
Network availability	99.999 availability
Network performance	On-the-fly updating/modification of network configuration
Effective network management	GUI-based network configuration
Enhanced network visibility	Network level visualization

4. *Changing traffic patterns*—Emergence of east-west (machine-to-machine traffic) and increasing conventional north-south (central location-to-user device) traffic pattern.

11.12 SDN: Conceptual Basis

SDN results in a network architecture that accords agility to enterprises to dynamically develop and deploy network services; reduce TCO, CAPEX, and OPEX; and provide enhanced visualization and network management capabilities with unmatched operational flexibility and scalability. SDN architecture delinks the control and forwarding functions of the network, abstracts the underlying infrastructure from the applications and services layers, and enhances service agility by making the control and program functions programmable. The following are some of the key features of SDN [6]:

1. Programmability and control plane abstraction: Direct programming of network control functions due to the delinking of the forwarding function.

2. Agility: Ability to dynamically configure and deploy services in tune with business needs. Control plane abstraction provides the capabilities to dynamically control and adjust traffic flows based on demand.

3. Centralized management: SDN controllers, with built-in network intelligence, provide a global view of the network. Abstraction layers

provide a common visualization to applications, services, and policy engines.

4. Dynamic configuration/reconfiguration: Automated programs provide on-demand capabilities for FCAPS support.

5. Use of open standards: Use of open standards simplifies network design.

Standardization

Two study groups, operating with the Standards Activities Council of the IEEE Communications Society, are responsible for examining standardization opportunities in SDN, NFV, and allied areas. The key IEEE standards related to SDN are presented in Table 11.3.

11.13 Architectural Framework for SDN

The architecture presented in this section and illustrated in Figure 11.7 is based on the model specified by ONF. The SDN architecture supports the use of open interfaces that facilitate the development of SDN applications that control connectivity between a set of network resources and the associated traffic flows [7]. The architecture consists of three interconnected layers or planes (illustrated in Figure 11.7), described as follows.

1. Application plane—Hosts SDN applications that interact with the control pane using northbound interfaces (NBIs).

Table 11.3
List of IEEE SDN Standards

Standard	Description
IEEE P1903.1	Standard for Content Delivery Protocols of Next Generation Service Overlay Network (NGSON)
IEEE P1913.1	Software-Defined Quantum Communication
IEEE P1915.1	Security for Virtualized Environments
IEEE P1916.1	Performance for Virtualized Environments
IEEE P1917.1	Reliability for Virtualized Environments
IEEE P1921.1	Software-Defined Networking Bootstrapping Procedures
IEEE P1930.1	SDN-based Middleware for Control and Management of Networks
IEEE P802.1CF	Recommended Practice for Network Reference Model and Functional Description of IEEE 802 Access Network

2. Control plane—The control plane consists of the SDN controller, the heart of the SDN architecture. The SDN control plane can have centralized, hierarchical, or decentralized design

3. Data plane—The data plane consists of the SDN datapaths that are the software representation of the network functions/elements. The datapaths interface with the control plane using the SDN control to data plane interface (CDPI).

11.13.1 Architectural Components

The primary components of an SDN framework are listed and described as follows.

1. SDN controller: The SDN controller, the primary entity (logical entity with centralized control) within an SDN network, is responsible for the following critical functions:
    a. Network topology information—Manual configuration or auto discovery;
    b. Abstraction—Provision of northbound interfaces for applications:
        i. Communicating forwarding instructions;
        ii. Communicating policy instructions;
        iii. Communicating configuration commands;
        iv. Network control and management.

The centralized SDN controller acts as an interface between the application layer and the datapaths, network abstraction, NBIs, control logic, and the interface between the control and data plane in the form of the CDPI driver. Figure 11.8 illustrates the functional architecture of an SDN controller.

Implementation of SDN controllers can be vendor-specific (proprietary) and/or based on open standards. It is also pertinent to note that while many use cases employ a centralized SDN controller, a distributed architecture is also feasible. In a distributed set-up the SDN controller function is distributed across multiple computing nodes for the purpose of capacity enhancement and/or redundancy.

Example 11.3—Mobile Broadband Networks
SDN implementations for telecom service providers are based on a distributed architecture with multiple domain-specific controllers. Probable domains include the following:

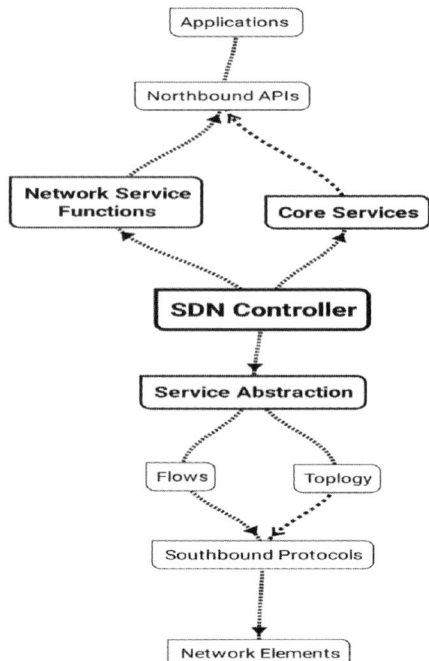

Figure 11.8 Functional blocks of an SDN controller.

1. RAN: RAN controller;
2. Core: Core network controller;
3. Data centers. The multiple SDN controllers, in a distributed architecture, are coordinated by an orchestration layer or controller of controllers. The SDN controller has an E2E network visibility.

2. SDN applications: The SDN applications are software programs that execute specific tasks within the SDN environment. Some of these applications replace the firmware present in the conventional devices. A typical SDN application consists of an application a logic, abstraction layer, and NBI agents. The underlying communication infrastructure is hidden (abstraction) from the applications by the SDN controller. The communication with the SDN controller is through the NBI. The applications could include network management applications, business applications, and analytics applications.

3. Datapath: The SDN datapath refers to the software representation of a network device (e.g., switches and routers). An SDN datapath is a logical device that provides access and control to the forwarding and data processing capabilities corresponding to the functional device being represented (e.g., the switch). The SDN datapath includes a CDPI

agent, a set traffic-forwarding engine (single or multiple), and traffic-processing functions (as required). The functionalities could include forwarding between external interfaces, traffic handling (processing), or termination functions. Single or multiple SDN fatapaths may be contained in a single (physical) network element and managed as a single entity or defined across multiple physical network elements.

4. SDN CDPI: The SDN CDPI is an open interface defined between an SDN controller and an SDN datapath. The following functionalities are provided by the interface:

   a. Programming control of forwarding operations;
   b. Capability advertisements;
   c. Reporting;
   d. Event notification.

5. SDN northbound interfaces: SDN NBIs are open interfaces positioned between SDN applications and SDN controllers to provide an abstract view of the network and facilitate direct expression of network behavior and other requirements. The NBI may be positioned at any level of abstraction (latitude) and different sets of functionality (longitude).

11.13.2 Protocol support

The communication protocols typically used in the telecommunication industry for SDN include the following:

1. OpenFlow: OpenFlow was the first standard communications interface defined between the control and the forwarding layers of SDN (in 2011). The OpenFlow standard is managed by the Open Networking Foundation (ONF), an organization is promoting the adoption of OpenFlow. OpenFlow provides access and control to the network device-forwarding plane (physical and virtual), thereby enhancing network agility and facilitating dynamic control of key network functions. OpenFlow sits on top of the transport layer (TCP) and uses the transport layer security (TLS) mechanisms. OpenFlow works with the device-routing tables (layer 3) to dynamically configure and later traffic flows through the network. Most vendors have products that work with OpenFlow.

2. Border gateway protocol: A hybrid version customized for SDH support.

3. NETCONF: A management protocol specified by the IETF that provides a secure mechanism to configure network elements like switches, routers, and firewalls. ONF has mandated the use of NETCONF for the configuration of OpenFlow-enabled devices (in a specification referred to as OS-CONFIG).
4. Extensible Messaging and Presence Protocol (XMPP): XMPP is a protocol based on the extensible mark-up language (EML), which provides instant messaging and online presence detection capability to SDN elements. The protocol provides servers and facilities with real-time operational capabilities and disseminates control and management plane information to the server endpoints.
5. Open vSwitch Database Management Protocol (OVSDB) : OVSDB is a virtual switch that supports standard management interfaces, protocols, and information dissemination across the server environment.
6. MPLS-TP: MPLS-TP is a layer 3 technology used in transport networks to extend the MPLS function to circuit-switched networks. MPLS-TP is explained in Chapter 10.
7. Packet cable multimedia (PCMM): PCMM is a cable-specific protocol for configuring dynamic DOCSIS service flows.
8. SNMP: SNMP (reconfigured to work for SDN) is a network-management protocol used to collect and configure information from network devices like servers, printers, switches, and routers.
9. Other common platforms include the open network environment (Cisco) and the network virtualization platform (Nicira).

11.13.3 Design and Deployment Considerations

Some of the key points to consider while planning for SDN deployments are detailed as follows:

1. Choice of architecture: Table 11.4 lists the available architectural options for SDN deployment.
2. Controller placement: The following are some of the important criteria for controller placement:
   a. The number of controllers and the placement of the control entities;
   b. The propagation delay (or the latency) between the control entities and the network devices;
   c. The reliability of the control path;

Table 11.4
SDN Architectural Choices

Architecture	Features	Drawback(s)
Centralized	Single-control entity Global network view Simplified control logic	Not easily scalable
Distributed	Distributed controllers Local views Support for adaptive SDN applications	Need for synchronization messages
Hierarchical	Centralized root Controller Distributed controllers Partitioned network view	Synchronization of root and distributed controllers

    d. The fault tolerance;

    e. The application-specific requirements.

3. Flow forwarding: OpenFlow uses ternary content addressable memory (TCAM) tables to route packet flows, described as follows:

    a. When a flow arrives at an ingress port, on a switch, a flow table lookup (software tables) is performed (depending on the type of switch—vSwitch or an ASIC-based switch).

    b. A request to the SDN controller is raised in case of no matching entries.

    c. The request can be processed (handled) using one of three available modes:

        i. *Reactive mode*—Controller creates and installs a rule in the flow table with the details of the flow under consideration.

        ii. *Proactive mode*—Controller populates flow table with all possible matches for the flow under consideration similar to routing table entries in hardware routers. This mode does not require a controller request to be raised by the function. Packets are forwarded at line rate with no delay.

        iii. *Hybrid mode*—Combines the benefits of the reactive and proactive modes (flexibility of the reactive mode and low latency forwarding of the proactive mode).

11.14 Management

A radically different approach is required for the management of SDN networks, as their architecture renders the traditional management techniques ineffective. In a traditional network any activity related to the deployment of new products and/or services takes a considerable amount of planning and a relatively long deployment time. In contrast SDN-enabled networks require only a fraction of the time to deploy and configure new services. The traditional networks are application-aware while in the case of SDN the applications are network-aware. The traditional networks are device-centric and manually configured while SDN-enabled networks are programmable and hence their behavior changes dynamically in response the changing network traffic flows.

The management of an SDN network mandates an automated framework. All critical areas need to be automated including (but not limited to) the following:

1. Capacity planning;
2. Monitoring;
3. Troubleshooting;
4. Security;
5. Route analysis.

Automated route analysis and update is one of the critical areas that impacts network availability and performance due to the ability to view the entire network topology coupled with real-time updates on traffic dynamics and route status.

11.15 OpenFlow

OpenFlow was the first standard communications interface defined between the control and the forwarding layers of SDN (in 2011). The OpenFlow protocol is layered on top of the TCP and enables SDN controllers to determine the path for network packets across the underlying physical infrastructure. The separation of the control and the forwarding plane facilitates optimal and effective traffic management in comparison to traditional networks that follow static routing. Some of the important features and benefits of OpenFlow are discussed in the following sections.

Features

1. Remote management of physical devices (with proprietary hardware, firmware, and software but standard network interfaces) from multiple vendors to be managed by a single open protocol.
2. Remote administration of layer 3 packet-forwarding tables—Addition, deletion, and/or modification of rules

3. Routing decisions to be made by the SDN controller with configurable life spans facilitating automatic discovery and management of routes.

Benefits
1. Abstraction;
2. Centralized intelligence;
3. Delink:
    a. Control and forwarding planes;
    b. Hardware and software;
    c. Physical and logical configuration.
4. Deployment of innovative services;
5. Granular policy management;
6. Network programmability;
7. Optimized performance;
8. Simplified provisioning.

In the initial days of SDN evolution OpenFlow was the only protocol used by SDN and virtualization technologies. However the last decade has seen the emergence and use of a number of other protocols as described in Section 11.9.2.

11.16 SDN Use Cases

Most of the functions of a converged triple-play telecommunication network are potential candidates for deployment of SDN and virtualization networks. In order to maintain continuity and to understand the nuances of NFV and SDN, we will revisit the example of the 4G network in the following use case and present the the SDN deployment details.

Use Case—Software Defined Mobile Networking

Software-defined mobile networking (SDMN) [7] is an innovative approach in the design and deployment of mobile broadband networks with software implementation of protocol-specific features in RAN and core networks and the use of generic and commodity hardware and software. The architecture of the conventional EPS network is illustrated in Figure 11.3, and the corresponding SDMN is illustrated in Figure 11.9.

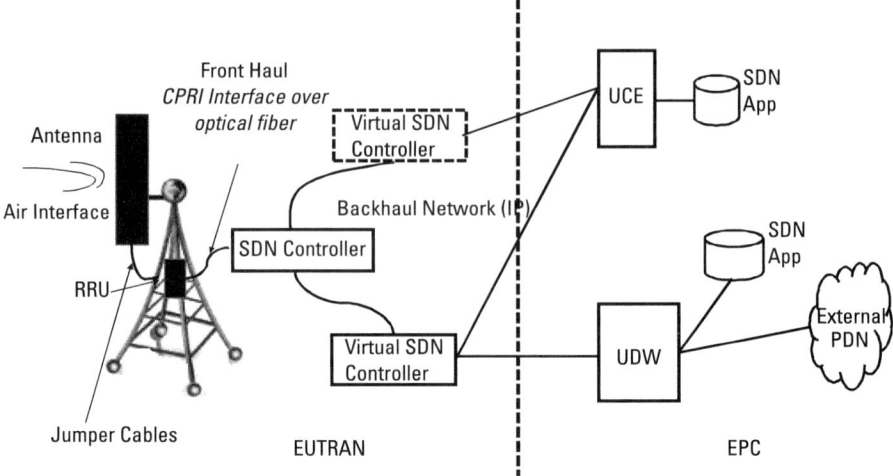

Figure 11.9 SDMN network architecture.

Existing Network Architecture

1. Air Interface: A UE (mobile) connects to the eNodeB over the air interface also referred to as the Uu interface.
2. EUTRAN:
    a. The EUTRAN consists of eNodeBs interconnected through an IP/MPLS backhaul network.
    b. The eNodeBs communicate with each other using the X2 interface.
    c. A fronthaul network connects the eNodeBs to the RRUs mounted with the RF antennas.
3. SAE architecture: The key system components of the SAE/EPC are outlined in Table 11.5.

SDMN Architecture

The SDMN architecture implementation of the EPC is illustrated in Figure 11.9, summarized in Table 11.6, and described as follows.

1. The key components of the SDMN architecture are the forwarding engine and the SDN controller.
2. The forwarding engines are connected through an IP/MPLS BH/transport network.
3. The separation of the control and the user plane is effected by the separation of the FE, representing the user plane, and the centralized

Table 11.5
EPS Components

Entity	Expanded Form	Interfaces
MME	Mobility management entity	S1 interface (eNB) S10 (MME) S11 (SGW) S6a (HSS)
HSS	Home subscriber server	S6a (MME)
SGW	Serving gateway	S1-U (eNB) S4 (SGSN) S5 (PGW) S11 (MME) S12 (UTRAN)
PSW	PDN gateway	S5 (SGW) Gx (PCRF) SGi (External PDN)
PCRF	Policy and charging rules function	Gx (PCRF) Rx (External PDN)

Table 11.6
SDMN Components

Old Entity	New Entity	Interfaces
MME (control plane)	SDN (flow) controller + SDN applications (MME, S-GW, P-GW...)	Swf (FE) Legacy support −S1 MME (eNB) −S10/S11 (Legacy MME) −S5/S8 (Legacy S/P GW)
HSS	SDN application	Legacy support −*S6a* (Legacy MME)
SGW (user plane)	SDN application	Smf Legacy support −*S1-U* (eNB) −S4 (SGSN) −S5 (PGW) o S10 −S11 (MME) −S12 (UTRAN)
PSW (user plane)	SDN application	Smf Legacy support −S5 (SGW) −Gx (PCRF) −SGi (external PDN)
PCRF	SDN application	Legacy support −Gx (PCRF) −Rx (external PDN)

SDN controller and the mobile applications (deep packet inspection (DPI), with video caching representing the control plane.

4. The forwarding can be virtualized for optimal utilization.

5. The forwarding engine applications are simpler in comparison to their physical implementation due to the separation of the control and user planes.
6. Traffic can be dynamically routed to either the forwarding engine or the underlying backhaul/transport network.
7. The forwarding engine functionalities include the following:

    a. Tunnel processing;

    b. GTP (GPRS tunneling protocol);

    c. GRE (generic routing encapsulation);

    d. Radio bearer management (using radio interfaces). Table 11.7 summarizes the key differences between 4G and 5G networks.

11.17 Summary

The primary objective of NVF is to separate the software and hardware components of a network node to facilitate their organic development, independent of each other. The decoupling also implies that dedicated resources can now

Table 11.7
Comparison of Key Parameters of Existing 4G and Proposed 5G Networks

Parameters	4G Networks	5G Network (Proposed)
Multiple access	DL- OFDM/OFDMA UL-SC-FDMA	OFDM
Frequency	700 MHz	3,100–3,550 MHz (and 3,700–4,200 MHz)
Bandwidth*	22 MHz, 34 MHz, 46 MHz	100 MHz
Spectral efficiency	2-5 times (3G) -16bits/Hz (Rel. 8.0), 30 bits/Hz (Rel.10)	30 bits/Hz
Scalable Spectrum Allocation	1.4, 3, 5, 10, 15, 20 MHz	100 MHz
Standard	3GPP, backward compatibility with non-3GPP networks	IMT 2020
Performance	All-IP network, low latency (user plane latency $\leq$ 5 ms, control plane latency $\leq$ 100 ms)	$\leq$ 1ms
Multiplexing	FDD/TDD	Dynamic TDD
Data coding	Turbo Code	LDPC Code
Throughput	1 Gbps DL/500MbpsUL**, 300 Mbps DL/150 Mbps (2*2 MIMO)	20 Gbps DL/10 Gbps UL(100 Mbps/User actual)

*Geography-dependent.
** LTE-A (Release 14) for stationary users (50% of the value for mobile users).

be shared or pooled amongst a larger set of entities resulting in elimination of redundancies and realizing economies of scale and process and product efficiencies. NFV and SDN are interrelated concepts that have value independently but that complement each other to bring out a radically new value proposition when deployed in tandem. The origins and the objectives of both these technologies are different with SDN being developed (by researchers) to introduce programmability to network devices, while NFV is a service provider–initiated concept developed to facilitate the rapid development and deployment of services.

The adoption of virtualization and SDN will result in incremental changes that will bring down the TCO and enhance revenue generation (the ability to dynamically provide and configure new services) for telcos. One of the earliest use cases for NFV was the virtualization of CPE, which led to logistic efficiencies and the associated reduction in OPEX.

SDN was necessitated by the ever-increasing bandwidth requirements of latency-sensitive real-time applications that support the dynamic deployment of services in tune with enterprise and industry requirements. SDN is another milestone in the movement toward the reduction of proprietary (albeit with standard interfaces) hardware and associated software systems in the ICT industry. The imminent arrival of the 5G systems and related growth in technologies like IoT and tactile internet, development of AI and related areas like MMI all contribute to the growth of SDN and virtualization technologies.

During the period of 2017 to 2020 service providers will enhance their portfolio of software-based services to customers using virtualized infrastructure, virtual customer premises equipment (vCPE), and enterprise vCPE. Spending on consumer and enterprise hardware and software is forecast to reach $1.5 billion worldwide by 2020.

11.18 Review

11.18.1 Review Questions

1. NFV will lead to the virtualization and consolidation of service provider NEs onto industry standard server, storage, and switching systems that can be shared with multiple users.

    a. True

    b. False

2. A VNF consists of one or more virtual machines that work on specific software and distinct processes that are hosted on high-configuration server farms with multiple, redundant high-capacity storage systems

connected to networks with high-speed switching devices and/or cloud infrastructure.

a. True

b. False

3. The two primary benefits of adopting virtualization technogies are: _____ and opportunity cost.

a. Bandwidth enhancement

b. Reduced TCO

c. Reduced FCAPS

d. Reduced OAM activities

4. The primary objective of NVF is to _____ the software and hardware components of a network node to facilitate their organic development, independent of each other.

a. Separate

b. Integrate

c. Eliminate

d. Combine

5. NFVI encompasses the set of hardware and software components that provide the environment for the deployment of VNFs.

a. True

b. False

6. A VNF service: implementation is through a _____ of network functions and end/terminal points.

a. Forwarding graph

b. Routing table

c. Flow table

d. Link

7. NFV and SDN are interrelated concepts that do not have value independently, but complement each other to bring out a radically new value proposition when deployed in tandem.

a. True

b. False

8. SDN results in a network architecture that accords agility to enterprises to dynamically develop and deploy network services; reduce TCO,

CAPEX, and OPEX; and provide enhanced visualization and network management capabilities with unmatched operational flexibility and scalability.

   a. True

   b. False

9. The SDN architecture consists of three interconnected layers or planes.

   a. True

   b. False

10. The SDN controller, the primary entity (logical entity with centralized control) within an SDN network, is responsible for the following critical functions: _____ and abstraction.

    a. Network topology information

    b. Storage

    c. Applications

    d. User plane

11.18.2 Exercises

1. Briefly describe NFV and SDN technologies and their interrelationship.

2. Explain the benefits of NFV and SDN deployment (separately) in the mobile broadband network environment (4G).

3. List and describe the compelling business case for the deployment of virtualization technologies.

11.18.3 Research Activities

1. Briefly describe the impact of virtualization in the reality technology domain. Justify your standpoint.

2. Will the impending arrival of 5G technologies have any impact on future developments in the field of NFV and/or SDN? Explain your viewpoint in detail.

3. Will NFV and/or SDN deployments have any impact on the development of IoT and tactile Internet? Provide the basis for your observations.

11.19 Case Study—Wireless Broadband Networks

Case—Verizon 4G Networks[7]

4G Deployment Objectives

1. Speed to support new-age applications (bandwidth-intensive);
2. Global standard—seamless roaming;
3. Backward compatibility and interoperability with other cellular technologies;
4. Low latency;
5. Security.

SDN/NFV-based Network Architecture[8]

1. Goals:
    a. Business transformation;
    b. Enhanced time-to-market;
    c. Fully automated OAM&P;
    d. Dynamic service provisioning and management.
2. Dynamic traffic management:
    a. Scaling up of operational efficiencies;
    b. Enhanced scalability and flexibility;
    c. Service agility.
    b. Features:
        a. Control and data plane separation;
        b. Efficient resource management;
        c. Automated resource orchestration;
        d. Automation of control, deployment, and business processes;
        e. Programmatic network control;
        f. Standards-based configuration protocols;
        g. NFV.

---

7. Refer to Section 9.3.3.
8. Verizon Network Infrastructure Planning, *SDN-NFV Reference Architecture v.1.0* (Feb 2016), Source URL<http://innovation.verizon.com/content/dam/vic/PDF/Verizon_SDN-NFV_Reference_Architecture. pdf>.

Network Architecture

Figure 11.10 illustrates the physical architecture of a 5G network. Table 11.8 lists the elements of a conventional 4G network and its SDN and SDN+NFV equivalence:

11.20 Referred Standards

ETSI GR NFV-TST 004 V1.1.1 (2017-05)—Network Functions Virtualisation (NFV); Testing; Guidelines for Test Plan on Path Implementation through NFVI

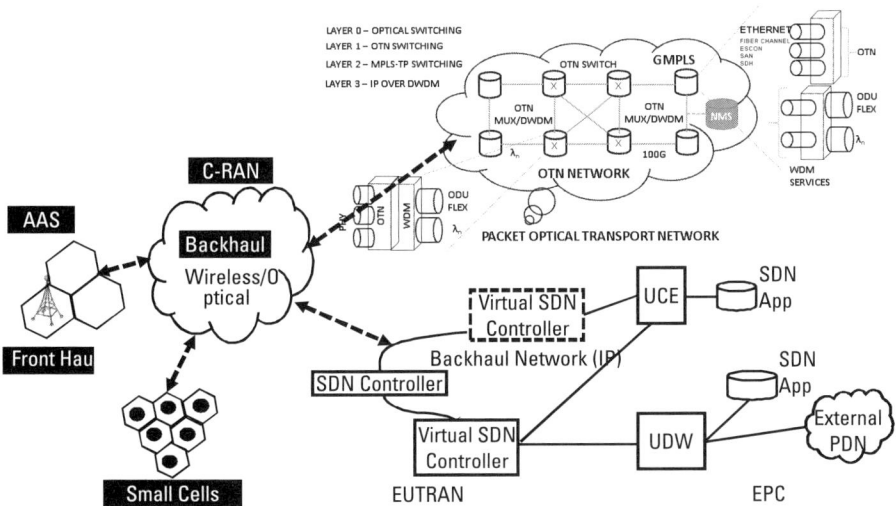

Figure 11.10 5G network architecture.

Table 11.8
Proposed SDN/NFV-Based Network Architecture

Existing 4G Architecture	Network Element	SDN-Based 5G Architecture	SDN + NFV-Based 5G Architecture
RAN	BBU	SDN controller	Virtual BBU
Core	MME	Unified control entity (UCE) (Control rule definition MME, S-gateway Control Plane, P-Gateway Control Plane)	Edge controller (EC)—Single-RAN domain
	HSS		+
	S-gateway		Global controller (GC)— Multiple RAN Domains
	P-gateway	+	
	PCRF	Unified data gateway (UDW) (data forwarding rules— S-gateway data plane, P-gateway data plane)	

ETSI GR NFV-SEC 009 V1.2.1 (2017-01)—Network Functions Virtualisation (NFV); NFV Security; Report on use cases and technical approaches for multi-layer host administration

ETSI GR NFV-SEC 003 V1.2.1 (2016-08)—Network Functions Virtualisation (NFV); NFV Security; Security and Trust Guidance

ETSI GS NFV-EVE 005 V1.1.1 (2015-12)—Network Functions Virtualisation (NFV); Ecosystem; Report on SDN Usage in NFV Architectural Framework

ETSI GS NFV-SWA 001 V1.1.1 (2014-12)—Network Functions Virtualisation (NFV); Virtual Network Functions Architecture

ETSI GS NFV-TST 008 V2.1.1 (2017-05)—Network Functions Virtualisation (NFV) Release 2; Testing; NFVI Compute and Network Metrics Specification

ETSI GS NFV-IFA 008 V2.1.1 (2016-10)—Network Functions Virtualisation (NFV); Management and Orchestration; Ve-Vnfm reference point—Interface and Information Model Specification

ETSI GS NFV-IFA 007 V2.1.1 (2016-10)—Network Functions Virtualisation (NFV); Management and Orchestration; Or-Vnfm reference point—Interface and Information Model Specification

ETSI GS NFV-IFA 013 V2.1.1 (2016-10)—Network Functions Virtualisation (NFV); Management and Orchestration; Os-Ma-Nfvo reference point—Interface and Information Model Specification

ETSI GS NFV-IFA 014 V2.1.1 (2016-10)—Network Functions Virtualisation (NFV); Management and Orchestration; Network Service Templates Specification

ETSI GS NFV-IFA 011 V2.1.1 (2016-10)—Network Functions Virtualisation (NFV); Management and Orchestration; VNF Packaging Specification

ETSI GS NFV-IFA 005 V2.1.1 (2016-04)—Network Functions Virtualisation (NFV); Management and Orchestration; Or-Vi reference point—Interface and Information Model Specification

ETSI GS NFV-IFA 004 V2.1.1 (2016-04)—Network Functions Virtualisation (NFV); Acceleration Technologies; Management Aspects Specification

ETSI GS NFV-IFA 006 V2.1.1 (2016-04)—Network Functions Virtualisation (NFV); Management and Orchestration; Vi-Vnfm reference point—Interface and Information Model Specification

ETSI GS NFV-IFA 003 V2.1.1 (2016-04)—Network Functions Virtualisation (NFV); Acceleration Technologies; vSwitch Benchmarking and Acceleration Specification

ETSI GS NFV-IFA 010 V2.1.1 (2016-04)—Network Functions Virtualisation (NFV); Management and Orchestration; Functional requirements specification

ETSI GS NFV-IFA 002 V2.1.1 (2016-03)—Network Functions Virtualisation (NFV); Acceleration Technologies; VNF Interfaces Specification

ETSI GS NFV-IFA 010 V2.2.1 (2016-09)— Network Functions Virtualisation (NFV); Management and Orchestration; Functional requirements specification

ETSI GS NFV-EVE 001 V3.1.1 (2017-07)—Network Functions Virtualisation (NFV); Virtualisation Technologies; Hypervisor Domain Requirements specification; Release 3

ETSI GS NFV-IFA 019 V3.1.1 (2017-07)—Network Functions Virtualisation (NFV); Acceleration Technologies; Acceleration Resource Management Interface Specification; Release 3

ETSI GS NFV-IFA 018 V3.1.1 (2017-07)—Network Functions Virtualisation (NFV); Acceleration Technologies; Network Acceleration Interface Specification; Release 3

ETSI GS NFV-EVE 007 V3.1.1 (2017-03)—Network Functions Virtualisation (NFV) Release 3; NFV Evolution and Ecosystem; Hardware Interoperability Requirements Specification

ETSI GS NFV-SEC 013 V3.1.1 (2017-02)—Network Functions Virtualisation (NFV) Release 3; Security ; Security Management and Monitoring specification

ETSI GS NFV-SEC 012 V3.1.1 (2017-01)—Network Functions Virtualisation (NFV) Release 3; Security; System architecture specification for execution of sensitive NFV components

ETSI GS NFV 001 V1.1.1 (2013-10)—Network Functions Virtualisation (NFV); Use Cases

ETSI GS NFV 002 V1.1.1 (2013-10)—Network Functions Virtualisation (NFV); Architectural Framework

ETSI GS NFV 002 V1.2.1 (2014-12)—Network Functions Virtualisation (NFV); Architectural Framework

ETSI GS NFV 003 V1.1.1 (2013-10)—Network Functions Virtualisation (NFV); Terminology for Main Concepts in NFV

ETSI GS NFV 003 V1.2.1 (2014-12)—Network Functions Virtualisation (NFV); Terminology for Main Concepts in NFV

ETSI GS NFV 004 V1.1.1 (2013-10)—Network Functions Virtualisation (NFV); Virtualisation Requirements

IEEE P1903.1—Standard for Content Delivery Protocols of Next Generation Service Overlay Network (NGSON)

IEEE P1913.1—Software-Defined Quantum Communication

IEEE P1915.1—Security for Virtualized Environments

IEEE P1916.1—Performance for Virtualized Environments

IEEE P1917.1—Reliability for Virtualized Environments

IEEE P1921.1—Software-Defined Networking Bootstrapping Procedures

IEEE P1930.1—SDN-based Middleware for Control and Management of Networks

IEEE P802.1CF—Recommended Practice for Network Reference Model and Functional Description of IEEE 802 Access Network

RFC 7665—Service Function Chaining (SFC) Architecture

11.21 Recommended Reading

11.21.1 Books

Doherty J., *SDN and NFV Simplified: A Visual Guide to Understanding Software Defined Networks and Network Function Virtualization,* Addison Wesley, 2016.

Gray, K., and T. D. Nadeau, *Network Function Virtualization*, Morgan Kaufmann, 2016.

Stallings, W., *Foundations of Modern Networking: SDN, NFV, QoE, IoT, and Cloud*, Pearson Education India, 2016.

Kapadia, K., and N. Chase, *Understanding OPNFV: Accelerate NFV Transformation Using OPNFV,* Mirantis, Inc., 2017.

Goransson, P., and C. Black *Software-Defined Networks: A Comprehensive Approach*

Monge, S. A., and S. G. Krzysztof, *MPLS in the SDN Era: Interoperable Scenarios to Make Networks Scale to New Services,* Morgan Kaufmann, 2014.

11.21.2 URLs

https://www.opnfv.org/.

http://searchsdn.techtarget.com/news/2240187268/Five-SDN-use-cases-From-video-to-service-orchestration.

https://www.sdxcentral.com/sdn/definitions/inside-sdn-architecture/.

11.21.3 Journals

Sher Decusatis, C. J., et al., "Communication Within Clouds: Open Standards and Proprietary Protocols for Data Center Networking," *IEEE Communications Mag.,* Vol. 50, No. 9, Sept. 2012.

Bari, M., et al., "Data Center Network Virtualization: A Survey," *IEEE Communications. Surveys and Tutorials,* Vol. PP, No. 99 (early access article).

Gibb, G., et al., "OpenPipes: Prototyping High-Speed Networking Systems," *Proc. SIGCOMM (Demo),* Barcelona, Spain, 2009.

Ali, I., et al., "Network-Based Mobility Management in the Evolved 3GPP Core Network," *IEEE Communications Mag.*, Vol. 47, No. 2, Feb. 2009.

Ekström, H., "QoS Control in the 3GPP Evolved Packet System," *IEEE Communications Mag.*, Vol. 47, No. 2, Feb. 2009.

References

[1] http://www.etsi.org/technologies-clusters/technologies/nfv.

[2] Holkkola, J., Transformation—The Real Business Case for NFV and SDN (2016). Retrieved from http://www.circleid.com/posts/20161005_transformation_the_real_business_case_for_nfv_and_sdn/.

[3] ETSI, "Network Functions Virtualization (NFV); Infrastructure Overview," NFV ISG, 2015.

[4] "Network Functions Virtualization—Introductory White Paper" (PDF), ETSI. 22 October 2012. Retrieved 20 June 2017.

[5] Ashton Metzler and Associates (2014). SDN LAN Business Case. Retrieved from Ashton-Metzler http://www.ashtonmetzler.com/SDN%20LAN%20Business %20 Case.pdf.

[6] Halpern, J., and C. Pignataro, "Service Function Chaining (SFC) Architecture," Internet-Draft draft-ietf-sfc-architecture-11, July 2015.

[7] Guo, C., et al., "BCube: A High-Performance, Server-Centric Network Architecture for Modular Data Centers," *SIGCOMM 2009*, 2009.

About the Author

Sudhir Warier, human capability management and leadership coach, author, speaker, poet, consultant, and freelance trainer, has over 23 years of corporate leadership exposure spanning the telecom and IT industries. He has had the opportunity to be associated with the entire telecom and IT value chain, both in the technology and management space. He is a fellow of the Institution of Electronics and Telecommunication Engineers (IETE), a chartered engineer (IETE), a graduate in electronics and telecommunication engineering (B.E.), a post-graduate in financial management (M.F.M.), and a master of philosophy—management. He is a member of the Board of Studies and an advisor to some of the leading management and engineering schools in India. He is also on the external examiner panel of the University of Mumbai and a reviewer for leading technology and management journals. In addition, he has authored/self-published 12 books and over 30 research papers. His book *Knowledge Management* is a reference text for several national and international universities for diverse fields including management, engineering, and science.

Index

Absorption, 57
Access layer, 162
Access routers, 314–15
Adaptive PCM (ADPCM)
    defined, 46
    example, 46–47
    predefined algorithm, 46
Add/drop multiplexers (ADMs)
    defined, 165
    in linear configuration, 166
    in ring configuration, 167
Advanced intelligent network (AIN), 10
AEL, 110
Aerial cables, 72–73
Agile optical networks, 300
Amplifiers, 173
Amplitude shift keying (ASK), 93–94
Analog-to-digital (A/D) conversion, 36
Analog transmission, 28
Angled physical contact (APC) connnectors, 100
APT wireless group (AWG), 283
Asynchronous networks
    bit stuffing, 127
    defined, 121
    limitations, 146–47
    OAM&P channels and, 149, 151, 156
    overview, 124–27
    T1 systems and, 121, 124
ATM PONs (APONs), 288–89
Automatically switched optical network (ASON)
    architecture specification, 204
    network reliability and, 210
    standards, 207
Automatically switched transport networks (ASTNs), 204
Automatic protection switching, 202

Back reflections, 89–91
Bending diameter, 70
Bidirectional connection, 169
Bit stuffing, 127
Border gateway protocol (BGP), 375
Border routers, 317
Bose-Chauduri-Hocquenghem codes, 213
Broadcast-and-select networks, 340–41, 342
Broadcast infrared systems, 19
Buried hetero (BH) lasers, 85

Cable preparation, 74
Capability set-1 (CS-1), 10
CAPEX, 156, 215, 233, 284, 357
Case studies
    DWDM deployment, 262–67
    next-generation access networks, 304–6
    OTN deployment, 223–28
    SDH network architecture, 179–84
    wireless broadband networks, 386–87
CCITT recommendations, 129–30
Centralized clock interface, 131–32
Channel coding. *See* FEC
Chip-based integrated optical circuits, 249
Chord coding, 42

Circuit switching, 6
Class 1 switch, 8
Class 2 switch, 8
Class 3 switch, 9
Class 4 switch, 9
Class 5 switch, 9
Clock recovery, 92
Coarse WDM (CWDM), 236, 239
   defined, 33
   frequencies, 248
   See also Wavelength division
      multiplexing (WDM)
Coaxial cable, 13–15
Codirectional interface, 131
Collection/aggregation layer, 162–63
Companding, 44–45
Connection management, OTN, 201–2
Connection-oriented packet-switched (CO-PS) network technology, 293–94
Connectors
   APC, 100
   E2000, 99
   lucent, 97–98
   MU, 98
   optical, 96–100
   SC/APC, 96–100
   SC/PC, 96–100
   subscriber, 97
   UPC, 100
Constrained shortest path first (CSPF)
   defined, 328
   enabling, 330
   functioning, 330–31
   shortest path computation, 330
Converged cable access platform (CCAP), 279–80, 302
Coordinated multipoint transmission/reception (CoMP), 362
Copper wires, 15
Core, 162–63
Core routers, 317
Couplers, 256
Cross-connects
   defined, 170
   functional blocks, 172
   key parameters, 171
   switching capacity, 170
   time slot interchange (TSI), 171
   types of, 171

Data over cable service interface specifications (DOCSIS), 20
   defined, 278
   full-duplex communication support, 279
   standards, 279
Dense WDM (DWDM)
   branched network, 262
   central frequencies, 240, 242
   channel bandwidth, 242
   channels, 237
   defined, 33, 236
   deployments, 243
   development of, 237
   grid, 243
   integrated systems, 234
   introduction to, 234–36
   IP over, 295, 296
   ITU-T 694.1 wave grid, 244–46
   links, 260–62
   multiplexers, 242
   transmission channels, 240
   wavelengths, 248
   See also Wavelength division
      multiplexing (WDM)
Device-to-device (D2D) communication, 284
Differential PCM (DPCM)
   code words, 45
   constant sampling frequency, 46
   defined, 45
Differential PSK (DPSK), 95
Differential quadrature PSK (DQPSK), 95
Digital subscriber line (DSL), 20
Digital-to-analog (D/A) conversion, 36
Digital wrapper technology, 197–98
Dimensioning, optical fiber, 69
Direct distance dialing (DDD), 8
Directly buried cables, 73
Direct sequence modulation, 18
Dispersion-shifted fiber (DSF), 62–63
Distributed feedback (DFB) lasers, 85, 86, 342
Distributed Raman amplifier (DRA), 257
DOCSIS/CCAP architecture, 280
DSL access multiplier (DSLAM), 20
DWDM deployment case study
   background, 262–64
   challenges, 264
   mesh topology, 267

network architecture illustration, 266
proposed solution, 264–68
requirements, 264
See also Dense WDM (DWDM)

E2000 connectors, 99
Edge routers/switches, 317
Electrical safety, 108
Electromagnetic interference (EMI), 14
eNodeB (eNB), 359
Epoxy and polish connectors, 96
Error detection/recovery, 93
Ethernet over PDH, 144
Extensible Messaging and Presence Protocol (XMPP), 376
External network-network interface (E-NNI), 208
Extinction ratio, 91–92

Fabry-Perot (FP) lasers, 85
Fabry-Perot amplifiers (FPAs), 257
Few-mode fibers, 64–65, 298
Fiber connecterization, 99–100
Fiber-handling techniques, 110–12
Fiber-laying techniques
    aerial cables, 72–73
    directly buried cables, 73
    horizontal directional drilling (HDD), 73
    illustrated, 71
    mauling, 73
    microtrenching, 72
    minitrenching, 72
    trenching and ducting (T&D), 71–72
Fiber management systems (FMSs), 76
Fiber-optic cables (FOCs), 14
Fiber-optic communication
    block diagram, 55
    cable preparation, splicing, and termination, 73–76
    fiber design specification, 65–70
    fiber-laying techniques, 71–73
    fundamentals, 53–81
    introduction to, 54–55
    light transmission and, 55–59
    optical fiber classification, 59–60
    optical fiber designs, 60–61
    optical fiber types, 62–65
    summary, 76–77

Fire safety, 107
5G mobile wireless networks
    antennas, 285
    application and business service plane, 286–87
    application types, 284
    architecture, 387
    backhaul, 285
    converged data plane, 286
    cutting-edge services, 284
    defined, 282
    hardware architecture, 285–86
    infrastructure control plane, 286
    integrated network management and operations plane, 287
    key features of, 282
    multiservice management plane, 287
    proposed spectrum allocation, 283
    software architecture, 286–87
    software plane, 286
    spectrum, 283
    U.S. frequency ranges, 283
Flexible WDM grids, 249–51
    arrangement, 250
    deployment of, 250
    illustrated, 250
    use of, 251
Flow forwarding, 377
Forward error correction (FEC)
    advantages of, 214–16
    in-band, 211
    Bose-Chauduri-Hocquenghem codes, 213
    channel code types, 211
    coding gain, 215–16
    defined, 211
    in dispersion-limited systems, 212
    hard-decision, 216
    LDPC block codes, 214
    in OSNR limited systems, 212
    Reed-Solomon codes, 212–13
    soft-decision, 216
4G mobile broadband networks (LTE), 175
4G network architecture, 359, 360
Frequency-division multiplexing (FDM), 30–31, 123
Frequency hopping, 18
Frequency shift keying (FSK), 95
FTTx networks
    architecture illustration, 282

FTTx networks (continued)
    defined, 281, 288
    design approaches, 288
    variants, 281
Full Service Access Network (FSAN)
    working group, 281, 289
Fusion-splicing, 74–75
Future packet-based networks (FPBNs), 347

G.652, 174
G.653, 174, 251
G.654, 174
G.655, 251
G.671, 237
G.691, 238
G.692, 238
G.694, 238
G.702, 129–30
G.703, 156
G.872, 197, 198
G.957, 238
G.982, 238
Generalized MPLS (GMPLS)
    classes of interfaces, 345
    defined, 345
    features not supported by, 345
    switching support, 297
Generic mapping procedure (GMP), 218

Hard-decision FEC, 216
HCFs, 298–99
Horizontal directional drilling (HDD), 73

In-band FEC, 211
Infrared, 19
Injury, common, 108, 109
In-line amplifiers (ILAs), 261
Inside plant (ISP) cables, 60
Intensity modulation (IM), 93
Interfaces, OTN, 200–201
Interleaving, 213
Internal network-network interface (I-NNI), 205
IP/MPLS optical core networks
    advantages of, 294–95
    defined, 294
    enhanced router interfaces, 295–96
    integration, 295

IP over DWDM architecture, 295, 296
IP over DWDM architecture, 295, 296
IP over MPLS, 334–36
ISPs, 319, 320, 321
ITU-T 694.1, 243, 244–46

Label distribution protocol (LDP), 333–34
Laser classification, 111
Light
    behavior on media, 57
    effects, 56–57
    modal propagation, 58–59
    propagation, 56, 57
    transmission basics, 55–59
Line decoding, 92
Link budgets
    defined, 100
    key parameters, 100–102
    power budget and margin
        computations, 101
    preparing, 100–102
    receiver sensitivity and dynamic range, 101
    transmit launch power, 100–101
Link loss, 100
Local loops, 11
Local network, 11–12
Logical interfaces
    control plane, 205–8
    OTN, 200–201
Long-distance network, 12
Longevity, optical fiber, 70
Loose-tube design, 60–61
Low-density parity check (LDPC) block
    codes, 214
Lucent connectors (LCs), 97–98

Material safety, 107
Mauling, 73
Maximum permissible exposure (MPE), 108–10
Mechanical splicing, 75
Microtrenching, 72
Microwave links, 175
Microwaves, 18–19
Minitrenching, 72
Modular optical interfaces, 86–89
MPLS
    benefits of, 323

elements (physical and logical), 324–26
forwarding table, 326
GMPLS and, 297–98, 345–46
IP over, 334–36
label description, 324
label format, 323
label switching process, 324–26
LSP types, 326–27
network architecture, 324
next-hop, 335–36
overview, 322–23
packet forwarding, 323–24
penultimate hop popping, 327
traffic engineering (TE), 327–34
tunnels, 334–35
MPLS-TP, 293–94, 346, 376
MU connectors, 98
Multiaccess edge computing (MEC), 284
Multicore fibers (MCFs), 299
Multi-mode fiber (MMF)
cables, 64–65
classes of, 64
core diameters of, 59, 298
cross-sectional view, 70
defined, 58
RI profiles, 63
types of, 64–65
Multimode graded index, 59
Multimode step index, 59
Multiplexers/demultiplexers, 164–70
functional architecture, 168
input channels, 168
Multiplexing
FDM, 30–31
overview, 29–30
SDM, 34–35
TDM, 31, 32
techniques, 30
WDM, 31–34
Multi-quantum well (MQW) lasers, 85
Multiservice provisionable platform (MSPP), 295
Multistage multiplexing, 218–19

NADH DS1 frame format, 134–35
NADH DS2 frame format, 135, 138, 139
NADH DS3 frame format, 135–37, 140
Narrowband WDM (NWDM), 236
NETCONF, 376
Network elements (NEs)

defined, 158
OTN, 192–93
in a ring, 166
Network function virtualization (NFV)
adoption of, 354
backward compatibility/interoperability, 365–66
business case, 357
defined, 355
design considerations, 365–67
distributed architecture, 367
in extending cutting-edge services, 284
flat structures and, 370
framework and services, 362–65
framework illustration, 363
functional blocks, 363–64
introduction to, 354–57
mobile broadband networks, 359–62
network management and orchestration, 369–70
network performance, 366
network reliability, 366
network security, 366
open-source platforms, 370
provisioning of services, 366–67
SDN and, 355, 368–69
service implementation, 364
summary, 382–83
use cases, 357–62
VNF services, 364
Network management and orchestration (MANO), 369–70
Next-generation access networks case study
challenges, 304
network architecture, 305
ODN, 306
overview, 304
solution, 305
wavelength provisioning, 306
Next-hop MPLS, 335–36
NG-PONs, 292–93
NG-POTN
agile optical networks, 300
architecture features, 297–301
architecture illustration, 297, 299
defined, 296
generalized MPLS (GMPLS), 297–98
optical fiber types, 298–99
P-OTS, 296
ROADMs, 300–301

NG-POTN (continued)
    SDM, 298
    transparent optical transport with traffic protection, 301
No-epoxy and no-polish connectors, 97
Nondispersion-shifted fibers (NDSF), 62–63
Non-return-to-zero/return-to-zero (NRZ/RZ) techniques, 94
Non-zero-dispersion-shifted fibers (NZ-DSFs), 63
North American Digital Hierarchy (NADH), 121, 124–27
Nyquist criteria, 36

ODN, 289, 291, 306
ODUflex, 218
O-E-E-O (OEO) conversion, 235, 261
OFDM PON, 292
OLT, 289
On-off keying (OOK), 93
ONU, 289
Opaque OXCs, 339
OpenFlow
    benefits, 378–79
    defined, 375
    features, 378–79
    flow forwarding, 377
    overview, 378
Open-source NFV platforms, 370
Open vSwitch Database Management Protocol (OVSDB), 376
Operations, administration, maintenance, and provisioning (OAM&P), 123, 137, 146, 161, 188, 192
OPEX, 156, 312, 338, 357
Optical access networks
    broadband, 278–80
    converged cable access platform (CCAP), 279–80
    DOCSIS, 278–79
    fiber, 280–82
    5G mobile wireless, 282–87
    overview, 277
    split-MAC, 280
Optical add/drop multiplexers (OADMs)
    deployment of, 96
    elements of, 259
    input to, 260
    wavelength multiplexing, 261
Optical amplifiers, 256–57

Optical burst switching (OBS)
    configuration, 344
    defined, 343
    differences, 343–44
    forwarding methods, 345
    mechanism, 344–45
Optical channel (OCh), 197
Optical connectors
    defined, 96
    E2000, 99
    epoxy and polish, 96
    LC, 97–98
    MU, 98
    no-epoxy and no-polish, 97
    preloaded epoxy/no-epoxy and polish, 96–97
    SC, 97
Optical control plane
    architecture, 208
    automatic discovery, 209
    call/connection management functions, 203, 209–10
    connections, 210
    defined, 203
    external network-network interface (E-NNI), 208
    functional entities, 208
    functions, 209–10
    illustrated, 203
    internal network-network interface (I-NNI), 205
    logical interfaces, 205–8
    protection, 204
    protection switching and restoration, 209–10
    standards, 204–5, 206–7
    user-network interface (UNI), 205
    See also Optical transport networks (OTN)
Optical data unit (ODU), 194
Optical distribution frames (ODFs), 75–76
Optical/electronic circuit switching, 339
Optical fibers
    attenuation characteristics, 66–67, 239
    bending diameter, 70
    cable preparation, 74
    classification, 59–60
    cost of laying, 53
    design parameters, 66
    design specifications, 65–70

dimensions, 69
dispersion types, 67
handling techniques, 110–12
inside plant (ISP) cables, 60
laying techniques, 71–73
longevity, 70
loose-tube design, 60–61
MMF, 58, 59, 64–65
operational wavelength for, 65
outside plant (OSP) cables, 59
overview of, 15
as preferred medium, 83
propagation modes, 68–69
propagation of light through, 57
sectional view of, 66
SMF, 58, 59, 62–63
splicing, 74–75
standard designs, 60–61
strength, 70
in synchronous optical networks, 173–74
termination, 75–76
tightly-buffered cable, 61
transmission losses, 65–67
transmission of light through, 58
transmission windows, 68
types of, 62–65
Optical interfaces, 90
Optical Internetworking Forum (OIF), 205, 243
Optical link design
link budgeting, 100–107
optical connectors, 96–100
optical modulation techniques, 93–96
optical receivers, 92–93
optical transmitters, 84–92
safety guidelines, 107–12
summary, 112–13
Optical modulation techniques
ASK, 93–94
FSK, 95
PolSK, 95
PSK, 95
types of, 94
Optical multiplex section (OMS), 197
Optical packet switching (OPS)
broadcast-and-select networks, 340–41, 342
defined, 340
functional blocks, 340

space switching, 343
wavelength routing, 341–43
WDM-based techniques, 340–43
Optical receivers
clock recovery, 92
defined, 92
error detection/recovery, 93
functions, 92–93
line decoding, 92
sensitivity and dynamic range, 101
WDM, 256
Optical safety, 108
Optical sources, 85–86
Optical switching
circuit, 339
overview, 338
techniques, 338
Optical transmission
absorption losses, 252–53
bands, 241, 247
challenges, 251–55
dispersion, 253
FWM, 254
linear characteristics, 252–53
nonlinearities, 253–54
SBS, 254
scattering losses, 252
spectral efficiency, 251
SRS, 254
waveguide losses, 252
windows, 240
XPM, 254
Optical transmission section (OTS), 197
Optical transmitters
back reflection, 89–91
defined, 84
design parameters, 89–92
extinction ratio, 91–92
functions, 84
launch power, 100–101
modular optical interfaces, 86–89
optical sources, 85–86
WDM, 256
Optical transport networks (OTN), 160–61, 187–228
automatic protection switching, 202
business imperatives, 189–92
case study, 223–28
connection management, 201–2
as digital wrapper technology, 197–98

Optical transport networks (continued)
    FEC, 211–16
    frame structure, 196
    generic mapping procedure (GMP), 218
    hierarchy, 195, 197
    interfaces, 200–201
    key features, 218–20
    layers, 195
    line rates, 196
    logical interfaces, 201
    multistage multiplexing, 218–19
    native format support, 198–200
    network architecture, 197–200
    network elements (NEs), 192–93
    network organization, 192–202
    next-generation architecture illustration, 199
    ODUflex, 218
    ODU information structure, 197
    optical channel (OCh), 197
    optical control plane, 202–11
    optical entities, 198–200
    optical multiplex section (OMS), 197
    optical transmission section (OTS), 197
    physical interfaces, 200–201
    protocol support, 198
    SDH/SONET network disadvantages and, 192
    standards, 192, 193
    summary, 220–21
    switching architecture, 217–18
    switch rates, 217
    tandem connection monitoring (TCM), 216–17
    TDM core networks and, 188–89
    transport hierarchy, 193–97
    transport mechanism, 195–96
    use in core network, 265
    user-defined path-layer monitoring, 216
    See also Transport networks
Optical transport unit (OTU), 194
OTN deployment case study
    background, 223
    challenges, 223
    proposed network architecture illustration, 226
    proposed solution, 225–27
    recommendations, 227–28
    requirements, 225
    See also Optical transport networks (OTN)
Outside plant (OSP) cables, 59

P2MP TE, 333
Packet cable multimedia (PCMM), 376
Packet optical networking platforms (PONPs), 293, 313
Packet optical networks (PONs)
    architecture and functioning, 289–92
    classification, 292
    components, 289
    connection solution, 291–92
    defined, 289
    for high-speed access networks, 288
    NG-PONs, 292–93
    ODN, 289, 291
    OLT, 289
    ONU, 289
    optical splitters, 289
    technologies, 277
    types of, 290
Packet optical transport service (P-OTS), 293
Packet-switched photonic networks
    access layer functionalities, 318
    access routers, 314–15
    architecture, 314–19
    backbone routers, 316
    border routers, 317
    burst, 343–45
    core routers, 317
    design considerations, 319–22
    distribution layer functionalities, 319
    distribution routers, 315–16
    edge routers/switches, 317
    GMPLS, 345
    imperatives, 312–13
    IP over MPLS, 334–36
    ISPs, 319, 320, 321
    MPLS, 322–34
    MPLS-TP, 346
    network elements (NEs), 314–18
    optical switching, 338–45
    packet, 340–43
    packet transport networks (PTNs), 346
    packet versus data flows, 346
    physical topology, 318–19
    service aggregation router, 317

summary, 347
virtual routing, 336–38
Packet switching, 6
Packet transport networks (PTNs), 175, 346
Passive routers, 258
PDH E1 frame format, 137–39, 141, 142
PDH E2 frame format, 139–44
PDH E3 frame format, 144, 145
PE function virtualization (vPE), 358
Penultimate hop popping, 327
Phase shift keying (PSK), 95, 251
Photonic circuit-switched networks
    architecture, 275–302
    case study, 304–6
    FTTx networks, 288–93
    introduction to, 276–77
    optical access networks, 277–87
    photonic core networks, 293–301
    summary, 301–2
Photonic core networks
    IP/MPLS, 294–96
    MPLS-TP, 293–94
    NG-POTN, 296–301
    overview, 293
    PBB-TE, 294
Physical interfaces, OTN, 200–201
Plesiochronous digital hierarchy (PDH), 121, 127–44
    12-frame multiframe alignment, 137
    24-frame multiframe alignment, 136
    CCITT recommendations, 129–30
    centralized clock interface, 131–32
    codirectional interface, 131
    data signal code conversion, 132
    defined, 127
    Ethernet over, 144
    first multiplexing order in, 128
    frame structures, 134
    G.702, 129–30
    G.703, 130
    interfaces, 130–33
    ITU multiplexing hierarchy, 125
    multiplexing hierarchy, 124
    multiplexing hierarchy (signaling structure), 141
    multiplexing hierarchy (simplified view), 129
    multiplexing hierarchy interface specifications, 131
    NADH DS1 frame format, 134–35
    NADH DS2 frame format, 135, 138, 139
    NADH DS3 frame format, 135–37, 140
    PDH E1 frame format, 137–39, 141, 142
    PDH E2 frame format, 139–44
    PDH E3 frame format, 144, 145
    timing concept, 128
Point-to-point infrared systems, 19
Polarity coding, 42
Polarization mode dispersion (PMD), 95
Polarization shift keying (PolSK), 95
Power budget, 101–2
Power margin (PM), 101–2
Preloaded epoxy/no-epoxy and polish connectors, 96–97
Protection switching, 202, 210–11
Provider backbone bridge traffic engineering (PBB-TE), 294
Public switched telephone networks (PSTNs)
    advancements, 19–20
    building blocks of, 5
    class 1 switch, 8
    class 2 switch, 8
    class 3 switch, 9
    class 4 switch, 9
    class 5 switch, 9
    components, 6–19
    defined, 4, 6
    illustrated elements of, 5
    local loops, 11
    local network, 11–12
    long-distance network, 12
    media types, 12
    networks, 6, 11–12
    switching equipment, 7–8
    switching hierarchy, 10
    switching offices, 6
    switch types, 8
    system hierarchy and call flow, 7
    terminology, 6
    transmission facilities, 6, 11
    transmission media and signals, 12–19
    trunks, 11
    wired media, 12–15
    wireless media, 15–19
Pulse-amplitude modulation (PAM), 36
Pulse-code modulation (PCM)
    adaptive (ADPCM), 46–47

Pulse-code modulation (continued)
    block diagram, 35
    companding, 44–45
    defined, 35
    differential (DPCM), 45–46
    quantization, 38–41
    quantization noise, 41–44
    sampling, 36–38
    value assignment, 43

Quadrature PSK (QPSK), 95
Quantization
    companding, 44–45
    defined, 38
    encoding/decoding stages and, 39
    input signal, 41
    intervals, increase in, 44
    levels, 41
    linear process, 44
    noise, 41–44
    process, 38–39
    process examples, 43
    uniform method, 44

Radio waves
    frequency spectrum, 17
    ISM bands, 16
    overview, 16
    single-frequency high-power, 16–17
    single-frequency low-power, 16
    spread spectrum, 17–18
Real-time virtualization, 361
Reconfigurable optical add/drop multiplexers (ROADMs)
    branching DWDM link and, 261
    CD/C, 260
    defined, 189
    NG-POTN, 300–301
Reed-Solomon codes
    coding gain, 212
    interleaving, 213
Reflection, 56
Refraction, 57
Regenerators, 173, 258
Ring topology, 167, 168
RSVP
    defined, 330
    functioning, 331–33
    message, 332

path message, 331–32
resource reservation, 332–33
signaled LSPs, 333
TE extensions for, 331

Safety guidelines
    AEL and, 110
    causes of injury and, 108, 109
    electrical safety, 108
    fiber-handling techniques, 110–12
    fire safety, 107
    material safety, 107
    maximum permissible exposure (MPE) and, 108–10
    optical safety, 108
    See also Optical link design
Sampling
    defined, 36
    frequency, 36
    illustrated, 37
Satellite microwave, 18–19
SC/APC connector, 76
Scattering, 57
SC/PC connector, 76
SDH network architecture case study
    background, 179
    key challenge, 181
    proposed, illustrated, 180
    proposed solution, 181–82
    recommendation, 182–83
    requirements, 179
SDH/SONET networks
    developing, 188
    drawbacks, 189–90
    evolution, 189
    legacy architecture, 191
    maximum bandwidth, 234
    NE switching, 219
    operational planes, 202–3
    OTN evolution to overcome disadvantages of, 192
    in ring topology, 189
    section overhead (SOH) bytes, 211
    switching layers, 217
    TDM-based, 235
    transport structure, 217
    WDM and, 235
Semiconductor optical amplifier (SOA), 257
Service-level agreements (SLAs), 156
SFP+ modules, 88–89

Index

Shielded twisted-pair (STP) cables, 13, 14
Shortest path first (SPF) algorithm, 328
Signaled LSPs, 327
Signal restoration at receiver, 42
Single-frequency high-power systems, 16–17
Single-frequency low-power systems, 16
Single-mode fiber (SMF)
    core diameters of, 59
    cross-sectional view, 70
    defined, 58
    RI profiles, 63
    types of, 62–63
Single-mode step index, 59
Small form-factor pluggable (SFP) modules, 86, 87
Soft-decision FEC, 216
Software-defined mobile networking (SDMN)
    architecture implementation, 380–82
    components, 381
    defined, 379
    forwarding, 381–82
    network architecture, 380
Software defined network (SDN)
    adoption of, 354
    applications, 374
    architectural choices, 377
    architectural components, 373–75
    architectural framework for, 372–77
    business case, 370–71
    CDPI, 374–75
    conceptual basis, 371–72
    controller, 373
    core network controller, 374
    datapath, 374–75
    design and deployment considerations, 376–77
    flat structures and, 370
    functional blocks of controller, 374
    IEEE standards, 372
    key features of, 371–72
    management, 377–78
    NFV and, 355, 368–69
    northbound interfaces, 375
    OpenFlow and, 378–79
    protocol support, 375–76
    RAN controller, 374
    software-defined mobile networking (SDMN), 379–82
    standardization, 372
    summary, 383
    technology, 356
    use cases, 379–82
SONET
    optical base carrier rate, 148, 160
    signals, 160
    standardization process, 147, 159
    standards, 147
    T1X1 subcommittee and, 159–60
    See also SDH/SONET networks
Space division multiplexing (SDM), 34–35, 298
Space switching, 343
Span analysis, 100, 104–5, 106
Span loss, 105
Splicing, 74–75
Split-MAC, 280
Spread spectrum, 17–18
Standard SMF (SSMF), 251
Static LSPs, 326
Steps coding, 43
STM-1 links, 103
Strength, optical fiber, 70
Subscriber connectors (SCs), 97
Switching architecture, OTN, 217–18
Switching equipment, 7–8
Switching offices, 6
Synchronous digital hierarchy (SDH) standards, 20
Synchronous optical networks
    abstract view of architecture, 161
    access layer, 162
    analogy, 162
    architecture, 161–64
    case study, 179–84
    collection/aggregation layer, 162–63
    components, 164–76
    connection types, 169
    core, 162–63
    cross-connects, 170–73
    evolution of, 159
    introduction to, 148
    key benefits, 159
    layer functions, 164
    layers, 161
    media, 173–75
    microwave links, 175
    multiplexers/demultiplexers, 164–70
    need, benefits, and function, 157–61
    network elements (NEs), 158

Synchronous optical networks (continued)
	network equipment, 164–73
	network topologies, 175–76
	optical fiber, 173–74
	optical transport networks, 160–61
	overview, 156–57
	summary, 176–77
	transport network topologies, 176
	universal standards, 158
Synchronous transport structure-level 1 (STS-1), 148, 160

T1 systems, 121, 124
Tandem connection monitoring (TCM)
	byte labeling, 216
	defined, 216
	operation principle, 217
	use of, 219
TDM core networks, 188–89
TDM PON, 292
Telecommunication, 4
Telecommunication networks, evolution of, 4
Ten-gigabit SFP (XFP) modules, 88
Terminal multiplexers (TMs)
	defined, 165
	in linear configuration, 166
	in point-to-point configuration, 166
	in ring configuration, 167
Terrestrial microwave, 18
Tightly-buffered cable, 61
Time-division multiplexing (TDM), 31, 32, 36
Time-slot assigner (TSA), 170
Time slot interchange (TSI), 171
Total cost of ownership, 357
Traffic engineering (TE)
	constrained shortest path first (CSPF), 328–31
	database, 328
	function of, 327
	label distribution protocol (LDP), 333–34
	in MPLS network, 329
	P2MP, 333
	paths, 328–30
	process, 327–28
	RSVP functioning, 331–33

Translucent OXCs, 339
Transmission facilities, 6, 11
Transmission windows, 68
Transparent OXCs, 339
Transport networks
	analogy, 162
	architecture, 161–64
	asynchronous, 124–27, 146–47
	components, 164–76
	Ethernet over PDH, 144–45
	evolution of, 123
	introduction to, 122–24, 156
	layer functions, 165
	need, benefits, and function, 157
	optical, 160–61, 187–228
	overview, 156–57
	PDH, 127–44
	prologue, 121–29
	SONET, 147–48
	summary, 148–49, 176–77
	topologies, 176
Traveling wave tubes (TWTs), 257
Trenching and ducting (T&D), 71–72
Trunks, 11
Two-fiber bidirectional ring, 175

Ultra-polished connectors (UPCs), 100
Unidirectional connection, 169
Unshielded twisted-pair (UTP) cables, 13
User-network interface (UNI), 205

Verizon 4G networks, 386–87
Vertical-cavity surface-emitting lasers (VCSELs), 86
Virtual customer premises equipment (vCPE), 353–54
Virtualization
	CPE, 358–59
	ecosystem, 356
	PE function, 358
	real-time, 361
	scope and function, 360–61
	See also Network function virtualization (NFV)
Virtual routing, 336–38
Voice and data transmission
	adaptive PCM (ADPCM), 46–47
	companding, 44–45

differential PCM (DPCM), 45–46
introduction to, 28–29
multiplexing, 29–35
PCM, 35–41
quantization noise, 41–44
summary, 47
Voice-over-IP (VoIP) services, 11

Wavelength converters/transponders/
    muxponders, 259
Wavelength division multiplexing (WDM)
    absorption losses, 252–53
    additional light ray transmission, 34
    classification, 33
    couplers, 256
    defined, 31–32, 236
    dispersion, 253
    evolution, 238
    evolution of networks, 246–51
    fiber bandwidth use, 32
    flexible grids, 249–51
    fundamentals, 238–46
    FWM, 254
    guard band, 246
    infrastructure integration, 292
    initial demonstration of, 33
    integration of legacy TDM, 235
    linear characteristics, 252–53
    link example, 32
    network components, 255–60
    nonlinearities, 253–54
    optical amplifiers, 256–57
    optical receivers, 256
    optical transmission bands, 247
    optical transmitters, 256
    passive routers, 258
    proprietary grids, 248
    regenerators, 258
    SBS, 254
    scattering losses, 252
    SDH/SONET networks and, 235
    SRS, 254
    standardization, 237–38
    summary, 267–68
    system classification, 236, 237
    transmission challenges, 251–55
    WADMs, 259–60
    waveguide losses, 252
    wavelength converters/transponders/
        muxponders, 259
    wavelength selective switch, 259
    XPM, 254
Wavelength provisioning, 306
Wavelength routing, 341–43
Wavelength selective switch, 259, 337
WDM PON, 292
Wideband WDM (WWDM), 236
Wired media, 12–15
Wireless broadband networks case study,
    386–87
Wireless media, 15–19